AP1000 核电厂
英汉常用技术词汇与术语

主　编　顾　军

副主编　缪亚民　王秀启　范福平
　　　　夏利明　金　湘　唐锡文

中国原子能出版传媒有限公司

图书在版编目(CIP)数据

AP1000 核电厂英汉常用技术词汇与术语/顾军主编.
—北京:中国原子能出版传媒有限公司,2011.6
ISBN 978-7-5022-5256-4

Ⅰ.①A… Ⅱ.①顾… Ⅲ.①核电厂—词汇—英、汉
②核电厂—名词术语—英、汉 Ⅳ.①TM623-61

中国版本图书馆 CIP 数据核字(2011)第 110804 号

内 容 简 介

《AP1000 核电厂英汉常用技术词汇与术语》包括两部分,第一部分是词汇与术语,约 12500 条;第二部分是术语缩写,约 5000 条,两部分共包括技术词汇与术语(及缩写)约 17500 条。此外,还把国际单位换算作为附录,方便读者工作使用。

本字典适用于在核电厂工作的人员使用,尤其适用于在 AP1000 核电厂工作的人员,以及刚刚进入 AP1000 核电厂工作的新员工,对他们快速掌握 AP1000 核电厂技术领域的专业词汇与术语,尽快适应岗位工作提供非常有益的帮助。同时,也为我们尽快消化、吸收 AP1000 核电厂技术,推动第三代压水反应堆核电厂在我国的建设,起到积极的促进作用。

AP1000 核电厂英汉常用技术词汇与术语

出版发行 中国原子能出版传媒有限公司(北京市海淀区阜成路 43 号 100048)
责任编辑 刘 岩 侯茸方
技术编辑 丁怀兰 王亚翠
责任印制 潘玉玲
印 刷 保定市中画美凯印刷有限公司
经 销 全国新华书店
开 本 850 mm×1168 mm 1/32
印 张 16.875 **字 数** 639 千字
版 次 2011 年 9 月第 1 版 2011 年 9 月第 1 次印刷
书 号 ISBN 978-7-5022-5256-4 **定 价** **90.00 元**

网址:http://www.aep.com.cn **E-mail:atomep123@126.com**
发行电话:010-68452845

《AP1000核电厂英汉常用技术词汇与术语》
编 辑 部

主　　编　顾　军

副 主 编　缪亚民　王秀启　范福平　夏利明　金　湘
　　　　　唐锡文

编　　著　唐锡文　胡兴强

校　　对　董赛君　杨小琴　叶菁菁

审　　核　曾　勋　裴卫农　陈　森　吴元明

集团审查

主　　审　张禄庆

统　　审　黄　芳　陈　森

总　序

核工业作为国家高科技战略性产业，是国家安全的重要基石、重要的清洁能源供应，以及综合国力和大国地位的重要标志。

1978 年以来，我国核工业第二次创业。中国核工业集团公司走出了一条以我为主发展民族核电的成功道路。在长期的核电设计、建造、运行和管理过程中，积累了丰富的实践和理论经验，在与国际同行合作过程中，实现了技术和管理与国际先进水平相接轨，取得了骄人的业绩。

中国核工业集团公司在三十多年的核电建设中，经历了起步、小批量建设、快速发展三个阶段。我国先后建成了秦山、大亚湾、田湾三大核电基地，实现了我国大陆核电“零”的突破、国产化的重大跨越、核电管理与国际接轨，走出了一条以我为主，发展民族核电的成功之路。在最近几年中，发展尤为迅猛。截至 2008 年底，核电运行机组 11 台，装机容量 907.82 万千瓦，全部稳定运行，态势良好。

进入 21 世纪，党中央、国务院和中央军委对核工业发展高度重视、极为关怀，对核工业做出了新的战略决策。胡锦涛总书记指出：“无论从促进经济社会发展看，还是从保障国家安全看，我们都必须切实把我国核事业发展好。”发展核电是优化能源结构、保障能源安全、满足经济社会发展需求的重要途径。2007 年 10 月，国务院正式颁布了《核电中长期发展规划(2005—2020 年)》。核电进入了快速、规模化、跨越式发展的新阶段。

在中国核电大发展之际，中国核工业集团公司继续以“核安全是核工业的生命线”的核安全文化理念和“透明、坦诚和开放”的企业管理心态，以推动核电又好又快又安全发展为己任，为加速培养核电发展所需的各类人才，组织核电领域专家，全面系统地对核电设计、工程建造、电站调试、生产准备和生产运营等各阶段的知识进行了梳理，构造了有逻辑性、系统性的核电知识体系，形成了覆盖核电各阶段的核电工程培训系列教材。

这套教材作为培养核电人才的重要工具，是国内目前第一套专业化、体系化、公开出版的核电人才培养系列教材，有助于开展培训工作，提高培训质量、节约培训成本，夯实核电发展基础。它集中了全集团的优势，突出高起点、实用性强，是集团化、专业化运作的又一次实践，是中国核工业 50 余年知识管理的积淀，是中国核工业 10 万人多年总结和实践经验的结晶。

21 世纪是“以人为本”的知识经济时代，拥有足够的优秀人才是企业持续发展的重要基础。中国核工业集团公司愿以这套教材为核电发展开路，为业界理论探讨、实践交流提供参考。

我们要继续以科学发展观为指导，认真贯彻落实党中央、国务院的指示精神，积极推进核电产业发展。特别是要把总结核电建设经验作为一项长期的工作来抓，不断更新和完善人才教育培训体系。

核电培训系列教材可广泛用于核电厂人员培训，也可用于核电管理者的学习工具书，对于有针对性地解决核电厂生产实践和管理问题具有重要的参考价值。

中国核工业集团公司总经理

2009 年 9 月 9 日

前 言

随着全球首台AP1000核电厂机组在我国的顺利开建，相关工程技术人员都希望尽快能够掌握有关AP1000英汉常用技术词汇与术语，以便于对核岛合同、AP1000专业技术资料等的消化、吸收，在这种背景下我们收集、整理了《AP1000核电厂英汉常用技术词汇与术语》字典，供核电厂相关人员工作参考。

全书分两部分，第一部分是词汇与术语，包含AP1000核电厂常用词汇与术语约10200条，设备采购常用词汇与术语800余条，设备监造常用词汇与术语1200余条，核电培训常用词汇与术语100余条和AP1000核电厂设备名称200余条，共约12500条。第二部分是术语缩写，包含AP1000核电厂常用缩写约600条，核电英语常用缩写约4000条，AP1000核电厂系统缩写100余条，设备监造专用缩写100余条，法规、标准缩写约50条和组织与机构缩写150余条，共约5000条。此外，还把国际单位换算作为附录，方便读者工作使用。

在本字典的编辑过程中，得到了三门核电有限公司编委会成员的大力支持，也得到了公司生产准备处、设计管理处等员工的积极协助，在此表示由衷的感谢！

特别需要提出的是集团公司主审张禄庆专家，对该字典的最终审查和定稿倾注了大量的心血，提出了许多好的修改意见和完善建议，对保证该字典的全面性、准确性发挥了巨大的作用。在此，对他付出的努力表示敬意和衷心感谢！

在本字典正式出版前夕，集团公司组织召开了专门的审查会，保证了内容的准确性和权威性。对组织审查会的中核集团核电部程慧平副主任，编委李苏甲，专家张禄庆、杨传德、左民、吴国安、石侠民，执行编委谢波、刘朔，以及编委会办公室成员姜福明、张红军、沈阳等表示感谢。

本字典在词汇的收集、翻译及整理的时间比较短，其中难免存在翻译不够贴切，内容不够全面等缺憾，恳请读者提出宝贵意见，以便我们未来升版，并为读者提供更加完美的工具书。

唐锡文

2010 年 10 月 25 日

目　　录

第一部分　词汇与术语

第二部分　术语缩写

第一部分　词汇与术语

Ⅰ. AP1000 核电厂常用词汇与术语

A

a matter of　有关…的事

a wide range of validity and high accuracy　宽幅有效和高精度

A/M transfer switch　自动/手动切换开关

abnormal condition　异常工况

abnormal load rejection　异常甩负荷

abnormal operating condition　异常运行工况

above grade　零标高以上,地面以上

above-mentioned　上述的

abrasive　磨料,研磨剂,有研磨作用的

absentee　不在者;缺席者,缺勤者

absorbed dose rate　吸收剂量率

absorbed energy　吸收能量

absorber chiller　吸收式制冷机

absorber cladding　吸收体包壳

absorber rod　吸收棒

absorption coefficient　吸收系数

absorption cross section　吸收截面

absorption ratio　吸收比

AC arc welding power source　交流弧焊电源

AC distribution and control system　交流配电和控制系统

AC important instrumentation power system　交流重要仪表电源系统

AC plasma arc welding　交流等离子弧焊

AC power　交流电,交流电源

AC/DC arc welding power source　交直流两用弧焊电源

accelerated corrosion　加速腐蚀

acceleration　加速

acceleration pressure drop　加速压降

acceleration pressure loss　加速压力损失

acceleration sensor　加速度传感器

acceleration voltage　加速电压

accept　验收,接受,认可

accept an offer　接受报盘

accept an order　接受订单

accept without modification　原样验收

acceptable daily intake　日容许摄入量

acceptable fuel design limits　燃料设计容许限值

acceptable heat transfer geometry　可接受的传热几何

acceptance　接受,验收

acceptance criterion　验收准则

acceptance limit　验收限值

acceptance report　验收报告

acceptance standard　验收标准
acceptance test　验收试验
access　通道,入口
access areas　可达区域
access control area　出入口控制区
access hatch　进出闸门,(人员通道)出入口
accessibility　可达性
accessory　附件
accident analysis　事故分析
accident conditions　事故工况
accident interlocking module　事故联锁组件,事故联锁模块
accident management　事故处理,事故管理
accident mitigation　事故缓解
accident prevention　事故预防
accident shutdown　事故停堆
accident source　事故源
accidental exposure　事故照射
accidental transient　事故瞬态
accident-initiated iodine spike　事故引发的碘释放峰
accommodate　顺应,调节;迁就,迎合,接受
accommodation　住宿设施
according to　照,依照,据,据…所说
accordingly　随着;相应地
account for　考虑
accumulated dose　累积剂量
accumulator　安全注射箱,蓄压箱
accumulator check valve　安全注射箱止回阀
accumulator nitrogen containment penetration　安全注射箱氮气管道安全壳贯穿件
accumulator nitrogen vent　安全注射箱氮气排放孔
accumulator pressure relief　安全注射箱压力释放
accumulator pressure transmitter isolation　安全注射箱压力变送器隔离
accumulator tank　安全注射箱
acetylene cylinder　乙炔瓶
acetylene generator　乙炔发生器
achieve consensus　达成一致,圆满解决
acicular ferrite　针状铁素体
acid　酸性的,酸
acid covering　酸性药皮
acid electrode　酸性焊条
acid resisting touchup paint for racks　台架用耐酸面漆
acid slag　酸性渣
acidity　酸度
acid-proof tile　耐酸瓷砖
acoustic booths　传音室
acoustic emission testing　声发射检测
acquiesce　默许,默认,勉强同意
actinide isotopic　锕系同位素的
activation　活化
activation product　活化产物
activation temperature range of flux　钎剂活性温度范围
active carbon filter　活性炭过滤器
active component　能动部件
active core　堆芯(活性区)
active core height　堆芯活性高度

active filler metal 活性钎料
active fuel length 燃料的活性区长度
active functions 能动功能
active high-pressure safety injection system 能动的高压安注系统
active metal brazing alloy 活性钎料
active power 能动功率
active soil pressure 主动土压力,实际土压力
active valve 能动阀门
active work order 处理中的工作票(未完工单)
activity 放射性,(放射性)活度
activity concentration 放射性浓度
activity level 放射性水平
actual capacity 实际能力,实际容量
actuate 驱动,触发
actuated equipment (被)驱动设备
actuation 驱动,触发
actuation devices 执行机构,驱动装置
actuation signal 触发信号、驱动信号
actuator 执行机构
acute radiological sickness 急性放射性疾病
adaptaters 转接件,转换件,适配件
adapter plate (燃料组件)管座格板
adaptive control welding 适应控制焊接
adaptor 贯穿件,适配器
addendum 附录,补遗
address 致力于
addressable heat detector 可寻址的热探测器
addressable ionization smoke detector 可寻址的电离烟雾探测器
addressable manual alarm pull station 可寻址的手动报警站
addressable monitor module 可寻址的监测模块
adhering slag 挂渣
adhesive strength 结合强度
adiabatic 绝热的,隔热的
adiabatic isochoric complete combustion 等容绝热完全燃烧
adjustable discharge gear pump 齿轮比例泵,可调排量齿轮泵
adjustable time delay actuating 时间可调的延迟驱动
adjustable time delay de-actuating 时间可调的延迟清除驱动
adjusting bolt 调节螺栓
adjusting ring 调节环
adjusting screw 调节螺钉,调整丝杠
adjusting valve 调节阀
administration building and emergency center water supply and drainage system 办公及应急中心给排水系统
administrative control 行政控制,行政管理
administrative program and procedure 管理大纲和程序
administrative withdrawal limit 撤销的行政限制
adoption by equation 等同采用
ADS blow down test ADS 回座试验,ADS 排放试验
ADS discharge header drain check valve ADS 排放集管疏水止回阀

ADS discharge header drain isolation valve　ADS排放集管疏水隔离阀
ADS discharge header vacuum relief　ADS排放集管真空释放
ADS stage 4 actuation　ADS第4级触发动作
advance payment　预付款
advanced　先进的，高级的
advanced alarm system　先进报警系统
advanced light water reactor evolutionary plant　渐进型先进轻水反应堆核电厂
advanced passive experiment　先进非能动试验
advanced plant experiment　先进电厂实验
adversary　对手，敌手
adverse　不利的，相反的，敌对的
adverse xenon distribution　不利的氙分布
aerial cable　架空电缆
aerodynamic behavior　空气动力特性
aerosol　气溶胶
aerosol activity　气溶胶活性
aerosol removal　气溶胶的去除
affected　有故障的，受影响的，受损伤的
affiliate　分支机构，分会；隶属企业
aforementioned　前面提到的
aforesaid　前面所说的；上述的
after cooler　后置冷却器
after filter　后置过滤器
after-heat　剩余释热，衰变热
after-power　剩余功率
against flotation　抗浮
against overturning　抗倾覆，抗侧翻
age hardening　时效硬化
ageing　老化
agenda　议事日程，会议事项
agglomerated flux　团块状焊剂，陶质焊剂，烧结焊剂
aggregate stockpile　石料堆场
aggressive element　侵蚀性元素
aging mechanism　老化机理
agree about　达成一致
agree to terms　达成协议
agreement　协议
air (circuit) breaker　空气（回路）断路器
air arc cutting　空气电弧切割
air assist check valve　气动止回阀
air baffle　（安全壳）空气导流板，空气挡板
air breathing source　呼吸空气源
air change rate　换气率
air compressor assemblies　空气压缩机组件
air compressor skid　空气压缩机成套设备
air cooler　空气冷却器
air delivery alternate isolation valve　供气备用隔离阀
air delivery isolation valve　供气隔离阀
air delivery line　供气管线
air delivery line pressure instrument isolation valve　供气管线压力仪表隔离阀
air delivery main isolation valve　供气

主隔离阀
air delivery pipe　供气管道
air diaphragm operator　气动隔膜膜片执行机构
air diaphragm operator with valve position　带阀位指示的气动膜片执行机构
air evacuation valve　放气阀,排气阀
air filter and absorber unit　空气过滤器与吸收体单元
air filter unit　空气过滤单元
air flow rate　空气流量,风流量
air flow ultra sonic air velocity monitor　气流超声气速监测仪
air handling unit　空气控制单元
air hose　通风软管
air ingestion　空气注入,空气摄入
air intake　吸风口,进风口,吸气管
air intake valve　进气阀
air lock　气闸,气塞,阻隔室
air plasma arc cutting　空气等离子弧切割
air plasma water injection arc cutting　水注入空气等离子弧切割
air pump　气泵
air receiver　压缩空气贮罐,压缩空气接收器
air receiver skid　压缩空气接受贮槽/组件
air release valve　排气阀
air sampling device　空气取样设备,空气取样装置
air self-cooling type　空气自冷式
air submersion dose　空气浸没剂量
air tank isolation valve　压缩空气贮罐隔离阀
air tank safety relief valve　压缩空气贮罐安全释放阀
air tight test　气密性检验
air valve　空气阀
air vent valve　放气阀
air-acetylene welding　空气乙炔焊
airborne　悬浮空气
airborne (radiation) monitor　悬浮空气(辐射)监测仪
airborne activity　悬浮空气活性
airborne monitoring　悬浮空气监测
airborne particulate　悬浮空气微粒
airborne particulate sample　悬浮空气微粒样本,悬浮空气微粒取样
air-cooled chiller　风冷冷冻机
air-operated double-diaphragm pump　气动双膜泵
air-operated valve　气动阀
air-powered piston pump　气动活塞泵
alarm　报警,警报
alarm(output) module　报警(输出)组件,报警(输出)模块
alarm check valve　报警止回阀
alarm guidelines　报警导则
alarm server　报警服务器
alarm signal　报警信号
alarm valve　报警阀
alarm window　报警窗
albedo　反照率
alert　警报
algae　藻类,海藻
algae growth　藻类生长

algicide　除藻剂
alienate　使疏远
aligning pin　定位销,准直销
alignment　定位,对中,对准,序列,准直
alignment guide stud　定位导向螺栓
alkali metal　碱金属
all in all　总的说来
all risks　一切险
all rods out　控制棒全部提出
all volatile treatment　全挥发处理
allotment　分批,分配
allowable stresses　许用应力
allowance　折扣
alloy　合金
alloying　渗合金
alloying constituent　合金剂
all-volatile chemical feed system　全挥发化学处理给水系统
along with…　随着…
alpha/beta counting system　α/β计数系统
alphanumeric keyboard　字母数字键盘
alternate　交替的,轮流的,预备的
alternate verified calculational method　交替验证过的计算方法
alternate wire-feed system　变速送丝方式
alternating current arc welding　交流电弧焊
alternating stress intensity value　交变应力强度值
alternation　交替,轮流,改造,预备
alternative　一种替代选择,代替品
alternative AC power source　备用交流电源
alumetized steel electrode　渗铝钢焊条
aluminium　铝
aluminium alloy arc welding electrode　铝合金焊条
ALWR Utility steering Committee　先进轻水堆用户指导委员会
ambient　周围的,周围环境的
ambient temperature　环境温度,周边温度
ambiguity　含糊;意义不明确
amendment ,modification ,alteration　变更,更改,修正
American society for training and development(ASTD)　美国训练发展协会
amicably　友好地
ammeter　电流表
amorphous filler metal　非晶态钎料
ampacity　载流量
amplifier　放大器
amplitude　振幅
an offer without engagement　虚盘
analog channel　模拟量通道
analog input　模拟量输入
analog link　模拟量连接线
analog output　模拟量输出
analog single　模拟信号
analog to digital converter　模数转换器
analog/digital converter　模拟/数字转换器
analogue link　模拟接续线
analyst　分析员,化验员
anchor bolt　地脚螺栓

anchorage　地脚螺栓,固定支座/装置,锚固
anchoring　抛锚效应
ancillary AC diesel generator　辅助交流柴油发电机组
ancillary diesel generator fuel oil storage tank　辅助柴油发电机燃油贮罐
ancillary diesel generator fuel oil supply system　辅助柴油机燃油供应系统
ancillary diesel generator packages　附属柴油机包
ancillary fans　辅助风机
ancillary fuel oil tank　辅助燃油箱
aneroid　膜盒,无液的气压计
angle beam method　斜射法超声波探伤
angle beam ultrasonic examination　斜波超声检测
angle joint　角接接头
angle leg　支撑腿,角钢支腿
angle stop check valve　角式截止止回阀
angle stop valve　角式截止阀
angle throttle valve　角式节流阀
angle type　角式
angle valve　角阀
angle-type axial flow pump　半贯流式轴流泵,弯管轴流泵
angular distortion　角变形
angular momentum　转动惯量,角动量
anion　阴离子
anion bed　阴离子床
anneal　退火
annealed austenitic stainless steel　退火奥氏体不锈钢
annex　附录;附属建筑
annex BLDG MCCs　附属厂房电机控制中心
annex building　附属厂房
annex building PWR XFMRs　附属厂房电力变压器
annihilation　灭绝,湮没,熄火
annual training plan　年度培训计划
annular fuel pellet　环状燃料芯块
annular valve　环形阀
annulus　环廊,环形空间
annulus drain　环形排水沟
annunciator panel　信号盘
anomaly　不规则,异常
anthropometric guidelines　人体测量学导则
anticipate　预期,指望
anticipated event　预期事件
anticipated operational occurrences　预期运行事件
anticipated operational transient　预期运行瞬态
anticipated transient　预期瞬态
anticipated transient without scram (ATWS)　未按预期停堆的瞬态
anti-form reagent　防沫剂
anti-rust agent　防锈剂
anti-seizer lubricant　抗咬死润滑剂
anti-vibration bar　抗震杆,防震棒,防震条
anti-whip device　管道防甩装置
AP1000 health physics program　AP1000 保健物理计划(大纲)

apart from　除了…以外,此外
apology　道歉
apparatus　器械,仪器,设备
apparent power　视在功率
appendix　附录(补充件)
applicability　适用性,应用范围
applicable　适用的
applicant　(许可证)申领者
application and computerized procedure server　计算机化应用规程服务器
application package　应用软件包
application program　应用程序
application server base software and tool　应用服务器基础软件和工具
application software　应用软件
application software-source code　应用软件—源程序
apply the same metaphor　用同样的比喻
approach　方法,手段
approach velocity　逼近速度,接近速度,入口流速
appropriate　恰当的,合适的
approval　批准,认可,审定
approximate　大概的
appurtenance　配件,附件
arbitration　仲裁
arc brazing　电弧钎焊
arc cutting　电弧切割
arc scratch　电弧擦伤
arc stabilizer　稳弧剂
arc surfacing　电弧堆焊
arc voltage feedback controlling arc welding　弧压反馈控制电弧焊
arc welding　电弧焊
arc welding generator　弧焊发电机
arc welding machine using a consumable electrode　熔化极弧焊机
arc welding machine using a non-consumable electrode　不熔化极弧焊机
arc welding power source　弧焊电源
arc welding rectifier　弧焊整流器
arc welding robot　弧焊机器人
arc welding transformer　弧焊变压器
arc welding with consumable　熔化电弧焊
arc welding with nonconsumable　不熔化极电弧焊
archive material　存档材料
arc-lamp　弧光灯
area radiation monitoring　区域辐射监测
area radiation monitoring system　区域辐射监测系统
argon　氩
argon arc welding machine　氩弧焊机
arm　摇杆,臂支架
armature　电枢,转子,衔铁
armored cable　铠装电缆
arrange a deal　达成交易
arrangement drawing　分布图,布置图
array　排列,矩阵
arrester　止挡,制动器,避雷器
artificial load　人工负载,模拟负载
as a gesture of　作为…表示
as already noted　如上所述
as appropriate　根据情况
as built　竣工

as built system　竣工系统
as far as…concerned　就…而言
as noted above　如上所述,正如(上面)所指出的那样
as-procured　按采购的
as regards　关于,至于
as requested　按照要求
as to　关于
ASB super element　ASB 超级单元
asbestos　石棉
as-built drawing　竣工图
as-built summary report　竣工汇总报告
ascension　上升,上到
ash valve　排灰阀
as-installed equipment　完工/竣工设备,已安装设备
ASME safety class 1,2 & 3 valve　ASME 1、2、3 级阀门
aspirating valve　吸(抽)气阀
assemble　装配,组装
assembler　汇编程序
assembly　燃料组件、装配
assembly drawing　组装图
assertive　断言的,肯定的
assessment　考核,评定,估定
assessment of exposure　照射评价
assessment of radiation protection　辐射防护评价
assessor　考核人员
assigner　委派者,指派者
assisted safety valve　带辅助装置的安全阀
associate　伙伴,同人
associate with　把…同…联系起来
associated apparatus　辅助仪器
assortment　种类
as-welded (AW)　焊态
asymmetrical　不对称的
asynchronous motor　异步电动机
at one's discretion　凭某人自行处理
at your earliest convenience　在你方便时,尽早
atmosphere　大气,空气
atmospheric blow down tank　常压排放/排污箱
atmospheric corrosion　大气腐蚀
atmospheric dispersion factor　大气弥散因子
atmospheric Dispersion Values-χ/Q　(放射性微粒的)大气弥散值-χ/Q
atmospheric over voltage　大气过电压
atmospheric pressure　大气压
atmospheric pressure tank　常压箱
atmospheric radiation monitoring apparatus　大气辐射监测仪
atmospheric stability　大气稳定度
atom absorption spectroscopy　原子吸收光谱
atom displacement　原子置换
atomic absorption spectrophotometer　原子吸收分光光度计
atomic hydrogen welding　原子氢焊
atomic hydrogen welding apparatus　原子氢焊机
attach great attention on　非常重视…
attached pump　附属泵
attachment　附属,附着物,附件
attachment lugs on the reactor vessel

head　压力容器顶盖吊耳
attack　腐蚀，侵蚀，攻击
attain，reach　达到
attendance　考勤，出席，到场
attendant　出席人，参加人
attitude　态度，行为
attorney　代理人，代言人
ATWS mitigation system　ATWS 缓解系统
ATWS mitigation system actuation circuitry　ATWS 缓解系统驱动回路
audible alarm　声响报警
audible count rate signal　计数率音响信号
audible local alarm　就地音响报警
audible signals　音响信号
audit　监查
audit follow-up　监查后续行动
audit plan　监察计划
audit record　监查记录
audit report　监查报告
audit team　监查组（队）
audit team leader　监查组（队）长
auditee　受监查方，受审核方
auditing　审计
auditor　监查员
augment　增加，提高，扩大
augmented in-service inspection　扩大的在役检查
austenitic electrode　奥氏体焊条
austenitic stainless steel　奥氏体不锈钢
austenitic stainless steel materials　奥氏体不锈钢材
authority　监管机构，官方，当局
authorization　授权
authorization to proceed　授权开工
authorization，permission，permit，release　认可，批准，许可
authorized approval　权威认证
authorized limit　授权（批准、允许）的限值
auto-drainer　自动排/疏水器
autogenously gas shielded tungsten arc-welding　自动气体保护钨极电弧焊
automatic actuation logic　自动驱动逻辑，自动触发逻辑
automatic analytical balance　全自动分析天平
automatic backwash strainer　自动反冲洗（海水）滤网
automatic battery monitoring system　自动蓄电池监控系统
automatic connection mode　自动连接方式
automatic control mode　自动控制模式
automatic depressurization　自动卸压
automatic depressurization system　自动卸压系统
automatic depressurization system actuation　自动卸压系统触发
automatic depressurization system valve　自动卸压系统阀门
automatic depressurization valve　自动卸压阀
automatic exchanger　自动交换机
automatic fire suppression system　自动灭火系统

automatic oxygen cutting 自动气割
automatic recirculation valve 自动再循环阀
automatic rod control system 自动棒控制系统
automatic rod withdrawal 自动提棒
automatic scalar 自动定标器
automatic selective call 自动选呼
automatic sprinkler system 自动喷淋系统
automatic start 自动启动
automatic surfacing 自动堆焊
automatic synchronization unit 自动同步装置
automatic water suppression system 自动水灭火系统
automatic welding 自动焊
autotitrator 自动滴定仪
autotransformer 自耦变压器
AUX BLDG MCCs 辅助厂房电机控制中心
AUX building power transformers 辅助厂房电力变压器
auxiliaries 辅助设备
auxiliary 辅助的
auxiliary boiler 辅助锅炉
auxiliary boiler de-aerator 辅助锅炉除气器/除氧器
auxiliary boiler feed water pump 辅助锅炉给水泵
auxiliary boiler makeup pump 辅助锅炉补水泵
auxiliary building 辅助厂房
auxiliary building common facility room ventilation 辅助厂房公共设施房间通风(系统)
auxiliary building cooler system 辅助厂房冷却系统
auxiliary building lighting transformers 辅助厂房照明变压器
auxiliary building water supply and drainage system 辅助厂房给排水系统
auxiliary component shielding 辅助设备屏蔽
auxiliary contact 辅助触点
auxiliary feed-water pump 辅助给水泵
auxiliary feed-water pump building ventilation system 辅助给水泵房通风系统
auxiliary feed-water system 辅助给水系统
auxiliary gutter (辅助)小线槽,辅助沟槽
auxiliary hoist 辅助提升机构,辅助吊车
auxiliary operator (AO) 辅助操纵员
auxiliary power supply 辅助厂用电源
auxiliary pressurizer spray line isolation (稳压器)辅助喷雾管隔离
auxiliary pump 辅泵,辅助泵
auxiliary relief valve 辅助卸压阀
auxiliary spray 辅助喷雾(流)
auxiliary spray and letdown purification line isolation 辅助喷雾和下泄净化管线隔离
auxiliary spray flow 辅助喷雾流
auxiliary spray line isolation valve 辅

助喷雾管道隔离阀
auxiliary steam distribution system 辅助蒸汽分配系统
auxiliary steam production system 辅助蒸汽生产系统
auxiliary steam supply system 辅助蒸汽供应系统
auxiliary steam system 辅助蒸汽系统
auxiliary stripping pump 辅助扫仓泵，辅助清仓泵
auxiliary transformer 辅助变压器，厂用变压器
auxiliary valve 辅助(副)阀
availability 供应能力，可用率
available 可得到的，有用的
available head 有效压头
avalanche 雪崩
average core outlet temperature 堆芯出口平均温度
average daily output 平均日供水量，平均日出力
average dust spot efficiencies 平均效率(比色法)，平均灰斑效率
avogadro's number 阿伏加德罗常数
avoid 避免
axial bend test 纵弯试验
axial blanket 轴向再生区，轴向转换区
axial flow pump with blades adjustable when stationary 半可调式轴流泵
axial flux difference 轴向(中子)注量率偏差
axial offset 轴向偏移
axial offset band 轴向偏移带
axial offset control rods 轴向偏移控制棒
axial peaking factor 轴向峰值因子
axial peaking hot spot factor 轴向峰值热点因子
axial power deviation 轴向功率偏离
axial power distribution 轴向功率分布
axial power factor 轴向功率峰因子
axial power imbalance 轴向功率不平衡
axial power shape 轴向功率分布
axial stiffness 轴向刚度
axis guide 轴套
azimuthal power oscillation 方位角功率振荡，象限功率振荡
azimuthal power tilt 象限功率倾斜

B

back flow preventer 回流挡板
back flow prevention check valve 防倒流止回阀
back lighted button 背光按钮
back pressure 背压
back pressure regulating valve 背压调节阀
back seal test 上密封试验
back seat 后座
back welding 封底焊
back-draft damper 逆风止回阀
backfill 回填
backfill soil 回填土
backfire 逆火
backflow prevention device 防倒流装置
background 背景，本底，后台

background survey　本底调查
backing　焊接衬垫
backing welding　打底焊
backseat　(阀门)上密封，回座
backup　后备，后援
backup group heater　(稳压器)备用组电加热器
backup heater　备用加热器
backup manual operator　备用手动操作器
backup power supply　备用电源
backward deformation　反变形
backward inclined blade　后倾式叶轮
bacteria　细菌
bacteriological　细菌学的，细菌学上的
baffle　围板，折流板，缓冲板
baffle assembly　(堆内构件)围板组件
baffle bolts　导流板螺栓，围板螺栓
baffle plate　隔板，挡板
balance　剩余部分
balance of plant (BOP)　核电厂配套设施
balance type　平衡式
balance valve　平衡阀
balanced beam　平衡梁
balanced pressure torch　等压式焊炬
balanced reducing valve　平衡式减压阀
balancing damper　调节阀，平衡阀/风门
ball drip valve　疏水球阀
ball float　浮球
ball float steam trap　浮球式蒸汽疏水器
ball valve　球阀
ballast　镇流器
ballast rheostat　镇定变阻器
ballistic research Laboratory (BRL)　弹道研究实验室
bank　(控制棒)组
bank demand position indication system　控制棒组需求棒位指示系统
bank insertion limit　(控制)棒组插入极限
bar　棒，杆
bar screen　(控制)棒栅
bar type cracking test　圆棒裂纹试验
bare　裸露的，无遮盖的
bare terminal (of an electrode)　焊条夹持端
bare weld filler metal　裸焊缝填充金属，光焊缝填充金属
barium group　钡系元素
barometer　气压计
barrel　吊篮
barrel base plate　吊篮底板
barrel outlet nozzle　吊篮出口接管
barrier　屏障
base　基座，底座
base load　基本负荷，基荷
base load operation　基本负荷运行，基荷运行
base metal　母材
base plate　底板，垫板
baseline design certification code　基准设计认证规范
baseline hammer set　基线锤装置
baseline success criterion　基准成功准则

baseline success sequence 基准成功序列
baseline test and operating data 基准试验和运行数据
basement 地下室,基础
basement foundation 基础底板,筏基
basic covering 碱性药皮
basic electrode 碱性焊条
basic limit 基本限值
basic of design 设计基准,设计依据
basic of purchase 采购依据
basic safety authorization training(BSA) 基本安全授权培训课程
basic slag 碱性渣
basic wind speed 基本风速
basicity 碱度
basin 集水池,(取样)水池
batch/lot 批,一次投料量;一次生产量
batching tank 配料箱,制备箱
battelle type underbead cracking test 巴特尔焊道下裂纹试验
batteries 蓄电池
battery charger 蓄电池充电器
battery monitors 蓄电池监测器
bay 海湾
be authorized to do 被授权,有权
be aware of 了解,知道,觉得
be bound by 受…的约束
be compatible with 与…协调的,相容的,可两立的
be concerned with 涉及,参与,干预
be consistent to(with) 一致的,协调的,相容的,不矛盾的
be corresponding to 相当的,对应的,符合…的
be due to do sth. 定于…做某事
be engaged in 从事
be engaged in a discussion 参与讨论
be entitled to 有…资格,被授予权利
be fully efficient 全面投产
be grateful to 感谢…
be greatly indebted to 非常感激…
be in a position to do sth. 能,有能力做某事
be in accord with 与…一致的,符合…的
be in production 投入生产
be in the position to do 有资格(权利)做决定,能够做
be in this line 从事这个行业
be liable for 对…应负(法律)责任,有义务的
be more likely to 很有可能
be not inclined to 不打算
be not prepared to 不准备
be not ready yet to 还不准备
be of great significance 具有重要意义
be operational 投产,投入使用
be referred to as 叫做,称为,被认为是
be responsible for 负责
be subject to 须经,受支配,从属于
be suitable to 合适的;适宜的,适当的;相当的
be under sb responsibility 由…负责
be unlikely to… 不太可能(做)…
be well under way 正在顺利进行
be well-versed in 熟悉,精通
bead 焊道

bead crack 焊道裂纹
beaker 烧瓶
beam current 电子束电流
beam power 电子束功率
beam power density 电子束功率密度
beam strip model 梁带模型
beam type bending 梁型弯曲
bearing alloy 轴承合金
bearing capacity 承载能力
bearing face 轴承支承面
bearing frame 轴承底座，轴承结构/框架
bearing housing 轴承座，轴承壳
bearing oil pump 轴承(润滑)油泵
bearing pad 轴承垫
bearing sleeve 轴承衬套
bearing swiveling ring 轴承回转(支承)环
bearing water 轴封水
becquerel (Bq) 贝克[勒尔](放射性活度单位)
bedplate 垫板
beforehand, in advance, upfront 事先，预先
beginning of cycle (燃料)循环初
beginning of life 寿期初
behavior 特性，性能，行为
behavioral 关于行为的
bellow seal 波纹管密封
bellow sealed valve 波纹管密封阀
bellows 波纹管
bellows pump 波纹管泵
bellows reducing valve 波纹管式减压阀
bellows seal balance safety valve 波纹管密封平衡式安全阀
bellows seal type 波纹管密封式
bellows valve 波纹管阀
belt-clip portable handset 便携式手持听筒
beltline (压力容器)筒体(尤指堆芯段筒体)
beltline forging 筒身锻件
beltline region (压力容器)筒身区
benchmark 基准，标杆
benchmark value 基准值
benchmarking 取经，对标
bend 弯头
bending strength 抗弯强度
bending stress 弯曲应力
bent frame 排架
berth 泊位
best estimate 最佳估算值
best estimate analyzer for core 堆芯最佳评估分析机
best estimate flow 最佳估算流量
bevel cutting 坡口切割
beveling 倒角，倒棱，坡口切
beyond design basis accident 超设计基准事故
beyond one's expectation 超出某人的预期
BFW pump 锅炉给水泵
bias 偏见，偏差，斜线
bias electrode 偏压电极
bid award 确定中标单位，发标
bid evaluation 评标
bidder 投标方

bidding　电厂招投标
bilateral tension stress　双向拉应力
bilge pump　仓底泵
billet　钢坯
bills of quantities　工程量清单
bimetal elements steam trap　双金属片式疏水阀
binder　粘接剂
bio-accumulation factor　生物积累因子
biocide　杀菌剂
biological half-life　生物半排期
biological shielding　生物屏蔽
bistable (module)　双稳态组件
bit　位
bite (control rods)　(控制棒)卡棒,咬住
bite position　(控制棒)卡棒位置
black & white camera with housing　带护套的黑白照相机
black & white monitor　黑白监视器
black RCCA　黑棒束控制组件
blackout　全厂断电
blade　刀刃,刀片
blank　毛坯,坯料
blanket　再生区,转换区
blast　爆破,毁灭,使枯萎,损害
blast gate damper　爆破闸阀
blender　混合器,混料机
blind flange　盲法兰,无孔法兰
block　块,滑轮(组),阻断,隔离
block diagram　方块图
block sequence welding　分段多层焊
block valve　切断阀,截止阀
blocking　垫块
blow pipe　焊炬
blow-down　排污,排放,放气
blow-down of a safety valve　安全阀排放
blow-down phase　喷放阶段
blow-down piping penetration　排污管贯穿件
blow-down pump　排污泵
blow-down rate　排污率
blow-down regenerative heat exchangers　排污再生式热交换器
blow-down valve　排污阀,排放阀
blower　风机,风箱
blowhole　气孔
blowout panel　排污控制屏
board　董事会
board in-word dialing (BID)　经话务台接入
body　阀体,物体
body stem seal　本体阀杆密封
body-bonnet connection　阀体与阀盖的连接
boiler　锅炉
boiler feed pump　锅炉给水泵
boiler return circulating pump　锅炉回水循环泵
boiling heat transfer　沸腾传热
bolt, bolting　螺栓
bolt tightener　螺栓紧定棘轮扳手
bolting material　螺栓件材料
bolting studs　柱头螺栓
bombardment　轰击
bond area　熔合区
bond coating　结合层

bonded flux 粘接焊剂
bonnet bush 阀盖衬套,阀帽螺纹
bonnet 阀盖,阀帽
bonnet gasket 阀盖密封垫片,阀盖密封衬垫
booster feedwater pump 给水增压泵
booster pump 增压泵,升压泵
bootstrap 引导,辅助程序
bootstrap loader 引导装入材料,引导装载程序
BOP 380 V AC power system BOP 380 V 交流电系统
BOP building radiation monitoring system BOP 厂房辐射监测系统
BOP communication system BOP 通讯系统
BOP lighting system BOP 照明系统
borated glass 硼玻璃
borated makeup 含硼补水,硼化补水
borated water 含硼水
boration 硼化
boration flow rate 硼化流量
bore scope with recorder 带记录仪的光学孔径探测仪
borehole shaft driven(centrifugal)pump 长轴深井(离心)泵
boric acid batching tank 硼酸制备箱
boric acid blender 硼酸混合器
boric acid concentration 硼酸浓度
boric acid filter 硼酸过滤器
boric acid mixing tank 硼酸混合箱
boric acid pump 硼酸泵
boric acid storage tank 硼酸储存箱
boric acid surge tank 硼酸波动箱
boric acid tank 硼酸箱
boric acid transfer pump 硼酸输送泵
boron coefficient 硼系数
boron concentration 硼浓度
boron concentration equalization 硼浓度平衡
boron concentration monitoring system 硼浓度监测系统
boron content 硼含量
boron dilution 硼稀释
boron dilution block 硼稀释隔离
boron dilution event 硼稀释事件
boron endpoint (燃料循环末期)末端硼浓度
boron equivalent 硼当量
boron lined counter 涂硼计数管
boron meter 硼浓度计,硼表
boron precipitation 硼沉淀,硼结晶
boron recycle system 硼回收系统
boron removal 去硼,除硼
boron worth 硼价值
borosilicate glass 硼硅酸盐玻璃
bottom grid 底部格架
bottom head 下封头
bottom head insulation 底封头保温层,底封头隔热层
bottom inlet pump 底吸泵
bottom line 底线
bottom nozzle 下管座
bottom plate 底板
bottom shell 下部筒体
bottom suction pump 底吸泵
bottom valve 底阀
bottom-end spacer (可燃毒物棒)

支撑管
bottoms pump　底料泵，塔底液泵
bounding analysis　边界分析，包络分析
bounding analysis curve (BAC)　包络分析曲线
bounding case　包络情况
bounding leak-before-break analysis　LBB 界值分析，LBB 包络分析
bow　弓形，弯曲
box foundation　箱形基础
box girder　箱形梁
brace metal　钎缝金属
braced joint　钎焊接头
bracket　牛腿，支撑架
braid　编织物
braided electrode　编织焊条
brake horsepower　制动马力
brake valve　制动阀
braking system　制动系统
branch connection　分支连接
branch line　支管
branch system　支状管网系统
branch T saddle joint　T 形管接头
brazability　钎焊性
braze cream　钎料膏
braze paste　钎料膏
braze welding　钎焊
brazed rate　钎着率
brazer　钎焊
braze-welding　钎接焊
brazing　钎焊
brazing (soldering) blowpipe　钎炬
brazing (soldering) process　钎焊过程
brazing filler metal　硬钎料
brazing filler metal erosion　钎料流失，钎料熔蚀
brazing flux　钎剂，焊剂
brazing in controlled atmosphere　保护气氛钎焊
brazing lamp　喷灯
brazing process variables　钎焊参数
brazing retort　钎焊盒
brazing seam　钎缝
brazing temperature　钎焊温度
brazing time　钎焊时间
breach　违反，破坏
breach of contract　违约
breach party　违约方
break　破裂，断裂，突变
break exclusion zone　破口排除区，破损排除区
break size　破口尺寸
breaker　断路器
breaking capacity　遮断容量，切断容量
breaking strength　断裂强度
break-point　断点
breakthrough　突破
breather　通气孔，呼吸阀
breather pipe　通气管
breather valve　通气阀
breathing air purifier　呼吸空气净化器
breathing apparatus　氧气呼吸器，吸氧器
bridge (over-head) crane　桥式起重机，桥式吊车
bridge span　桥架跨距
brine　浓盐水

bring about　引起
bring the price down　降价
brittle　易破裂的,易碎的,脆性的
brittle fracture　脆性断裂
brittle temperature range　脆性温度区间
broadcasting control panel　广播控制盘
broaden,enlarge　加宽,放阔,使扩大
broadly,widely,extensively　广泛地
bronze fitted pump　衬青铜泵
brush　电刷,刷子
bucket　吊桶,桶状物,铲斗
bucket float　浮桶
buckling　曲率,屈曲,拱设
building boundary　厂房边界
building lighting fixtures　厂房照明装置
building power transformer　厂房电源变压器
building site pump　工程用泵
buildup　局部堆焊
build-up factor　积累因子
buildup of activity　放射性的累积
build-up sequence　(堆焊或多层焊)熔敷顺序
buildup welding　堆焊
built-in　固有的,内置的,内含的
built-in reactivity　(初始)后背反应性
built-up member　组装件,装配件
bulk coolant temperature　冷却剂整体温度
bulk density　容积密度,体密度
bulk-boiling　整体沸腾
bulkhead mounted pump　舱壁泵
bulletin　通告,公告,报告
bumper　缓冲器
bundled 3rd party software　连带的第三方软件
buoyancy　浮力
buoyant flow　浮力流
buoyant force　浮力
burden　焊剂层
burial　掩埋,埋藏(放射性废物)
buried cable　地下电缆
burn through　烧穿
burnable absorber　可燃毒物吸收体
burnable absorber material　可燃毒物材料
burnable absorber rod　可燃毒物棒
burnable poison　可燃毒物
burnable poison rod　可燃毒物棒
burnable poison rod assembly handling tool　可燃毒物组件操作工具
burnout　烧毁,熄火
burnout correlation　烧毁关系式
burnout ratio　烧毁比
burn-up　燃耗
burring　去毛刺
burst　突然爆发,猝增
burst pressure　冲破压力
bus　总线,母线
bus ground fault detection　母线接地故障监测
bushing　套管,衬套
business intelligence　商业智能
business process reengineering (BPR)　商业重组过程

bus-way 母线槽,母线通道
butt and lap joint 搭接对接接头
butt weld 对接焊缝
butt welding 对焊
butterfly swing check valve 旋启式蝶式止回阀
butterfly type non-slam check valve 蝶式缓冲止回阀
butterfly valve 蝶阀,瓣阀
butterfly valve with gear actuator 机械驱动蝶阀
buttering 隔离层堆焊,预堆边焊
buttering parts 打底焊件
buttress 扶壁,水管支墩
buttwelding valve 对焊连接阀
buyer 需方,买方
buzzer 蜂鸣器
BWRA cracking test for austenitie steel BWRA奥氏体钢裂纹试验
by and large 从全体看来
by virtue of 凭借,因为,由于
bypass 旁通,旁路
bypass feature 旁路特性
bypass flow 旁通流量,旁流
bypass logic 旁通逻辑
by-pass region 旁流区
bypass valve 旁路阀,旁通阀
bypassed and inoperable status indication application 旁路和不可运行状态显示的应用
bypassed and operable status 旁通及可运行状态
byte 字节

C

C&F(cost and freight) 到岸价(成本+运费)
C/C 每隔,中心矩
cab (起重机)驾驶室
cabinet 设备柜
cable box 电缆箱,电缆盒
cable connector 接线盒,电缆接头盒
cable end fitting 电缆终端配件
cable fill ratio 电缆填充率
cable flat/cable layer 电缆层
cable gallery 电缆廊道
cable list 电缆清册
cable pulling vault 电缆井
cable spreading 电缆敷设
cable supply 电缆供电,电缆供应
cable termination 电缆头
cable tray 电缆托架,电缆桥架
cable tray fill 电缆桥架填充
cable tray support 电缆桥架支吊架
cache memory (高速)缓冲存储器
calculated peak pressure 计算峰值压力
calculation uncertainty 计算不确定性
calculator module 计算机组件
calendar plan 日历计划
calibration 标定
calibration of instrumentation 仪表标定/校准
calorimeter 量热计
calorimetric power measurement system 量热式功率测量系统
cam 凸轮
cam-operated limit switch 凸轮式限位

开关
cancel an order 取消订单
cancellation 取消
candidate high-energy lines 待选高能管道
candling 蜡烛状熔化
canned (reactor coolant) pump （反应堆冷却剂）屏蔽泵
canned high-inertia reactor coolant pump （反应堆冷却剂）大惯量屏蔽泵
canned motor pump 屏蔽电机泵
canned motor reactor coolant pump 反应堆冷却剂屏蔽泵
canned-motor RCP 屏蔽电机式反应堆冷却泵
canopy seal 卡努比密封
cans of anti-oxidation grease 听装抗氧化油脂
capable fault （地质）能动断层
capacitor 电容器
capacity （起重机）起重量，容量
capped end 加帽端
capsule 辐照监测管，胶囊
capsules holder （辐照监督）管固定架
capture cross section 俘获截面
capture gamma ray 俘获 γ 射线
carbon 碳
carbon arc air gouging 碳弧气割
carbon arc cutting 碳弧切割
carbon arc welding 碳弧焊
carbon dioxide electric vaporizer 二氧化碳（干冰）电加热气化器
carbon dioxide skid 二氧化碳（干冰）成套装置
carbon dioxide suppression system 二氧化碳抑制系统
carbon equivalent 碳当量
carbon-dioxide arc welding 二氧化碳气体保护电弧焊
carburization 渗碳处理，碳化
carburizing 渗碳处理，碳化
carburizing flame 碳化焰
card 插件，卡片
cardinal point 基点，坐标原点
carriage return 回车
carry on 从事
carry over （汽）带水，带水率，夹带
carry under （水）带汽
carrying rail 行走轨道
carryover moisture （带水）湿蒸汽
cartridge(filter) （过滤器）滤芯
cartridge fuse 熔丝管
cascade 级联，串级
case 箱，盒，柜（仪表），案例
case-by-case 逐个案例
cash 现金
casing adapter （主泵）吸水导向管
cask 乏燃料组件运输容器
cask handling machine(crane) 乏燃料运输容器吊车
cask loading pit(CLP) 乏燃料运输容器装料池，装罐井
cask shipping rail and rail car 乏燃料运输容器运输轨道和轨道车
cask trunnion 乏燃料运输容器吊耳
cask wash down pit (CWP) 乏燃料运输容器清洗池

cask wash down pit drain isolation　乏燃料运输容器清洗井排水隔离
cask wash down pit level instrument　运输容器冲洗池液位仪表
cask wash down pit level instrument isolation　乏燃料运输容器清洗池液位仪表隔离
cassette tape　盒带,盒式磁带
cast　铸造,塑造
cast austenitic stainless steel　铸造奥氏体不锈钢
cast iron　铸铁
cast iron electrode with steel core　钢芯铸铁焊条
cast nodular　球墨铸铁
casting part　铸件
cast-in-place　现场浇铸
cast-in-site concrete　现场浇灌混凝土
casual　偶然的,碰巧的;临时的
catalytic hydrogen recombiner　催化氢复合器
catalytic oxygen removal system (CORS)　催化除氧系统
catalytic oxygen removal system (CORS) package　催化除氧系统成套设备
catch basin　雨水井,集水井
catchment area　汇水区域
category　种类,类别,范畴
cathode ray tube　阴极射线管
cathodic protection　阴极保护
cathodic protection system　阴极保护系统
cation　阳离子
cation bed　阳离子床
cation bed demineralizer　阳离子交换床,阳离子除盐器
cation conductivity　阳离子电导率
cavitation　气蚀,气穴现象
cavity　堆腔,腔,洞,孔,空穴
cavity liner seal ring　堆腔衬里密封环
ceiling grating　天花板网格,吊顶隔栅
ceiling light　天花板照明,顶灯
ceiling speaker　装在天花板上的扬声器
cell　电池,小室
cell battery　单元电池,盒式电池
cellar drainage pump　地下室排水泵
cement solidification　水泥固化
center of gravity　重心
central equipment administration system　主设备管理系统
central processing unit(CPU)　中央处理器
centralized control　集中控制
centralized sampling　集中取样
centrifugal　离心(式)的
centrifugal canned-motor pump　离心屏蔽泵
centrifugal charging pump　离心式上充泵
centrifugal fan　离心式风机
centrifugal pump with semienclosed impeller　半开式叶轮离心泵
centrifugal pump with shrouded impeller　闭式叶轮离心泵
ceramic pellet　陶瓷芯块
ceramic spraying　陶瓷喷涂

cerium group 铈系元素
certificate 证书,发证
certificate of approval 批准证书
certificate of compliance 合格证
certification 毕业
certified capacity 额定排量,确认容量
certified design 经确认的设计
certify, attest to, demonstrate, prove 证明
challenge 威胁,挑战
chamber 腔,小室
chamber pressure 燃烧室压力
chamfer 倒角,修切边缘
change control 变更控制
change notice 变更通知
change out 替换,代替
change request 变更申请
change request form 变更申请表
change review board 变更审查委员会
changeover 转向,转接
changeover valve 切换阀,多向阀
changing fixtures 控制棒与阻力塞等的抽插机
channel 通道
channel deviation 通道偏差,通道偏离
channel flow area 通道流通面积
channel head 蒸汽发生器下封头
channel integral model 通道积分模型
channel operational test 通道运行试验
channel spacing 通道间隔,通道格架
channeling wall 通道壁
character 字符,特性点
character printer 字符打印机
character string 字符串
characteristics curve 特性曲线
characterize 表现…的特征,刻画…的性格
charcoal absorber 活性炭吸附器
charcoal deep bed filter vessel 活性炭深床过滤器壳体
charcoal filter 活性炭过滤器
charcoal guard and delay bed 活性炭保护和延迟床
charge ionizing particle 带电电离粒子
charge pump 给料泵,上充泵
charge-coupled device 电荷耦合器件
charged particle 带电粒子
charger 充电装置
charging path 上充通道
charging pump 上充泵
Charpy upper-shelf energy 夏氏上平台能量
Charpy V test 夏氏 V 型试验
Charpy V notch 夏氏 V 型缺口
Charpy V notch test 夏氏 V 型缺口试验
charter a ship 租船
chatter 频跳
check 阻止,制止,控制,校验
check source 校验用放射源
check valve 逆止阀,止回阀,单向阀
check-hole 检查口,检查孔
check-list 校核清单
chemical addition pot 化学物添加箱
chemical addition tank drain isolation valve 化学物添加箱排水隔离阀
chemical addition tank fill isolation valve 化学物添加箱进料隔离阀

chemical addition tank vent isolation valve　化学物添加箱排气隔离阀
chemical additive　化学添加物
chemical analysis　化学分析
chemical and volume control　化学和容积控制
chemical and volume control system isolation　化学和容积控制系统隔离
chemical and volume control system letdown isolation　化学和容积控制系统下泄隔离
chemical and volume control system makeup subsystem　化学和容积控制系统补给子系统
chemical cleaning　化学清洗
chemical interaction　化学相互作用
chemical mixing tank　化学混合箱
chemical oxygen demand　化学需氧量
chemical poison　化学毒物
chemical reagent injection　化学试剂注入
chemical seal　化学密封
chemical shim　(反应性)化学补偿
chemical shim reactivity control　化学补偿反应性控制
chemical treatment tank　加药箱
chemical waste tank　化学废液箱
chemistry element　化学元素
chemistry laboratory and its drainage system　化学实验室及其疏水系统
chequered plate　花纹(格)钢板
chest　箱,柜
chevron cracking　锯齿形裂纹
chevron plate　波纹板
chilled　冷冻的
chilled unit　冷冻机组
chilled water　冷冻水
chilled water circulation pump　冷冻水循环泵
chiller　冷冻机,制冷机
chipping　碎片
chipping hammer　敲渣锤
chloride　氯化物
chloride concentration　氯离子浓度
chloride content　氯离子含量
chlorinator　氯化装置,加氯器
chlorine　氯
chocked flow　壅塞流
choke valve　节流阀
chokeless pump　无堵塞泵
chop　截断,切割
chopped cosine profile distribution　截余弦分布
chromatograph　色层分离,色谱
chrome　铬,铬合金
chroming　镀铬
chromium　铬
chromium depletion　贫铬
chromium equivalent　铬当量
chronological　按时间顺序的
CIF(cost,insurance and freight)　到岸价(成本+保险+运费)
circuit breaker　(电路)断路器
circuit breaker test cabinet with test coupler　带连接器的电路断路器试验柜
circuit diagram　电路图
circuiting cooler　循环冷风机

circuiting cooling water energy dispersion system　循环水(排放口)消能系统

circuiting cooling water pumping station system　循环(冷却)水泵站系统

circuiting cooling water treatment system　循环(冷却)水处理系统

circuiting water pump　循环水泵

circular　圆周的，环形的

circular magnetization　(磁粉检验)周向磁化

circular railway　环形轨道

circular runway rail　环行轨道

circular-groove cracking test　环形槽裂纹试验

circular-path cracking test　圆形镶块裂纹试验

circulate　流转，流行，传播

circulating capital　流动资金

circulating cooler　循环冷风机

circulating cooling water　循环冷却水

circulating cooling water drainage system　循环(冷却)水排放系统

circulating cooling water intake system　循环(冷却)水吸入系统

circulating cooling water yard pipe network system　循环(冷却)水厂区管网系统

circulating water pump　循环水泵

circulating water trash screen　循环水(拦污)滤网

circulation ratio　循环倍率

circumferential　环形的，周向的

circumferential ridge　周向环脊

circumferential weld　环焊

civil works　土建工程

clad creep　包壳蠕变

clad flattening　包壳压塌，包壳压扁

clad pellet gap　包壳与芯块之间的间隙

clad pressure boundary　(燃料棒)包壳压力边界

clad rupture　包壳破裂

cladding　包壳，包覆

cladding collapse　包壳坍塌

cladding defect　包壳破损，包壳失效

cladding part　堆焊件

cladding strain　包壳应变

claim　索赔，要求，请求

clamp valve　对夹式阀门

clamps　夹，夹具，夹紧

clarification　澄清，净化

clarifier　澄清槽

clarity　清晰

clarify misunderstanding　消除误解

class 1E　1E 级

class 1E DC　1E 级直流电

class 1E DC and uninterruptible power supply　1E 级直流不间断电源系统

class 1E DC switchboard voltage　1E 级直流配电盘电压

class 1E division　1E 级序列

class 1E fused transfer switch box　1E 级装有保险丝的转换开关箱

class 1E radiation monitor　1E(安全)级辐射监测仪

classification of records　记录分类

classroom　教室

classroom training 课堂培训
clause 条款
clay 黏土,泥土
cleaning 清洗
cleaning and flushing 清洁和冲洗
cleanliness 清洁
cleanliness program 清洁程序
clean-out 清扫
clear up,clarify,make sure,identify 澄清,弄清
clear well 澄清池,澄清井
clearance 间隙,清除
clearance between focal point and (plate) surface 离焦量
clearance fillability test 填缝性试验
clear-water reservation 清水池
cleavage fracture 解理断口
cleave insert (堆内构件)镶块,楔块
clevis U形夹,U形接头
clevis insert U形镶块,U形楔块
climatology 气候学
clock system 时钟系统
clockwise 顺时针方向
clogging 堵塞
close coupled pump 单体泵
close this deal 做成这笔生意
closed circuit TV system 闭路电视系统
closed cycle cooling water system 闭式循环冷却水系统
closed peripheral pump 闭式旋涡泵
closed sprinkler 闭式喷头
closed system 封闭系统
closed-circulating cooling water 闭路循环冷却水
closed-loop mode 闭环模式
closing (circuit breaker) (断路器)合闸
closing damper 密闭阀,关闭风门
closing valve temperature 关阀温度
closure (压力容器顶盖)封闭
closure bolting 压力容器顶盖螺栓
closure flange 压力容器顶盖法兰
closure head 压力容器顶盖
closure nut 主螺母,顶盖螺母
closure stud 主螺栓
cloud point 始凝点
clutch 离合器
CMT discharge isolation valve 堆芯补水箱出口隔离阀
CMT fill isolation 堆芯补水箱充水隔离
CMT inlet isolation valve 堆芯补水箱进口隔离阀
CO_2 arc welding 二氧化碳气体保护电弧焊
CO_2 arc welding machine 二氧化碳弧焊机
CO_2 arc welding with thick wire 粗丝二氧化碳亚弧焊
CO_2 arc welding with thin wire 细丝二氧化碳亚弧焊
coagulant 凝聚剂
coarse grained region 粗晶区
coarse sand 粗砂
coast down 惰转
coast down flow 惰转流量
coated electrode 药皮焊条,涂层焊条

coating applicator　涂层施工员
coating debris　涂层碎片
coating flux　涂料
coating material　涂层材料，包覆材料
coaxial cable　同轴电缆
cobalt　钴
cobalt-base alloy　钴基合金
cobalt-base alloy surfacing electrode　钴基合金堆焊焊条
cock　旋塞阀
code　规则，法规
code and standards　规范和标准
code case　规范案例
code controller　编码控制器
Code of Federal Regulation　联邦管理法规
code plate　(ASME)规范钢印/铭牌
coefficient of discharge　排量系数
coefficient of heat flow concentration　热流集中系数
coefficient of thermal expansion　热膨胀系数
coil　线圈
coil housing　磁轭
coil stack assembly　磁轭线圈部件
coincide with　与…一致，相合，符合，相符，相巧合
coincidence logic　符合逻辑
cold condensate capacity　水冷凝能力
cold conditions　冷态
cold crack　冷裂纹
cold cracking susceptibity coefficient　冷裂敏感系数
cold differential test pressure　(安全阀)冷态试验差压
cold formed　冷成形的
cold junction box　(热电偶)冷端补偿箱
cold leg　(管道)冷段
cold leg balance line　冷段平衡管线
cold leg bend instrument root　冷段弯管处仪表根部
cold leg insulation package　冷段保温包
cold leg nozzle　冷段接管
cold overpressure　冷态超压
cold performance test　冷态性能试验
cold pressing　冷态加压，冷态升压
cold shutdown　冷停堆
cold soldered joint　虚钎，虚焊
cold startup　冷态启动
cold test　冷态试验
cold void volume　冷空腔体积
cold wall type vacuum brazing　冷壁真空钎焊炉
cold work　冷加工，冷作
cold-wall effect　冷壁效应
cold-worked　冷加工的，冷作的
collaboration　协作，合作
collapse　压扁
collapse level　坍塌水位
collapsed liquid level　坍塌液位，净液位
collar　焊筋
collection　托收
collection duct　进风管
collection trough　收集槽
collective dose　累积剂量

collective dose equivalent　集体剂量当量
colophony flux　松香钎剂
color code　色标,色码,彩色代码
color graphic display　彩色图形显示器
color laserjet　彩色激光打印机
column　柱,圆柱
combination fire/smoke dampers　防火/防烟组合风门
combination of type test and analyses　型式试验和分析的组合
combination valve　组合阀
combined cutting and welding torch　焊割两用炬
combined frequency　组合频率
combined license　综合执照取证,一步法执照申请
combined license(COL) applicant　综合执照申请者,一步法执照申请者
combined license holder　综合执照持有人,一步法执照持有者
combined license information　综合执照信息,一步法取照信息
combined normal and seismic design basis loads　正常与地震情况下合并的设计基准载荷
combined operating license applicant　综合运行许可证申请者,一步法运行执照申请者
combined plasma arc　联合型等离子弧
combined stress　复合应力
combined total leak rate　汇总泄漏率
combustible air intake and engine exhaust subsystem　助燃空气入口和发动机排气子系统
combustible gas　可燃气体
combustible material　可燃物
combustion　燃烧
combustion intensity　燃烧强度
combustion rate　燃烧速度,燃烧率
come into effect,come into force　被实施,生效
come into operation　投入运行
come quite a long way　进展不小
come to a compromise　达成妥协
come to an agreement　达成协议
come to terms with　与…达成协议
come up　的确出现,发生
come up with,put forward,bring up　提出
command　命令
commencement of lift　开始起吊
commencement of plant commercial operation　电厂开始投入商业运行
comment　评语,评论
commercial operation　商业运行
commission　佣金
commissioning　调试,试运行
commissioning program(CP)　调试大纲
commit to　承诺
commitment　许诺,诺言
committed dose equivalent　待积剂量当量
common base plate　公共底板(座)
common cause failure　共因失效,共因/故障
common ground　共同立场,一致点

common mode failure　共模故障，共模/失效

common point　公共点

common primary sample line penetration　一回路取样管共用贯穿件

common Qualified Platform　Common Q 安全级仪控平台

common spray line　喷雾母管

common switch board　公共配电盘

commonplace　平常事情

communication　交流，联络，通信

communication/CCTV　通信/闭路电视

communication link　通信连接

communication station　通信站

communication subsystem　通信子系统

communication system　通信系统

commuting allowance　交通补助

compact tension fracture mechanics test　紧凑拉伸断裂力学试验

company standard　企业标准

compartment　隔间，小室

compartment isolation　舱室隔离

compatibility　兼容性

compensated ionization chamber　补偿电离室

compensation　补偿

compensation cable　补偿电缆

competence　工作能力

competitive　有竞争性的

compiler　编程者，编译器

complement，replenish　补充

complete logbook　全日志

completion certificate　竣工证书

completion of erection release　安装完工证书

compliance　一致性

compliance with requirements　符合要求

complication　麻烦情况

comply with　符合，与…一致，依从，照办

comply with，meet　满足

component　部件，设备，成分，分量

component cooling heat exchanger bolt tightener　设备冷却水热交换器螺栓紧定棘轮扳手

component cooling water　设备冷却水

component cooling water heat exchanger　设备冷却水热交换器

component cooling water pump　设备冷却水泵

component cooling water surge tank　设备冷却水波动箱

component-level actuation　设备级触发，设备水位触发

composite　合成的，复合的，合成物

composite structure　混合结构，组合结构

composure　冷静，沉着

compound　混合物，复合物，化合物

comprehensive antenna system　全向天线系统

comprehensive vibration assessment　综合振动评定

comprehensive，overall　全面的

compressed　压缩的

compressed air dryer skid　压缩空气干

燥装置
compressed air system　压缩空气系统
compressible flow　可压缩流
compressive strength　抗压强度
compressor　压缩机
compromise　妥协，调和，和解
computer aided design for welding structure　计算机辅助焊接结构设计
computer aided welding process programming　计算机辅助焊接工艺设计
computer network　计算机网络
computer with software for iltr test　带 iltr 试验软件的计算机
computer-based learning (CBL)　计算机化学习
computerized procedure system　计算机化的规程系统
computerized procedure system engineer　计算机化程序系统工程师
concave　凹面
conceal　隐藏
concede　让步，允许，承认
concentrate storage tank　浓缩液储存箱
concentration　集中，专心，浓缩，浓度
concentrations load　集中荷载
concentrations of radioactive material　放射性物质浓度
concentric reducer　同心大小头，同心管接头
conceptual design　方案设计，概念设计
concession　让步，妥协，特许
conciliation　调解
conciseness　简明
conclude　(条约等的)缔结，议定
conclude a contract　缔结合同
conclude the business　达成交易
concrete compressive strength　混凝土抗压强度
concrete containing fly ash　粉煤灰混凝土
concrete structure　混凝土结构
concrete-filled structural modules　填充混凝土的结构模块
concreteness　具体
condensate　凝结水
condensate and feedwater system　凝结水及给水系统
condensate cooler　凝结水冷却器
condensate pot　凝结水罐，冷凝液罐
condensate pump　凝结水泵
condensate storage tank　凝结水箱，冷凝液储存箱
condensate transfer pump　凝结水输运泵
condenser　(汽轮机)凝汽器，冷凝器
condenser hotwell　凝汽器热阱
condenser steam dump valve　凝汽器蒸汽排放阀
condenser vacuum breaker　凝汽器真空破坏器
condenser vacuum pump　凝汽器真空泵
condenser vacuum system　凝汽器抽真空系统
condition　条件，状况，工况

condition adverse to quality 不利于质量的条件/状况
conducing wire 引出线
conductance 导热
conduction pump 传导泵,直流电磁泵
conductive integral 积分热导率
conductivity meter (benchtop) 电导计(台式)
conductivity meter (portable) 电导计(便携式)
conductor 导体,导线
conduit 电缆管,电线套,管道
cone 锥形物,圆锥体
confidence 心,自信心,确信
confidence level 置信度
confidential 保密的
configuration 配置,构造,形状,外貌,轮廓
confined point 约束点
confirm 证实,证明(陈述、主张等)有根据
confirm an order 确认订单
confirmation,acknowledgement 确认
confirmatory 确认的,证实的
confirmatory open signal 确认开信号
conform to 与…一致
conformatory test 验证试验
conformity 合格,符合
conical 锥形的
conical gear actuator 锥齿轮传动装置
conical head 锥形封头
conical shell 锥形筒体
conical valve 锥形阀,翼形阀
connection (隔离阀试验)接口,连接
connection box 连接盒,接线盒(箱)
connection diagram 接线图
connection dimension 连接尺寸
connection isolation 接口隔离
connection point 连接点,结点
connector 接头
connector plate 连接板
consent 同意
consequence of accident 事故后果
consequently 所以,因此
conservation 守衡
conservation of energy 能量守衡
conservation of mass 质量守恒
conservation of momentum 动量守衡
conservative 保守的
considerable 明显的,值得考虑的
consideration 体贴,考虑,报酬
consignee 收货人
consigner,consignor 发货人
consignment 交运货物
consignment note 发货通知书
console 控制台
consolidate 巩固
constant axial offset control (CAOC) 恒定/常轴向偏移控制
constant current power supply 恒流电源
constant pressure pump 定压泵
constant wire-feed system 等速送丝方式
constant-current arc welding power source 恒流弧焊电源
constantly 不断地,经常地
constant-voltage arc welding power

source 恒压弧焊电源
constituted component 组装件
constraint stress 约束应力
constricting nozzle 压缩喷嘴
construction 施工,建造
construction activity 施工,建造活动
construction and installation test 建造和安装试验,建安试验
construction funds 建设资金
construction inspection 施工检查
construction manager 工程经理,施工经理
construction permit 建造许可证
construction quality 施工质量
construction testing 构造试验、结构试验
construction tolerance 施工允许偏差
consult with 商量
consumable electrode 可熔电极
consumable insert 可熔性嵌条
consumable nozzle 熔嘴
consumable wire 熔丝
cont. recirc. isolation valve 连续循环隔离阀
contact conductance 接触导热
contact welding 接触焊
contactless gear pump 非接触齿轮泵
contactor 接触器
contact-reaction brazing 接触反应钎焊
contained source 内装源
containment 安全壳
containment air baffle 安全壳空气导流板
containment air cleanup system 安全壳空气净化系统
containment air filtration exhaust 安全壳空气过滤排风
containment air filtration isolation 安全壳空气过滤隔离
containment air filtration system 安全壳空气过滤系统
containment air filtration system isolation 安全壳空气过滤系统隔离
containment air supply containment isolation IRC 安全壳送风(系统)壳内隔离
containment air supply debris screen 安全壳送风滤网
containment area high-range radiation level 安全壳区域高放射性水平
containment atmosphere 安全壳大气,安全壳气体
containment atmosphere radiation level 安全壳气体辐射水平
containment atmosphere radiation monitor skid 安全壳气体辐射监测装置
containment atmosphere sample line penetration 安全壳气体取样管贯穿件
containment back pressure 安全壳背压
containment building 安全壳厂房
containment bypass 安全壳旁通
containment chilled water supply penetration 安全壳冷冻水供水管贯穿件
containment cooling 安全壳冷却

containment demineralized water supply penetration 安全壳除盐水供水管贯穿件

containment dome 安全壳穹顶

containment effectiveness 安全壳有效性

containment electrical penetration assembly 安全壳电气贯穿件

containment exhaust duct penetration 安全壳排风贯穿件

containment failure early (CFE) 安全壳早期失效

containment failure intermediate (CFI) 安全壳中期失效

containment failure late(CFL) 安全壳晚期失效

containment flood up conditions 安全壳淹没工况

containment flood up level 安全壳淹没高度

containment free volume 安全壳自由容积

containment hydrogen concentration instruments 安全壳氢气浓度仪表

containment hydrogen control system 安全壳氢气控制系统

containment hydrogen igniter 安全壳氢点火器

containment hydrogen mixing system 安全壳氢气混合系统

containment hydrogen recombiner system 安全壳氢复合系统

containment hydrogen ventilation system 安全壳氢气排风系统

containment instrument air inlet penetration 安全壳仪表空气入口贯穿件

containment instrumentation system 安全壳仪表系统

containment integrated leak rate testing 安全壳整体泄漏率试验

containment integrity 安全壳完整性

containment internal structures 安全壳内部结构

containment isolation 安全壳隔离

containment isolation failure 安全壳隔离失效

containment isolation monitoring application software 安全壳隔离监测应用软件

containment isolation signal 安全壳隔离信号

containment isolation system 安全壳隔离系统

containment isolation test 安全壳隔离试验

containment isolation test connection isolation 安全壳隔离试验接口隔离

containment isolation valve 安全壳隔离阀

containment isolation valve test connection 安全壳隔离阀试验接口

containment leak rate test system 安全壳泄漏率试验系统

containment lighting XFMRs 安全壳照明变压器

containment MCCs 安全壳电机控制中心

containment mixing　安全壳气体混合
containment penetration　安全壳贯穿件
containment penetration test connection　安全壳贯穿件试验接口
containment piping penetrations with flued heads　带管接头的安全壳管道贯穿件
containment polar crane　安全壳环吊
containment power transformers　安全壳电力变压器
containment pressure　安全壳压力
containment pressure instruments　安全壳压力仪表
containment purge　安全壳吹扫
containment purge discharge containment isolation valve　安全壳吹扫排气管安全壳隔离阀
containment purge exhaust air　安全壳吹扫排气
containment purge supply containment isolation valve　安全壳吹扫进气管安全壳隔离阀
containment purge ventilation system　安全壳吹扫通风系统
containment reactor coolant drain system　安全壳反应堆冷却剂疏排系统
containment recirculation　安全壳再循环
containment recirculation cooling system　安全壳再循环冷却系统
containment recirculation fan coil units　安全壳再循环风机盘管机组
containment recirculation isolation　安全壳再循环隔离
containment recirculation isolation valve　安全壳再循环隔离阀
containment recirculation screen　安全壳再循环滤网
containment recirculation subsystem　安全壳再循环子系统
containment return header penetration　安全壳返回集管贯穿件
containment service air inlet penetration　安全壳厂用压缩空气入口贯穿件
containment spray flow path　安全壳喷淋流道
containment spray heat exchanger　安全壳喷淋热交换器
containment spray pump　安全壳喷淋泵
containment spray ring headers　安全壳喷琳环集管
containment spray signal　安全壳喷淋信号
containment spray system　安全壳喷淋系统
containment spray system heat exchanger　安全壳喷淋系统热交换器
containment structural acceptance test　安全壳结构验收试验
containment structural integrity test　安全壳结构完整性试验
containment sump　安全壳地坑
containment sump level　安全壳地坑液位
containment sump pump　安全壳地坑泵

containment supply duct penetration 安全壳送风风道贯穿件
containment supply header penetration 安全壳送风集管贯穿件
containment testing boundary isolation valve 安全壳试验边界隔离阀
containment vessel 安全壳容器
contaminated 被污染的
contemplate 考虑,思考
contentious 有争议的
contingency 意外事件
continuity 连续性
continuous cooling transformation 连续冷却转变图
continuous deep flexural member 深度柔性构件
continuous laser welding 连续激光焊
continuous welding 连续焊
continuously manned alarm stations 连续值班报警站
contract 合同
contraction coefficient 收缩系数
contraction deformation 收缩变形
contraction pressure loss 收缩压力损失
contraction stress 收缩应力
contractor 承包商,订约人
contractor's pump 施工用泵
contractual party 缔约方
contribute 出力,做出贡献
control 控制,调节
control board 控制盘
control cable 控制电缆
control console 操控台,控制台
control damper 控制风阀
control group heater (稳压器)控制组电加热器
control interface 操作界面,控制界面
control interlocks 控制联锁
control level of surface contamination 表面污染控制水平
control logic cabinet 控制逻辑柜
control of items 物项管理/控制
control of purchased items 采购物项管理/控制
control rod 控制棒
control rod assembly 控制棒组件
control rod bank 控制棒组
control rod bank worth 控制棒组价值
control rod calibration 控制棒刻度
control rod cluster 控制棒束
control rod control system 控制棒控制系统
control rod drive line 控制棒驱动线
control rod drive line alignment 控制棒驱动线对中
control rod drive mechanism 控制棒驱动机构
control rod drive mechanism cooling system 控制棒驱动机构冷却系统
control rod drive mechanism housing 控制棒驱动机构承压壳体
control rod drive mechanism position 控制棒驱动机构位置
control rod drive mechanism seismic support 控制棒驱动机构抗震支撑
control rod drive mechanism support 控制棒驱动机构支撑

control rod drive shaft　控制棒驱动轴
control rod drive shaft handing tool　控制棒驱动轴装卸工具
control rod drive shaft storage racks　控制棒驱动轴贮存架
control rod drive shaft unlatching tool　控制棒驱动轴拆卸工具
control rod drive storage rack　控制棒驱动机构贮存格架
control rod drive system (CRDS)　控制棒驱动系统
control rod drive unlatch tool　控制棒驱动拔闩工具
control rod drop　控制棒落棒,落棒
control rod dropping time　控制棒落棒时间
control rod ejection　控制棒弹棒
control rod ejection accident　弹棒事故
control rod group　控制棒组,棒组
control rod guide thimble　控制棒导向套筒
control rod guide tube　控制棒导向管
control rod insertion limit　控制棒插入极限
control rod misalignment　控制棒失错对中/未对中
control rod movement speed　控制棒移动速度
control rod position indication　控制棒棒位指示
control rod withdrawal　控制棒提升/抽提
control room access door　主控制室出入门
control room isolation and air supply initiation　主控制室隔离和送风启动
control switch　控制开关
control valve　调节阀,控制阀
control volume　控制容积
controlled area　控制区
controlled atmosphere arc welding　充气室中电弧焊
controlled release　受控释放(排放)
controller area access personal dose monitoring　控制区出入口剂量监测
controversial　争议的
convection　对流
convection boiling　对流沸腾
conventional island　常规岛
conventional island DC & AC UPS system　常规岛直流和交流不间断电源系统
conventional load voltage　常负载电压
converter　转换器
conveyer　转送者,让与人,输送机
conveyer belt　传送带,运输皮带
conveyer car　运输小车
cool down　冷却
cool down and depressurization　降温降压
cool down dump valve　冷却排放阀
coolant channel　冷却剂通道(流道)
coolant chemistry　冷却剂化学
coolant flow blockage effect　断流效应,冷却剂流动堵塞效应
coolant flow channel　冷却剂流道
coolant flow coast down　冷却剂流量下降/惰转

coolant flow rate 冷却剂流量
coolant jacket 冷却剂夹套,冷却套
coolant mass flow rate 冷却剂质量流量
coolant mixing 冷却剂交混
coolant pressure boundary 冷却剂压力边界
coolant specific enthalpy 冷却剂比焓
cooler 冷却器
cooling coil 冷却盘管
cooling ducts and fans 冷却风管和风机
cooling season 需冷却的季节
cooling shroud 冷却风罩,冷却屏蔽
cooling time 冷却时间
cooling tower 冷却塔
cooling tower cell 冷却塔单元
cooling tower fan 冷却塔风机
cooling water 冷却水
cooperative 合作的
coordinates exchange 坐标变转换/变换
coordinator 协调员
copper 铜
copper shoe 铜滑板
copper-alloy arc welding electrode 铜合金弧焊条
copper-plating welding wire 镀铜焊丝
cord 线,绳,电线
core alteration 堆芯更换,堆芯改造
core area 堆芯区
core average heat flux (density) 堆芯平均热流密度
core barrel 堆芯吊篮,堆芯吊篮筒体
core barrel assembly 堆芯吊篮组件
core barrel flange 堆芯吊篮筒体法兰
core barrel hold down spring 堆芯吊篮压紧弹簧
core barrel nozzle 堆芯吊篮管嘴/接管
core barrel shell 堆芯吊篮筒体
core bypass flow 堆芯旁通流量,堆芯旁流
core condition 堆芯状况/状态
core configuration 堆芯布置/结构
core cooling and boration 堆芯冷却和硼化
core damage frequency 堆芯损坏频率,堆芯损坏概率
core debris relocation 堆芯碎片移位
core equivalent diameter 堆芯等效直径
core exit temperature 堆芯出口温度
core exit thermocouple 堆芯出口热电偶
core feedback 堆芯(反应性)反馈
core flow subfactor 堆芯流量分因子,堆芯流量子因子
core heat flux 堆芯热流密度
core heat removal trip 堆芯导热失效/中断
core instrument assembly 堆芯测量组件
core integrity 堆芯完整性
core life 堆芯寿期,堆芯寿命
core lifetime extension 堆芯延寿
core limit 堆芯限值
core loading 堆芯装料
core makeup tank 堆芯补水箱

core makeup tank heated recirculation test　堆芯补水箱热态再循环试验
core makeup tank injection　堆芯补水箱(向堆芯)注水
core mapping　堆芯燃料装载图
core maximum heat flux　堆芯最大热流密度
core melt scenario　堆芯熔化状况
core meltdown　堆芯熔化
core midplane　堆芯中分平面
core mixing sub factor　堆芯交混子因子
core multiplication factor　堆芯增殖因子
core neutron moderator　堆芯中子慢化剂
core operating limits report　堆芯运行限值报告
core pad　堆芯支撑块
core physics　堆芯物理
core power　堆芯功率
core power distribution　堆芯功率分布
core radial power distribution　堆芯径向功率分布
core reactivity　堆芯反应性
core reactivity coefficient　堆芯反应性系数
core reflooding　堆芯再淹没
core region　堆芯区域
core release　堆芯(熔融物)喷放
core relocation　堆芯(熔融物)移位
core shell　(反应堆容器)堆芯段筒体
core shroud　堆芯围筒,堆芯围板
core shroud cooling flow　堆芯围筒冷却流量
core shutdown margin　堆芯停堆裕量,堆芯停堆深度
core specific enthalpy rise　堆芯比焓升
core stability　堆芯稳定性
core support block　堆芯支撑块
core support pad　堆芯支承垫(块)
core thermal design　堆芯热工设计
core thermal shield　堆芯热屏蔽
core uncover　堆芯裸露
core unloading　堆芯卸料,堆芯未装料
core void fraction　堆芯空泡份额,堆芯汽泡份额
corner double seam　锁缝角接头,双缝角
corporate learning strategy　企业的学习策略
corrective action　纠正措施
corrective action processing and tracking　纠正行动处理和跟踪
correspond to　对应,与…一致,符合
correspondence　商函
corridor　通道,过道,廊道
corrosion　腐蚀
corrosion fatigue　腐蚀疲劳
corrosion inhibitor　缓蚀剂
corrosion inhibitor addition tank　缓蚀剂添加箱
corrosion medium　腐蚀性介质
corrosion product　腐蚀产物
corrosion protection　腐蚀防护
corrosion rate　腐蚀速率
corrosion resistance　抗腐蚀强度
corrosion resistant material　抗腐蚀材

料，耐蚀材料
corrosion resistant stainless steel 耐腐蚀不锈钢
cosmic radiation 宇宙辐射
cost benefit analysis 成本效益分析
cost management 费用管理，成本管理
cost threshold 成本阈值
count on 指望，依靠
count rate 计数率
counter area monitor 计数管区域，监测仪
counter offer 还盘（买方），买方还价
counter proposal 反建议
counter-counter offer 再换盘（卖方）
counterflow 逆流，逆向流动
counterpart 对方，对手
counter-sign 会签
counting room 计数室
coupling 接头，耦合
coupling breaker 耦合开关
coupon 试块，试样
course 课程
courseware 教材
courtesy, etiquette, protocol 礼仪，得体
cover 盖板
cover, insure 保险
cover flange 盖板法兰
cover for valve box 阀箱盖
cover glass 防护镜片
cover quite a lot of ground 取得了不小的进展
coverage 险别
covering letter 伴函
CPS application and computerized procedure CPS 应用和微机化程序
CQ thread ball valve CQ 螺纹球阀
crack 裂纹
crack propagation rate 裂纹扩展速率
cracking sensibility 裂纹敏感性
cracking test 裂纹试验
crane/elevator/lift 吊车/电梯/起吊设备
crank 曲柄，转动曲柄
crater 弧坑
crater crack 弧坑裂纹
CRDM cooling system 控制棒驱动机构冷却系统
CRDM external cables 控制棒驱动机构外部电缆
CRDM power supply system 控制棒驱动机构电源系统
CRDM rod travel housing 控制棒驱动机构棒行程套管，控制棒驱动机构耐压壳
CRDM ventilation shroud 驱动机构通风罩
CRDM ventilation system 控制棒驱动机构通风系统
creativity 创新性
credible event 可信事件
credit terms 信贷条件
creep 蠕变
creep collapse 蠕变坍塌
creep down 向内蠕变
creep fatigue 蠕变疲劳
creepage distance 表面漏电距离
crevice corrosion 缝隙腐蚀

criteria, code and standard 准则、规范和标准
criterion 标准,准则
criterion of cold cracking 冷裂判据
critical boron concentration 临界硼浓度
critical concentration 临界浓度
critical crack size 临界裂纹尺寸
critical current of double arcing 双弧临界电流
critical damping value 临界阻尼值
critical experiment 临界实验
critical exposure pathway 关键照射途径
critical flaw size 临界裂纹尺寸
critical flow 临界流
critical heat flux modified cold wall factor 临界热流密度冷壁修正因子
critical heat flux modified shape factor 临界热流密度形状修正因子
critical heat flux modified spacer factor 临界热流密度格架修正因子
critical heat flux ratio 临界热流密度比
critical human action 关键的人员动作(行为)
critical kinetic energy 临界动能
critical load 临界载荷,危险载荷
critical mass 临界质量
critical mass flow 临界质量流量
critical mass flux 临界质量流密度
critical nuclide 关键核素
critical path outage model 大修关键路径模型
critical safety function 关键安全功能
critical size (core) (堆芯)临界尺寸
critical stress intensity 临界应力强度
critical stress intensity factor 临界应力强度因子
critical transfer pathway 关键转移途径
critical volume 临界体积
criticality 临界,临界度
criticality accident 临界事故
criticality alarm system 临界报警系统
criticality analysis 临界分析
criticality approach curve 临界逼近曲线
criticality monitoring 临界监测
criticality safety 临界安全
cross connect valve 桥管阀门
cross connection 十字接头
cross section 截面,横断面
cross-connected 交叉连接
cross-flow 横向流,交叉流动
cross-sectional area 横截面,交叉区域
crucial 关键的,极紧要的
cruciform 十字形
cruciform cracking test 十字接头裂纹试验
cruciform weld 十字焊缝
crud 垢污,杂质,污垢
crud deposit 垢污物沉积
crud trap area 污垢物/杂质聚集区
crud-induced power shift (CIPS) 污垢物导致的功率偏移
crush resistance 抗压性,抗压强度
crux 要点,难点,要害

cryogenic nitrogen　低温氮气
cryogenic valve　低温阀门
crystal structure　晶体结构
cubicle　小室
cu-cooling plate　铜冷却板
cultivated　有修养的，文雅的
culvert　涵洞，管路
cumulative absolute velocity　累积绝对速度
curb　阻挡栏，路缘
cured　治愈的，凝固
current　电流，流
current carrying capacity　载流能力
current curve versus time　电流波形图，电流一时间曲线
current density　电流密度
current generator-DCS testing　发电机-DCS 试验
current limiting　限流，电流限制
current transformer　电流互感器
curriculum　课程
cursor　光标，指针
curtain area　帘面积，帘区，屏障
curvature　坪曲线，弯曲，曲率
curved vane　曲面叶片
cushion　衬垫，垫层，加衬垫
customary　习惯的，通常的
cut a deal　达成协议
cut face　切割面
cut line　切割线
cut set　割集
cut-off energy　切割能，截断能
cut-off frequency　截止频率，截断频率
cut-off valve　断流阀，截断阀
cutting blowpipe　割炬
cutting gun　割枪
cutting nozzle　割嘴
cutting oxygen　切割氧
cutting robot　切割机器人
cutting shoulder　切口上缘
cutting speed　切割速度
cutting stream　切割氧
cutting tip　割嘴
cutting torch　割炬
cycle life　循环寿期
cyclic fatigue　循环疲劳
cyclic load　循环载荷
cyclic stress　循环应力
cycloidal screw pump　摆旋线螺杠泵
cyclone separator　旋风分离器
cyclone translational speed　（热带气旋）平移速度
cylinder cap　阀罩
cylinder manifold　汇流排
cylinder valve　气瓶阀
cylindrical gear actuator　（圆）柱齿轮传动装置
cylindrical shell　圆柱形筒体

D

D/A(document against acceptance)　承兑交单
D/D (remittance by bank ' s demand draft)　票汇
D/P(document against payment)　付款交单
dado　护壁板，墙裙
daily load follow　日负荷跟踪
daily variation coefficient　日变化系数

dam　闸,坝,障碍
dam failure　溃坝
damaged fuel rod　破损燃料棒
damped natural frequence　有阻尼的固有频率,阻尼固有振动频率
damper　风阀,阻尼器,节气闸
damper actuator　挡板控制器
damping　阻尼
damping ratio　阻尼比
DAS processor cabinet　DAS 处理器机柜
DAS squib valve controller cabinet　数据采集系统爆破阀控制器机柜
dashpot action　缓冲作用,阻尼作用
dashpot characteristics　缓冲特性,阻尼特性
dashpot drop time　(控制棒)缓冲落棒时间
dashpot entry　(控制棒)缓冲段入口
dashpot region　缓冲区
dashpot section　缓冲段
dashpot tube　缓冲管
data acquisition　数据采集
data acquisition and processing system　数据采集和处理系统
data acquisition hardware　数据采集硬件
data base　数据库
data bus　数据总线
data channel　数据通道
data communication system (DCS)　数据传输系统
data file　数据文件
data highway　数据总线
data link　数据链,数据链接
data link server　数据链(接)服务器
data logging system　数据登录系统
data organization　数据结构
data processing and display function　数据处理和显示功能
data processing module　数据处理模块
data reduction analysis method　数据简化分析方法
date of issue　开证日期
date of standard implementation　标准实施日期
daughter product　子体产物
DC & AC UPS system　直交流不间断电源系统
DC arc welding power source　直流弧焊电源
DC hold cabinet　直流保持柜
DC stabilized power supply　直流稳压电源
DC uninterruptible power system　直流不间断电源系统
DCS support equipment　DCS 支持设备
DDS data highway　设计开发系统数据总线
DDS processor cabinet　设计开发系统处理器机柜
de-soldering　脱钎,脱焊
dead band　死区
dead leg　死水段
dead load　静载荷
dead pocket　死区
dead-end compartment　死角隔间

deadline 截止时间,工期
deadweight 自重
deadweight of the mat 底板自重
deaerator 除氧器
deaerator lift pump 除氧器供给泵
deaerator recirculating pump 除氧器再循环泵
deaerator storage tank 除氧器储水箱
dealing 生意,业务
deborating demineralizer 除硼床
de-brazing 脱钎,脱焊
debris 碎片
debris filter bottom nozzle (DFBN) (燃料组件)防异物下管座
debug 查错
deburring 去毛刺
decade boxes-DCS testing DCS 试验用十进制电阻箱
decay 衰变
decay chain 衰变链
decay constant 衰变常数
decay heat 衰变热
decay heat rate 衰变热率
decay tank 衰变箱
deceleration valve 减速阀
deck 甲板,舰板,覆盖物
decline,reject,turn down 拒绝
decommissioning 退役
decompression valve 卸压阀
decontamination 去污
decontamination basin 去污盆
decontamination factor (DF) 去污因子,去污系数
decontamination solution drain 去污废液疏排水
decontamination system 去污系统
decontamination water 去污水
dedicate 专门用于
dedicated containment penetration 安全壳专设贯穿件
dedicated controls and displays 专用控制和显示盘
dedicated safety panel ♯1 1 号专用安全控制盘
dedicated safety panel ♯2 2 号专用安全控制盘
dedication 鉴定,见证
dedication entity 鉴定/见证实体
deenergize 失电,断电
de-energized state 失电状态
defect 缺陷
defective fuel 破损燃料组件
defense in depth 纵深防御
defense-in-depth control 纵深防御控制
defense-in-depth function 纵深防御功能
deferment 延期
deficiency 不足,亏空
define 规定
defined base 确定的基础,基本定义
definitely 的确,一定,绝对
deflagration 爆燃
deflagration to detonation transition 爆燃向爆炸的过渡
deflection 挠度,变形,偏转,偏斜
deflection deformation 挠曲变形
deflector 偏转板,导流板

deformation　变形
degasifier　脱气剂
degasifier column　脱气柱
degasifier separator　脱气分离器
degasifier separator pump　脱气分离器泵
degasifier vacuum pump　脱气塔真空泵
degradation　退化，降低，降解，降质
degradation mechanism　劣化机理，降质机理
degraded core　降级堆芯
degree of demonstration　证实程度，验证度
degree of enrichment　富集度
degree of free for robot　机器人运动自由度
degree of redundancy　冗余度，多重度
degree of subcooling　欠热度，过冷度
degree of superheat　过热度
dehydrogenation　去氢
delay　延迟，延时，缓发
delay circuit module　延迟电路组件
delay on actuating　延迟驱动，延迟触发
delayed crack　延迟裂纹
delayed critical　缓发临界
delayed neutron　缓发中子
delayed neutron fraction　缓发中子份额
delineate　描写，描绘
delithium demineralizer　除锂床
delivery　运送，投递，传送
delivery valve guard　导出阀限位器，排气阀防护器
delong's diagram　德龙组织图
delta connection　三角形连接
deluge sprinkler system　喷淋系统
deluge valve　喷淋阀
demand position system　需求棒位系统
demarcate　画界，区别
demineralization　除盐
demineralize　去矿化，去矿物质
demineralized pre-filter　（去离子）床前置过滤器
demineralized water　除盐水
demineralized water isolation valve　除盐水隔离阀
demineralized water pump　除盐水泵
demineralized water storage tank　除盐水贮存箱
demineralized water supply containment　除盐水供水安全壳外隔离(阀)
demineralized water system isolation　除盐水系统隔离
demineralized water tank　除盐水箱
demineralized water transfer and storage system　除盐水输送和储存系统
demineralizer　除盐装置(器)
demineralizer after-filter　（除盐）床后置过滤器
demineralizer resin flush line containment penetration　除盐床树脂冲洗管安全壳贯穿件
demography　人口统计学
demonstration　技术示范
demonstration run　验证运行，示范

运行
densification (燃料)密实化
densification effect (燃料)密实化效应
densimeter 密度计
density wave 密度波
denting 凹坑
deoxidizer 脱氧剂
deoxygenated water 除氧水
deoxygenated water cooler 除氧水冷却器
deoxygenated water pump 除氧水泵
deoxygenated water tank 除氧水箱
departure 背离,违背
departure from nucleate boiling (DNB) 偏离泡核沸腾
departure from nucleate boiling ratio (DNBR) 偏离泡核沸腾比
dependability 可信性,可信赖度
dephosphorization 脱磷
depleted fuel 乏燃料
depletion 燃耗,烧乏
depletion chain 消耗链
depletion rate 消耗速率
depletion worth curve 反应性价值消耗曲线
deposit (买方支付的)定金,保证金
deposit corrosion 沉积腐蚀
deposited corrosion product 沉积的腐蚀产物
deposited metal 熔敷金属
deposition 沉降,沉积量
deposition efficiency 沉积效率
depressurization 降压
depressurization sparger 降压喷头,卸压鼓泡管
depth of corrosion 腐蚀深度
depth of cutting vein 割纹深度
depth of foundation 基础埋置深度
depth of slag bath 渣池深度
derated coefficient of discharge 降低的排放系数
derating 降级,减量
derivation 导出,来历,出处
derived air concentration (DAC) 导出的空气浓度
derived limit 导出的限值
descaling 去氧化皮
descaling pump 除锈泵
description for drawing up standard 标准编制说明
deserve 值得
desiccant 干燥剂
desiccant dryer 干燥器
desiccation 干燥作用
desiccator 干燥器
design assumption 设计假设
design basis 设计基准/基础,设计依据
design basis flood 设计基准淹没
design basis ground motions 设计基准地面运动
design basis network load follow pattern 设计基准网络负荷跟踪图
design basis source terms 设计基准源项
design basis tornado 设计基准龙卷风
design change 设计变更
design code 设计规范

design commitment 设计承诺,设计委托,设计执行
design condition 设计工况
design control 设计管理,设计控制
design criteria 设计准则
design deficiencies 设计缺陷
design for reactor core safety 堆芯安全设计
design guidelines 设计导则
design implementation 设计实施
design interface control 设计接口管理
design limit 设计限值
design load 设计载荷
design objective limit 设计目标限值
design overspeed condition 超设计转速工况
design parameter 设计参数
design pressure 设计压力
design report 设计报告
design response spectra 设计响应谱
design review 设计审查
design specification 设计技术规格书
design step load change 设计阶跃负荷变化
design summary report 设计总结报告
design temperature 设计温度
design tolerance 设计允许误差
design transient 设计瞬态
design verification 设计验证
design wind 设计风荷载
design wing velocity 设计飞行速度
designate, appoint, assign 指定,规定,限定
designation 名称,指定,选派
designation of system 系统名称
designed basis flooding (DBF) 设计基准淹没
designed basis low water level (DBL) 设计基准低水位
desilting basin 清淤池
de-soldering 脱钎,脱焊
desorption 解吸附,脱吸,解析
desoxydation 脱氧
desperate 急切需要的
despite 尽管
destine 预定,注定
destructive examination 破坏性检验
destructive test 破坏性试验
desulphurization 脱硫
desuperheater 过热蒸汽减温器
detached part 脱落零件
detail 细节,详告
detection limit 探测限值
detector 探头,探测器
detector channel 探测通道
detector push-pull device 探测器推拉装置
detector storage tube 探测器存放管
detector well 探测器孔道
detergent waste 洗涤剂废水
detergent waste decontamination factor 洗涤废液去污因子
deterioration 退化,劣化
deterministic safety criteria 确定论安全准则
detonation 爆炸
devastating 破坏性的
developed spray pattern 完善的喷淋

模式
developer 显影剂
development, format and review （询价单的）编制，格式和审核
deviation 偏离，偏差，背离
deviation disposition requestion(DDR) 偏差处理要求
deviation permit 偏离许可
devote one's attention to 集中精力…
dew point temperature 露点温度
de watering pump （地下水）排水泵
dewatering, de-watering 脱水，去水
diagnostic 诊断的，判断的
diagnostic program 诊断程序
diagnostic self testing 自诊断试验
dialog keyboard 对话键盘
diameter 直径
diametral power oscillation 径向功率振荡
diaphragm 隔膜，膜片
diaphragm metering pump 膜片式计量泵
diaphragm operated control valve 膜动式控制阀
diaphragm piston pump 膜片活塞泵
diaphragm plunger pump 隔膜柱塞泵
diaphragm pump 隔膜泵
diaphragm reducing valve 薄膜式减压阀
diaphragm valve 隔膜阀
die formed 冲制成形
diesel 柴油机
diesel engine 柴油机
diesel engine starting subsystem 柴油机启动子系统
diesel fuel day tank vault 柴油机日用油箱存放间
diesel GEN BLDG lighting XFMRs 柴油发电机厂房照明变压器
diesel GEN BLDG MCCs 柴油发电机厂房电机控制中心
diesel generator 柴油发电机
diesel generator building 柴油发电机厂房
diesel generator building power transformers 柴油发电机厂房电力变压器
diesel generator fuel oil day tank 柴油发电机日用油箱
diesel generator set 柴油发电机组
diesel load sequencer 柴油机带载顺序
diesel oil transfer module 柴油机燃油输送模块
diesel oil transfer module enclosure exhaust fan 柴油机燃油输送隔间排风机
diesel oil transfer package 柴油输送成套设备
diesel pump 柴油机驱动泵
differ from 与…不同，与…有区别
differential piston pump 差动活塞泵
differential pressure 差压
differential pressure controller 差压调节器
differential pressure gauge 差压计
differential pressure instrument line isolation valve 差压仪表管线隔离阀
differential pressure loading 差压荷载

differential pressure sensor 差压传感器
differential pressure transmitter 差压变送器
differential protection 差动保护
differential rod worth 控制棒微分价值
differential worth 微分价值
differentiator module 微分器组件
diffuser 扩散器，导叶
diffuser plate (堆内构件)流量分配板
diffuser pump 导叶泵，扩散泵
diffusible desoxydation 扩散脱氧
diffusible hydrogen 扩散氢
diffusible hydrogen remained at 100 ℃ 100 ℃残余扩散氢
diffusible oxydation 扩散氧化
diffusion 扩散
diffusion brazing 扩散钎焊
diffusion coefficient 扩散系数
diffusion flame 扩散火焰
diffusion phoresis 扩散电泳现象，离子扩散
diffusion theory 扩散理论
digital/analog (D/A) converter 数/模转换器
digital circuit board 数字电路板
digital electro-pneumatic calibrator 数字式电-气动标定仪/校准器
digital filtering 数字滤波
digital heat prober thermometer 数字式热探针温度计
digital high precision multimeter 数字式高精度万用表
digital input 数字输入
digital metal impact monitoring system 数字化金属松动部件监测系统
digital misalignment alarm module 数字化失调报警模块
digital multimeter 数字式万用表
digital rod position indication (DRPI) system 数字式棒位指示系统
digital voltmeter module 数字电压表组件
dilution 稀释
dilution factor 稀释因子
dimension 尺寸，尺度，维，度，元
dimension of connecting channel 连接槽尺寸
dimension of valve stem end 阀杆端部尺寸
dimension of valve stem head 阀杆头部尺寸
dimensional diagram 尺寸图
dimensional variation 尺寸改变，尺寸误差
dimple (燃料组件格架上的)搅混叶
dimple fracture 韧窝断口
diode gun 二极电子枪
dip brazing 浸渍钎焊
dip soldering 浸渍钎焊
direct acting reducing valve 直接作用式/弹簧式卸压阀
direct burial 直接埋葬
direct contact method (磁粉检验)直接接触法
direct containment heat 安全壳直接加热

direct current(DC) 直流电
direct current arc welding 直流电弧焊
direct current suppressor 直流分量抑制器
direct dose 直接剂量
direct grounding 直接接地
direct inward dialing (DID) 直接接入拨号
direct loaded safety valve 直接加载式安全阀
direct outward dialing 直接外拨
direct spring loaded safety valve 直接弹簧式安全阀
direct vessel injection line 压力容器直接注射管
direct vessel injection nozzle 压力容器直接注射管嘴/接管
direct vessel injection valve 压力容器直接注射阀
direct-acting reducing valve 直接作用/弹簧式卸压阀
direction interlocking circuit 定向联锁电路
direction of welding 焊接方向
directory 目录
disadvantage factor 不利因子
disc face friction 阀瓣表面摩擦
disc valve 圆板阀,圆盘阀,瓣阀
disc 蝶板,阀瓣
discharge 卸料,排放,放电
discharge burn up 卸料燃耗
discharge isolation valve 排放隔离阀
discharge piping 排放管线
discharge pump 排放泵
discharge valve 排放阀,出口阀
disconnect 断开,切断,断路
disconnect button 切断按钮
disconnect plug 可拆接头
disconnect rod 拆卸杆,切断杆
discontinuity 不连续,间断
discount rate 贴现率
discrepancy 不一致,不符,偏差,违背
discrete 离散的
discrete burnable absorber 离散型可燃毒物
discrete burnable absorber rod 离散型可燃毒物棒
discrete level 离散液位
discrete time sample data storage system 离散时间采样数据存储系统
discretion 判断力
discuss,talk over 讨论,磋商
dished head 碟形封头
dished pellet 碟形(燃料)芯块
dishing 冲压
disinfection 消毒
disk 磁盘
disk file 磁盘文件
disk operation system 磁盘操作系统
disk storage 磁盘存储
dislocation 错边
dismantle 拆除,拆卸,解体
dispersant 分散剂
dispersed flow 弥散流
dispersion 弥散
displacement 位移
display 显示
display frame 显示结构

display guideline 显示(画面)导则
display operation station (显示)操作台站
disposal 支配权,(自由)处置权
disposition 处理,安排
disposition of nonconforming items 不符合项处理
dispute 争议
dissolution 溶解
dissolved boron concentration 可溶硼浓度
dissolved oxygen (DO) 溶解氧
distance between welding wires 焊丝间距
distance learning 远距/程学习
distinct 明显的,明确的,显著的
distinction,difference 差别
distinctly 独特地
distinguish 区别,辨别,识别
distributed architecture 分布式结构
distributed computer processor 分布式计算机处理器
distributed computer system 分布式计算机系统
distributed controller 分布式控制器
distributed network 分布式网络
distributed processing cabinet 分布式处理机柜
distribution 分配,分布
distribution box 配电柜,配电箱
distribution bucket (安全壳冷却水)分配桶/盘
distribution cabinet 分配柜
distribution list 分发清单
distribution panel 配电盘
distribution system 配水管网,分配系统
distribution weir (水)分配围堰
district heating pump 地区供暖泵
ditch 地沟,沟渠,壕沟
divergent nozzle 快速割嘴
diverse 不同的,多样的
diverse actuation 多样性驱动
diverse actuation system panel 多样化驱动系统控制盘
diverse mitigation function 多样缓解功能
diverse parameter 多样性参数
diversified 多种形式的
diversity 多样性
divert into 转化到,转移到
diverter pump 分流泵
divider plate 隔板,分流板
division 篇,序列
DNB heat flux 偏离泡核沸腾热流密度
do something reciprocal in return 礼尚往来,有来有往
dock pump 船坞泵
document approval 文件批准
document change 文件变更
document control 文件管理
document control manager 文件管理经理
document distribution 文件分发
document preparation 文件编制,文件准备
document retention 文件保存

document review 文件审查
document submittals 文件提交
document submittals, approvals and tracking 文件提交,批准和跟踪
documentary draft 跟单汇票
domain 领域,范围,领土
domain controller 域控制器
domestic 国内的,生活中的
domestic sewage 生活污水
domestic water 生活用水
domestic water yard pipe network system 生活水厂区管网系统
dominant 支配性的,占优势的
dominant cut set 支配性割集
door seal 门密封
doors and hatch 门及闸门
Doppler broadening 多普勒展宽
Doppler coefficient 多普勒系数
Doppler effect 多普勒效应
Doppler feedback 多普勒反馈
Doppler power coefficient 多普勒功率系数
Doppler reactivity defect 多普勒反应性亏损
Doppler weighting factor 多普勒权重因子
dose 剂量
dose assessment 剂量评价
dose commitment 剂量负担
dose constraint 剂量约束
dose equivalent 剂量当量
dose equivalent commitment 剂量当量负担
dose equivalent limit 剂量当量限值
dose rate conversion factor 剂量率转换因子
dose risk 剂量风险
dosimeter 剂量计
dosimeter reader 剂量计读取器
dosing pump 比例泵,定量给料泵
double arcing 双弧现象
double coated electrode 双层药皮焊条
double diaphragm pump 双隔膜泵,双膜片泵
double disc flat gate valve 双闸板平板闸阀
double disc swing foot valve 双瓣摆动式底阀
double disk parallel gate valve 双闸板平行式闸阀
double ended break (管道)双端断裂
double fume chamber and hood 双位通风柜
double gaskes 双重衬垫
double gate disc 双闸板
double hook 双钩
double opening exhaust valve 双口排气阀
double period 倍增周期
double pole double throw transfer switch 双极双掷转换开关
double reeling system 双卷绕系统
double sleeve encapsulation 双套管密封,双套管封装
double tension test 双重拉伸试验
double width double inlet (DWDI) 双进风
double-contingency principle 双重偶然

性原则
double-ended cold-leg guillotine　冷段双端剪切断裂
double-ended guillotine (DEG)　双端剪切断裂
double-ended rupture　双端破裂
double-pole single-throw (DPST) switch　双极单掷开关
double-pulse generation　双脉冲发生(器)
down payment　定金,入门费
downcomer　下降环腔,下降段,下降管
downcomer region　下降区
down-hand welding　平焊
downhill welding　俯焊
downstream　下游
draft at sight　即期汇票
draft contract　合同草案
draft standard for approval　标准报批稿
draft standard for examination　标准送审稿
draft, draw up　草稿,草案,草图,起草
drag on, prolong, make longer　拖延,持续
drag　拖,拖拉
drain　排水沟,排水,疏水
drain flash tank　疏水闪蒸箱
drain heat exchanger　疏排水箱热交换器
drain isolation valve　疏水隔离阀
drain line　疏水管线
drain path　排水路径
drain recovery tank　疏水回收箱
drain sump　排水地坑,疏水地坑
drain system　排水系统,疏水系统
drain tank　疏水箱
drain traps　排水存水弯
drain valve　疏水阀
drainage　排水,放水
drainage capacity　排水能力
drainage system　排水系统
drainage valve　排水阀
draindown mode　疏排模式
draw up a contract　起草合同
drawer　抽屉
drawing　图纸
drawing list　图纸目录
drawing volume list　图册目录
drawled part　拉制件
drier　干燥器
drift　冲洗,漂流
drift flux　漂移流
drinking　饮用的
drip　水滴,使滴下
drip pan　集油盘
drip pot　疏水槽
dripping　滴漏
drive line drop testing　(控制棒)驱动线落棒试验
drive rod　驱动杆
drive rod assembly　驱动杆部件
drive shaft　驱动轴
drive shaft assembly　驱动轴部件
drive unit　驱动机构
driveline　驱动线
driving force　驱动力
drop size　下落尺寸/高度,液滴尺寸

drop test 落棒试验
drop time 落棒时间
droplet 液滴
droplet entrainment 液滴夹带
dropping characteristic arc welding power source 下降特性弧焊电源
drop-weight fracture toughness test 落锤法断裂韧性试验
dross 结瘤,浮渣
drum 卷筒
dry bulb temperature 干球温度
dry deposition 干沉降
dry dock pump 干船坞泵
dry fire protection piping 干式消防管道
dry flashback arrestor 干式回火防止器
dry method underwater welding 干式水下焊
dry pipe 干管
dry pipe valve 过热蒸汽阀,干管阀
dry pit pump 干坑泵
dry powder coating (polyesters) (聚酯)干粉涂层/包覆
dry self-priming pump 干式自吸泵
dry sump pump 干坑泵
dry type transformers for hydrogen igniter 用于氢点火装置的干式变压器
dry vacuum pump 干式真空泵
dryer 干燥器
drying oven 烘干箱,烘箱,干燥炉
dry-insulated transformer 干式绝缘变压器
dryout 干涸
dry-type transformer 干式变压器
dual gas plasma arc cutting 双层气体等离子弧切割
dual range 双量程
dual trace oscilloscope 双踪示波器
dual-frequency simplex operation 共频单工操作,双频单一操作
dual-tone multi-frequency 双音多频
duct 风管
ductile fracture 延性断裂
ductility 延性
ductility factor 延性因子
ductility of steel structure 钢结构的延性
ductility-dip crack 失延裂纹
ductwork 管网,管道系统
due 预定的,到期的
dummy fuel element 假(燃料)元件
dump valve 事故排放阀
duplex filter 双联滤器
duplex safety valve 双联(弹簧式)安全阀
duplex transceiver 双方收发信机
duplication 复制,转储
durability 耐久性
duration time 持续时间
dust spot efficiency 比色法效率,灰斑效应
duty cycle 负载持续率
dye 染,染色
dye penetrant examination 着色渗透检查
dye penetrant inspection 着色探伤

dye penetrant test 着色渗透检验
dynamic 动力的,动力学的,动态的
dynamic(static)pressure 动(静)压
dynamic active earth pressure 动态主动土压力
dynamic analysis 动态分析
dynamic balance 动态平衡
dynamic characteristics 动力学特性
dynamic effect 动态效应
dynamic equilibrium 动态平衡
dynamic head 动压头
dynamic instability 动态不稳定性
dynamic lateral force 动态侧向力
dynamic load factor 动态载荷因子
dynamic parameter 动态参数
dynamic poisson ratio 动泊松比
dynamic range 动态范围
dynamic response 动态响应
dynamic shear modulus 动剪切模量
dynamic soil pressure 动态土压力
dynamic testing and analysis 动态试验和分析
dynamic wind pressure 动风压
dynamic young's modulus 动弹性杨氏模量
dynamics 动力学

E

earned value 赢得值
earth clamp 地线夹头
earth leakage protection 接地保护
earth return 大地回馈
earthing 接地
earthing resistance 接地电阻
earthing system 接地系统
earthquake 地震
earthquake loadings 地震荷载
earthquake subsidence 震陷,地震下沉
ease of construction 便于施工,易于施工
easy-going 轻松的,舒适的
eccentric 偏心的,古怪的
eccentric reducer 偏心大小头
eccentric rotary pump with dividing plate 带隔片的偏心转子泵
eccentric rotary pump with elastic ring 带柔性环的偏心转子泵
eccentric rotary pump with sliding sleeve 带滑动套筒的偏心转子泵
eccentricity (of an electrode) 焊条偏心度
ECS switchgear buses 发电机控制系统开关母线
eddy current 涡流
eddy current examination 涡流检测
eddy current inspection 涡流检查
editor 程序编辑
educational administration 教务处
educational improvements 教学改进
eductor 喷射器
eductor water supply skid 喷射器供水装置,射流供水装置
eductor water supply tank skid 喷射器供水箱,射流供水箱
effect a contract 实施合同,合同生效
effective delayed neutron fraction 有效缓发中子份额
effective dose equivalent 有效剂量当量,有效外照射剂量当量

effective dynamic pressure 有效动压力
effective flow area 有效流动面积
effective flow rate 有效流量
effective fuel temperature 燃料有效温度
effective full power day 等效满功率天
effective full power year 等效满功率年
effective multiplication factor (k_{eff}) 有效增殖因数
effective pressure 有效压力
effectiveness 有效性
effectuation 发生,生效
effluent 流出物,废水
effluent concentration limit 流出物浓度限值
effluent holdup tank 排出流滞留箱
effluent management software 流出物管理软件
effluent monitoring 排出流(放射性)监测
effluent radiation monitoring system 排出流辐射监测系统
egg crate grid (蒸汽发生器)蛋篓型格架
ejected rod accident 弹棒事故
ejected rod worth 弹棒价值
ejection 弹出,排出,射出
ejection accidents of low worth rod 低价值棒弹棒事故
ejector 喷射器,排放管
elastic scattering 弹性散射
elastoplastic fracture mechanics 弹塑性断裂力学
elbows 弯头
e-learning 电子化学习
electret microphone 驻极体传声器,电介体话筒
electric actuated stop valve 电动截止阀
electric actuated wedge gate valve 电动楔式闸阀
electric actuated 电动的
electric actuator 电动装置
electric arc spraying 电弧喷涂
electric building 电气厂房
electric building cooler system 电气厂房循环冷风系统
electric building smoker exhaust system 电气厂房排烟系统
electric building ventilation system 电气厂房通风系统
electric building water supply and drainage system 电气厂房给排水系统
electric dispatch network 电力调度网
electric double disk parallel gate valve 电动平行式双闸板闸阀
electric furnace 电炉
electric heater 电加热器
electric horn 电喇叭
electric monorail hoist 单轨电动葫芦
electric operated valve 电动阀
electric unit heater 电加热器组
electric welder 电焊机
electric welding machine 电焊机
electrical and mechanical equipment 电气和机械设备

electrical conductivity　电导率
electrical conduit　引出线管
electrical panel　配电屏
electrical penetration　电气贯穿件
electrical penetration test isolation valve　电气贯穿件试验隔离阀
electrical power system　电气供电系统
electrical test　电气试验
electrically driven feedpump　电动给水泵
electrically driven pump　电动泵
electrically isolate　电气隔离
electrically operated brake　电动制动器
electric-hydraulic actuating brake　电动液压制动器
electricity grid　输电网
electricity rate　电费
electric-powered hoist　电动起升机构,电动葫芦
electro impulse pump　电脉冲泵
electroacoustic component　电声元件
electrode　电极,焊条
electrode for under-water cutting　水下切割用焊条
electrode for welding spheroidal graphite cast iron　球墨铸铁焊接用焊条
electrode holder　焊钳,电极夹
electrode keep time of slider　焊丝停摆时间
electrode setback　电极内缩长度
electrode with cast core wire　铸芯焊条
electrodeionization　电去电离作用
electrodeionization demineralizing unit (EDI)　电离除盐装置
electrodeionization waste brine　电离除盐废液,电离除盐浓缩液
electro-gas (arc) welding　气电弧立式焊
electro-gas (arc) welding machine　气弧立式电焊机
electro-hydraulic　电动液压,电一液
electro-hydraulic servo valve　电液伺服阀
electro-hydraulic turbine control system　电动液压汽轮机控制系统
electroless nickel plated　化学镀镍的
electromagnetic braker　电磁制动器
electromagnetic diaphragm pump　电磁隔膜泵
electromagnetic flowmeter　电磁流量计
electromagnetic inspection　电磁探伤
electromagnetic interference (EMI)　电磁干扰
electro-magnetic jet pump　电磁喷射泵,电磁泵
electromagnetic lens　电磁透镜
electromagnetic metering pump　电磁计量泵
electromagnetic proportioning pump　电磁式比例泵
electromagnetic pump　电磁泵
electromagnetically coupled pump　电磁偶合泵
electron beam brazing　电子束钎焊
electron beam cutting　电子束切割
electron beam deflection coils　电子束

偏转线圈
electron beam welding (EBW) 电子束焊
electron beam welding machine 电子束焊机
electron gun 电子枪
electronic bistable 电子双稳态定值器
electronic cabinet 电子机柜
electronic pocket dosimeter 电子式袖珍剂量计
electroslag process stability 电渣工艺稳定性
electro-slag remelting 电渣重熔
electro-slag remelting process 电渣重熔工艺
electroslag surfacing 电渣堆焊
electroslag welding (ESW) 电渣焊
electroslag welding machine 电渣焊机
electroslag welding with consumable nozzle 熔嘴电渣焊
electroslag welding with plate electrode 板极电渣焊
electroslag welding with tube electrode 管极电渣焊
electroslag welding with wire electrode 丝极电渣焊
electro-sound telephone system 电声电话系统
electrostatic discharge 静电释放
electro-submersible drainage pump 电动潜水排水泵
electro-submersible pump 电动潜水泵
electrotechnical box 电气箱
element 要素,纲要,成分
elemental iodine 元素碘,碘元素
elements real num 按实常数显示单元
eletro hydraulic actuator 电子液压装置
eletro magnetic actuator 电磁驱动装置
eletro magnetic hydraulic actuator 电磁液压驱动装置
elevated nozzle 高位接管口,提升段管嘴
elevation 立面图,高程,提升
elevator 电梯
elevator pit 电梯井底
elite course 精品课程
ellipse head 椭球形封头
elongated cavity 长气孔,长空腔
elucidate 阐明,说明
embark on 着手,开始从事
embedded 内含的,植入的
embedded part 预埋件
embrittlement 脆化
emergency 应急,紧急情况
emergency air storage tank (主控制室应急居住系统)应急空气储罐
emergency announcement 紧急通知,紧急广播
emergency boration 紧急硼化
emergency conditions 紧急工况,应急工况
emergency control room system 应急控制室系统
emergency core cooling system (ECCS) 应急堆芯冷却系统
emergency core decay heat removal 应

急堆芯衰变热去除
emergency cut-off valve　紧急截断阀
emergency diesel generation system　应急柴油发电机系统
emergency disk brake　应急盘式制动器
emergency drain　应急疏水
emergency exposure　应急照射
emergency feedwater tank　应急给水箱
emergency habitability　应急可居留性
emergency letdown　紧急排放
emergency makeup to the spent fuel pool isolation valve　乏燃料水池应急补水隔离阀
emergency offsite communication　应急对外通信
emergency oil pump　应急油泵
emergency operating instruction　应急操作说明
emergency operating procedure (EOP)　应急运行规程
emergency operation　应急运行
emergency outlet　安全出口,紧急出口
emergency passive cooling heat sink　应急非能动冷却热阱
emergency power source　应急电源
emergency power system　应急电力系统
emergency response guideline　应急响应导则
emergency shutdown　紧急停堆
emergency telephone system　应急电话系统
emergency transfer equipment　应急(电源)转换设备
emergency trip time　紧急停堆时间
emphasize　强调
emptying valve　放空阀
enabling objective　分解目标
encapsulation　封装,包装
enclose　随信付上,付入
enclosed motor　封闭式电动机,防水电动机
enclosed please find　随函附上
enclosure　附件
encounter　遇到
end beam　端梁
end of cycle　循环末期
end of erection state report　安装完工报告
end of life　寿期末
end of manufacturing report　完工报告
end plug　端塞
endeavor to　全力以赴
end-of travel limit switch　行程终端限位开关
endowment insurance　养老保险
end-packed type piston pump　端部密封式活塞泵
end-to-end dimension　结构长度,全长
end-window counter　钟罩形计数管
energization　通电
energize　激励
energized　通电的
energy group　能群
energy level　能级
energy resolution　能量分辨率
energy response　能量响应

energy spectrum　能谱
energy window　(辐射)能区
enforcement　强制执行
engage in　经营,从事
engagement　约定,承诺,婚约
engine cooling subsystem　柴油发电机冷却子系统
engine lubricating oil subsystem　柴油发电机润滑油子系统
engine speed control subsystem　柴油发电机转速控制子系统
engineer station　(安全)工程师站
engineered safety features actuation (ESFA)　专设安全设施驱动
engineered safety features actuation subsystem　专设安全设施驱动子系统
engineered safety features coincidence logic　专设安全设施符合逻辑
engineering company　工程公司
engineering enthalpy rise factor　工程焓升因子
engineering factor　工程因子
engineering heat flux hot channel factor　热流密度工程热管因子
engineering hot channel factor　工程热管因子
engineering manual　工程手册
engineering practice and lesson　工程实践和经验教训
engineering safety feature actuation　专设安全设施驱动
engineering station　工程管理站
enhance, strengthen　加强
enriched　富集的,加浓的,浓缩的
enriched fuel　富集燃料
enriched uranium　富集铀
enriched uranium reactor　富集铀反应堆
enrichment　富集度
enter into an agreement　签置协议
enter into/establish business relations　建交
enterprise　企业集团
enterprise data server　企业数据服务器
enthalpy　焓
enthalpy rise engineering hot channel factor　焓升工程热管因子
enthalpy rise factor　焓升因子
enthalpy rise uncertainty factor　焓升不确定因子
entity　实体
entrainment　夹带
entrainment rate　夹带率
entrust　委托
entry　登录,进入,入口,报关手续
entry level requirement　入学水平要求
environment monitoring vehicle　环境监测车
environment qualification　环境鉴定
environmental assessment　环境评价
environmental condition　环境条件
environmental effect　环境效应
environmental monitoring　环境监测
environmental qualification program　环境评定大纲
environmental quality (EQ)　环境质量

environmentally friendly 环保的
epicenter (地震)震中
epitaxial solidification 联生结晶
epithermal neutron 超热中子
epoxy 环氧的,环氧树脂
epoxy sealer 环氧密封剂
epoxy top coat 环氧面漆
equality 平等
equalization 均衡,平等
equalization heater 均衡加热器
equalizing voltage 均衡电压
equilibrium 平衡,均衡
equilibrium cycle 平衡循环
equilibrium enrichment (燃料)平衡富集度
equilibrium no-load conditions 平衡零负载工况
equilibrium poisoning 平衡中毒
equilibrium power distribution 平衡功率分布
equilibrium samarium 平衡钐
equilibrium xenon 平衡氙
equipment classification 设备分级
equipment hatch 设备闸门
equipment hatch hoist 设备闸门吊车
equipment leakoff 设备引漏(水)
equipment list 设备明细表
equipment module 设备模块
equipment name 设备名称
equipment qualification 设备鉴定/评定
equipotential connection 等电压连接
equivalent 相当的,等量的,等效的
equivalent calculated capacity 当量计算排量
equivalent core diameter 堆芯等效直径
equivalent diameter 当量直径
equivalent dose 当量剂量
equivalent static load 等效静载荷
erection drawing 安装图
erosion 溶蚀
erosion/corrosion resistance 抗磨蚀/腐蚀
erosion-corrosion 磨蚀—腐蚀
erosion-corrosion-induced 磨蚀腐蚀导致的
escape 泄漏,逃脱
escape rate coefficient 逃脱率系数,泄漏系数
escape valve 泄放阀
especially 尤其
essential chilled water system 重要冷冻水系统
essential service water intake filteration system 重要厂用水取水过滤系统
essential service water pump 重要厂用水泵
essential service water pumping station system 重要厂用水泵房系统
essential service water yard pipe network system 重要厂用水厂区管路系统
estimate assumption 评估假设
estimated critical position(ECP) 预计临界控制棒位置
evaluate 对…估价,对…作评价
evaluation 评价

evaluation report 评估报告,评价报告
evaluation tool 评价工具
evaporation 蒸发(作用)
evaporative air cooling (EAC) unit 蒸发式空冷机组
evaporator 蒸发器
evenness of cutting surface 切割面平面度
event 事件
event oriented emergency operational procedure 事件导向应急操作规程
event plan 事件计划
event sequence 事件顺序
events of moderate frequency (condition Ⅱ) 中等频率事件(工况Ⅱ,Ⅱ类工况)
eventually 最后,终于
evolutionary 渐进的,改良的,进化的
exaggerate 夸大,浮夸
examination 检验
examination & approval 审查/批复
examination category 检查范畴,检查类别
excavation 开挖,负挖
exceptionally 特别地
excess flow check valve 限流止回阀
excess letdown 过剩下泄
excess letdown heat exchanger 过剩下泄热交换器
excess power 超功率
excess reactivity 剩余反应性,过剩反应性
excessive feedwater flow 给水过流量
excessive penetration 塌陷
excessive slumping 过度坍塌
exchanger 交换器
excitation 励磁
excitement 趣味性
exciter 励磁机
exciter air cooler 励磁机空气冷却器
exclusion area boundary 非居住区边界
exclusive agent 独家代理
exclusively 专门地,排他地
exclusivity clause 排他条款
excore detector 堆外探测器
ex-core instrumentation 堆外测量
ex-core long ion chamber 堆外长电离室
ex-core nuclear instrumentation 堆外核测仪器(仪表)
ex-core surveillance 堆外监测(指示)
excursion 骤增,猝增,漂移
execute a contract 履行合同
execute an order 执行订单
executive 总经理,董事
exemption 豁免
exercisability 可操作性
exercise testing (阀门)动作试验
exhaust 出口排气
exhaust cleanup unit 排风净化机组
exhaust fan 排风机
exhaust head 排气集管
exhaust hole 放气孔
exhaust hood 排气罩
exhaust valve 排气阀
exhausting smoke fan 排烟风机
exhaustiveness (论述等)详尽,彻底

exit steam pressure 蒸汽出口压力
expand 扩大,扩展
expanded part 胀件
expansion 膨胀
expansion bellow 膨胀波纹管
expansion joint 膨胀结合,伸缩缝
expansion ring 膨胀环
expansion tank 膨胀水箱
expansion, vibration and dynamic effects testing 膨胀、振动和动态效应试验
expected site boundary concentration limit 预计厂址边界浓度限值
expediting plan 催交计划
expediting techniques and tool 催货的技巧和工具
experimental stress analysis 试验应力分析
expert opinion 专家意见
expert report 专家报告
expertise 评定,鉴定(报告),专业鉴定
explicit arrangement 详细布置
explosive welding 爆炸焊接
explosively opening valve 爆破阀
export-oriented 出口型的
exposure 照射,曝射,曝光
exposure category 照射等级
exposure effect 辐照影响,辐照效应
exposure rate 辐照率
extend 拓宽,拓展
extend a warm welcome to 向…致以热烈欢迎
extend an offer 延长报盘
extend business 扩大业务,扩大经营
extend/renew an agreement (协议)延续/续约
extended flange 沿伸法兰
extended range 宽量程
extension stem (reach rod) thru shield wall 延伸穿屏蔽墙的阀杆
extensively 广泛地
exterior flooding 外部水淹
external 外部
external characteristic (电源)外特性
external contamination 外污染
external event 外部事件
external exposure 外照射
external interface 外部接口
external load 外负荷
external pressure 外压
external screw 外螺纹
external wall 外墙
extinction 消灭,消失,废除
extinguisher 灭火器
extraction 取出,抽出,拔除
extreme environmental loads 极限环境荷载
extremely 相当
ex-vessel cooling 压力容器外部冷却
ex-vessel core catcher 压力容器外堆芯捕集器

F

fabrication isometric 三维等轴加工图
fabrication weldability 工艺焊接性
face and bypass dampers 迎风阀和旁通阀
face bend testing 面弯试验
face of cut 切割面

face of weld　焊缝正面
face plate　面板
face-to-face dimension　结构长度
facilitate　使容易,使顺当,助长,促进
facility specialties　设施专用产品
factor of safety　安全因素
faculty　师资
fail closed　失效关闭
fail safe　失效安全
fail safety operation　失效安全操作
fail to honor the contract　不履行合同
failed fuel assembly location detecting system　破损燃料定位检测系统
failed fuel fraction　破损燃料份额
failure　失效,故障,损坏
failure analysis of weldments　焊件失效分析
failure detector module　故障检测模块
failure mechanism　失效机理
failure mode　故障模式
failure modes and effects analysis　故障模式和影响分析
fall into　分成,落入
fall short of　没有达到
fall within the scope of　属于…范围
fallout　辐射微尘,放射性沉降(物)
familiarize　熟悉
fan　风机
fan blower　风机
fan cooler　风机冷却器
fan coolers return containment isolation　风机冷却器回风管安全壳隔离
fans　风扇
fashion　方式,式样,风尚
fast connector　快拆接头
fast fission　快中子裂变
fast fission factor　快中子裂变因数
fast neutron flux　快中子注量率
fastener　紧固件,扣件
fastener locking device　紧固件锁紧装置
fast-neutron bombardment　快中子轰击
fatigue　疲劳
fatigue fracture　疲劳断口
fatigue strength　疲劳强度
fault　故障,事故,错误
fault current　故障电流
fault report　缺陷报告,故障报告
fault tree　故障树
faulted condition　事故工况,故障工况
faults of moderate frequency　中等频率故障
faying face　钎焊面
feasibility　可行性
feasibility study　可行性研究
feature　特色,特点
fecal pump　粪便泵
feed pump　给料泵
feed ring　给水环
feed water　给水
feed water heater　给水加热器
feed water pump　给水泵
feed water storage tank　给水贮存箱
feed-and-bleed　补给和排放,充排
feedback　反馈
feeder　馈电线,送水管
feedhead　(铸件的)冒口
feedwater booster pump　给水增压泵

feedwater check valve　给水止回阀
feedwater control　给水控制
feedwater control system　给水控制系统
feedwater control valve（FCV）　给水调节阀
feedwater cycling　给水循环
feedwater flow　给水流量
feedwater flow controller　给水流量控制器
feedwater flow demand signal　给水流量需求信号
feedwater inlet nozzle　给水进口管嘴
feedwater isolation　给水隔离
feedwater isolation trip　给水隔离停堆
feedwater isolation valve　给水隔离阀门
feedwater pump　给水泵
feedwater ring　(SG)给水环
feedwater valve　给水阀
ferreous　铁的
ferrite　铁素体
ferrite content　铁素体含量
ferrite side-plate　铁素体侧板条
ferritic　铁素体的
ferritic base material　铁素体母材
ferritic electrode　铁素体焊条
ferritic low-alloy　铁素体低合金钢
ferritic stainless steel　铁素体不锈钢
ferrule　套环
fertile nuclide　可转换核素
fettling　去毛刺
few-group parameter　少群(截面)参数
FHS fixed underwater light　燃料操作系统固定水下照明灯
FHS portable underwater light　燃料操作系统移动式水下照明灯
fiberglass　玻璃纤维
fiber-optic link　光纤链路,光纤连接
fibrous　纤维性的,含纤维的
fibrous debris　纤维碎片
fibrous zone　纤维区
field change　现场修改(设计)
field inspection　现场巡回检查
field mounted instrument　现场安装仪表
field of weld temperature　焊接温度场
field welding　现场焊
file　文件
filing　归档
fill　充注,填料
filler　装填物,充填物
filler block　端塞,衬块
filler material　焊接填充材料
filler metal　填充金属,焊条
fillet　钎角
fillet welding　角焊,条焊
film badge　胶片剂量计
film boiling　膜态沸腾
film boiling coefficient　膜态沸腾传热系数
film coefficient　膜层传热系数
film heat transfer　膜态传热
filter　过滤器,滤池
filter cartridge　过滤器滤芯
filter cartridge replacement　更换过滤器滤芯
filter glass　滤光镜片

filter pump 过滤泵
filter transfer cask 过滤器运输容器
filtered containment vent 安全壳过滤排放
filtration 过滤
filtration supply and exhaust system 过滤进排风系统
filtration units 过滤单元
fin 散热片
final acceptance 最终验收
final acceptance report 最终验收报告
final design 最终设计
final inspection 最后检查
final product pump 成品泵
final safety analysis report (FSAR) 最终安全分析报告
final storage of document 文件最终储存
finance 财务
financial position, financial standing 财务状况
financing 财务
findings 发现的问题,调查结论
fine 优良的,精美的,精细
fine grained region 细晶区
fine sand 细沙
finger 指状管
finger (cracking) test 指形裂纹试验
finishing 精加工,完工
finite element model 有限元模型
finned slab 翼缘楼板
fins 肋板,翅片
fire alarm 失火报警
fire and smoke damper 防火和防烟阀
fire annihilation protecting equipment 灭火保护装置
fire area 防火分区
fire barrier 防火屏障
fire break 防火间距,防火线
fire brigade 消防队
fire control cabinet 消防控制机柜
fire cracker welding 躺焊
fire damper 防火阀
fire detector 火灾探测器
fire extinction 灭火
fire extinction agent 灭火剂(器)
fire extinction pump 消防泵
fire fighting foam system 泡沫消防系统
fire hose 消防水龙带
fire hose station 室内消火栓(箱)
fire hydrant 室外消火栓
fire main 消防总管
fire proof installation 防火设施
fire protection 消防
fire protection analysis 防火分析
fire protection containment penetration 消防系统安全壳贯穿件
fire protection program 防火大纲
fire protection water system 消防水系统
fire pump 消防泵
fire rating 耐火等级
fire resistant 耐火的
fire secondary storage tank 辅助消防水储存箱
fire stop 阻火件
fire stopping wall 挡火墙

fire suppressant 灭火剂
fire suppression system 灭火系统
fire water 消防水
fire water containment test connection isolation 消防水安全壳试验接口隔离
fire water storage tank 消防水储存箱
fire water supply containment isolation 消防水供水管安全壳隔离
fire zone 防火区
fire-fighting water 消防水
firm 公司
first (fuel) cycle (燃料)第一循环
first bay 第一跨
first collision probability 首次碰撞概率
first core 初始堆芯
first out alarm 首发报警
first stage ADS isolation ADS 第一级隔离
FISCO (type) cracking test FISCO 裂纹试验
fish eye 白点
fishbone cracking test 鱼骨形裂纹试验
fissile nucleus 易裂变核素
fission and corrosion product activity 裂变和腐蚀产物活度
fission chain 裂变链
fission chamber 裂变(电离)室
fission cross section 裂变截面
fission energy 裂变能
fission fragment 裂变碎片
fission gas 裂变气体
fission gas release 裂变气体释放
fission neutron 裂变中子
fission product 裂变产物
fission product barrier 裂变产物屏障
fission product control system 裂变产物控制系统
fission product poison/poisoning 裂变产物中毒
fission product removal and control system 裂变产物去除和控制系统
fission rate 裂变率
fission source 裂变源
fission spectrum 裂变谱
fission yield 裂变产额
fissionable 可裂变的
fissure 裂纹
fittings 装配
fixed ball valve 固定式球阀
fixed differential reducing valve 定差减压阀
fixed in-core detector 固定式堆内探测器
fixed incore flux detector 固定式堆芯中子注量率探测器
fixed incore instrument 固定式堆内测量仪
fixed louver 固定百叶窗
fixed neutron detector 固定式中子探测器
fixed point 固定点
fixed position control 固定位置控制
fixed pressure reducing valve 定值卸压阀
fixed radio station 无线固定电台

fixed tube　固定管
fixed underwater light　固定式水下照明灯
fixture　固定设备,装配器,夹具
flake　白点
flame arrester　阻火器,火焰消除器
flame cone　焰芯
flame cutting　氧气切割
flame envelope　外焰
flame front propagation　火焰传播,火势蔓延
flame gouging　火焰气刨
flame heating efficiency　火焰热效率
flame planer　门式切割机
flame resistant　耐火烟的
flame spraying　火焰喷涂
flame straightening　火焰矫正
flame welding　卷边焊
flammability　易燃性,易燃度
flammable　易燃的
flange　法兰
flange ball valve　法兰式球阀
flange bolting　凸缘螺栓
flange end　法兰端
flange facing　法兰密封面,法兰面
flange facing finish　法兰面加工
flange joint　法兰连接
flanged connection　法兰连接
flanged coupling　法兰连接
flanged motor driven pump　法兰式电动泵
flanged overhung pump　法兰悬吊泵
flap valve　片状阀,瓣阀,拍门,铰链阀
flapper　(内置换料水箱)排气挡板
flapper valve　铰链阀,片状阀
flare groove welding　喇叭形坡口焊,胀口焊
flare test　(传热管)扩口试验
flare tube fitting　外喇叭口套管接头
flaring　扩口
flash　闪蒸
flash point(open cup and closed cup)　(开式杯和闭式杯)闪点测定计
flashback　回火
flashback arrestor　回火防止器
flashing　闪蒸汽化,闪蒸,闪烁
flashing alarm　闪光报警
flashing fraction　闪蒸份额
flask　长颈瓶,曲颈瓶,屏蔽容器
flat head　平封头
flat lock seam joint　平面锁缝接头
flat position weld　平焊
flattening　压扁,压平
flaw　缺陷
flaw characterization　缺陷特性
flaw detection sensitivity　探伤灵敏度
flexibility　灵活性
flexible connection　柔性连接
flexible cord　柔线
flexible coupling　挠性联轴器,挠性接头
flexible gate disc pump　弹性刮片泵
flexible pipe　软管,柔性管
flexural　弯曲的
flexural strength　抗弯强度
float ball valve　浮动式球阀
float charge　浮充电
float trap　浮动阀,浮动汽门

float type flowmeter　浮子流量计
float valve　浮球阀
float voltage　浮充电压
floatation tank　浮选池
floating dock pump　浮船坞泵
floating point　浮点
floating pump　浮动泵
flood level　洪水位,水淹水位
flood light　泛光灯,探照灯,泛光照明
flood source　洪水水源
flooding　淹没
flood-up　水淹,淹没,洪水上涨
flood-up level　淹没水位,涨水水位
floor contamination monitor　地面污染监测仪
floor drain　楼面疏水,地面疏水
floor drain tank　楼面疏水箱
floor module　地板模块
floor opening　楼板开孔
floor response spectra　楼面响应谱
floor seals　地板密封
floor sleeve　楼板套筒
floppy disk　软盘
flow & level corrections application software　流量和液位修正应用软件
flow area　流道面积
flow characteristics　流量特性,流动特性
flow characteristics derivation　流量特性推导
flow coastdown　流量惰走,流量惯性下降
flow control valve　流量控制阀
flow diagram　工艺流程图
flow diameter　流道直径
flow distribution　流量分配
flow distribution baffle　流量分配板
flow distribution plate　流量分配板
flow disturbance　流量扰动,流体扰动
flow induced vibration　流致振动
flow instability　流动不稳定性
flow lift force　流动升力
flow metering orifice　流量测量孔板
flow orifice plate　流量孔板
flow path　流道
flow rate　流量
flow rate rotometers　转子流量计
flow redistribution　流量再分配
flow resistance　流阻
flow restricting orifice　节流孔板
flow restrictor　限流器,流量限制器
flow sensor　流量传感器
flow soldering　波峰钎焊
flow transmitter　流量变送器
flow transmitter root valve　流量变送器根阀
flow-head curve　(泵)流量—扬程曲线
flow-induced　流致
flow-induced vibration　流致振动
flowmeter method　流量测量方法
flow-passing　流通,流经
flow-tuning orifice　流量调节孔板
fluctuation　波动,起伏
flue　烟囱,通风管,烟道
fluid load　流体荷载
fluid pressure　流体压力
fluid process monitor　流体工艺监测仪
fluidity of the slag　熔渣流动性

fluorescent lamp 荧光灯
fluorescent light 荧光,荧光灯
fluorescent penetrant inspection 荧光渗透探伤
fluoride 氟化物
flush 冲洗,使活力旺盛
flush lap joint 平面搭接头
flush line containment isolation relief 冲洗管安全壳隔离释放
flush pump 冲洗泵
flushing 冲洗
flushing and hydrostatic testing 清洗和水压试验
flush-mounted 镶嵌,暗装
flutter 颤振,摆动
flux 焊药,焊剂
flux activity 钎剂活性
flux backing 焊剂垫
flux coated brazing rod 药皮钎料,药皮焊条
flux density 通量密度,注量率密度
flux distribution 注量率分布,通量分布
flux flattening 注量率展平
flux for narrow-gap submerged arc welding 窄间隙埋弧焊焊剂
flux gradient 注量率梯度,通量梯度
flux inclusion 钎剂夹杂
flux instrumentation guide tube 注量率测量导管
flux layer 焊剂层
flux map 中子注量率分布图
flux paste 钎剂膏
flux shape consideration 注量率分布形状影响因素
flux thimble and mandrel 注量率测量子套管和芯棒
flux-cored colder wire 药芯钎料丝
flux-cored electrode 管状焊条,药芯焊条
fluxer 钎剂涂敷器
flux-hopper 焊剂漏斗
flux-shielded welding 熔渣保护焊接
flux-volume weighing 注量率—体积权重
flywheel failure 飞轮失效,飞轮故障
focal length 焦距
focal spot 焦点
focussed diameter of the beam 焦斑直径
folded joint 锁缝接头
folded type 折叠型
follow up action 后续行动
food chain 食物链
foot mounted pump 地脚固定泵
foot valve 底阀
forced circulation 强迫循环
forced circulation cooling 强迫循环冷却
forced convection 强制对流,强迫对流
forced flow 强制流动,强迫流动
forced reactor coolant flow 反应堆冷却剂强制流动
forced-air cooling 强制风冷
forcing function 力函数
foreground 前台
foreign direct investment(FDI) 外国直接投资

foreign elements　杂质元素
foreign material　异物，外来物
forenamed　前面举出的，前面所举的
forged part　锻制件
forging　锻件，锻造
forging part　锻件
forklift　叉式升降机，铲车
formalize　使成正式，正式化
format　格式，形式
formation　格式，形式，成型性
former　辐板（堆芯围板的径向支撑板）
forming　形成，组成，构成
forming part　成型件
forthcoming　即将来到的
forthwith　立即，立刻
foster　促进（发育），鼓励，扶植
foster cooperation and trust　促进发展合作与互信
fouling　污垢
fouling coefficient　（热交换器）污垢系数
foundation medium　基础介质
foundation ring　基础环
foundation settlement monitoring　基础沉降监测
four way redundant　四通道冗余
fourth stage ADS valve　第 4 级自动泄压系统阀
fractography　断口金相
fracture　断口
fracture analysis diagram　断裂分析图
fracture apperance　断口形貌
fracture mechanics　断裂力学
fracture test　断口试验
fracture toughness　断裂韧度
fracture toughness property　断裂韧性
fracture toughness test　断裂韧性试验
frame　框架，构架，机架
framework　框架，构架
free deformation　自由变形
free field　自由场
free float type steam trap　浮球式疏水阀
free from particular average　全损险（平安险）
free liquids　游离液体，自由液体
free on board　离岸价（船上交货价）
free-running oscillation test　自由振荡试验
freestanding　独立的
free-standing cladding　自立式包壳
freeze seal　冰冻密封
freezing　冻结
freight cost　运费
freon pump　氟利昂泵
frequency analysis　频谱分析
frequency counter　频率计
frequency meter　频率计，周波表
frequency range　频率范围
fresh fuel　新燃料
fresh water　淡水
fresh water pump　淡水泵
fretting　磨蚀
fretting corrosion　摩振腐蚀，激振腐蚀
fretting wear　微振磨损
friction coefficient　摩擦系数
friction factor　摩擦因子
friction loss　摩擦损失

fringe air pressure （热带气旋）边缘气压，外围气压
front computer 前置计算机
front panel 前面板，前仪表盘
fruitful 富有成效的
fuel 燃料
fuel/clad chemical interaction 燃料/包壳化学相互作用
fuel/clad mechanical interaction 燃料/包壳机械相互作用
fuel（rod）temperature coefficient 燃料(棒)温度系数
fuel additive 燃油添加剂
fuel and control rod 燃料和控制棒
fuel assembly 燃料组件
fuel assembly handling 燃料组件操作，燃料组件装卸
fuel assembly sipping device 燃料组件啜吸装置
fuel assembly visual inspection equipment 燃料组件目视检查装置
fuel basket 燃料篮，燃料筐
fuel building 燃料厂房
fuel building bridge 燃料厂房桥架
fuel building crane 燃料厂房吊车
fuel building ventilation system 燃料厂房通风系统
fuel building water supply and drainage system 燃料厂房给排水系统
fuel bundle 燃料棒束
fuel burn up 燃料燃耗
fuel centerline melt 燃料中心熔化
fuel cladding 燃料包壳
fuel cladding breach 燃料包壳破损
fuel cladding gap coefficient 燃料包壳间隙传热系数
fuel cladding temperature 燃料包壳温度
fuel cluster 燃料棒束
fuel column 燃料柱
fuel cycle 燃料循环
fuel cycle lifetime 燃料循环寿期
fuel damage 燃料破损，燃料损坏
fuel densification 燃料密实化
fuel depletion 燃料燃耗，燃料贫化
fuel element 燃料元件
fuel element burnout 燃料元件烧毁
fuel element failure accident 燃料元件破损事故
fuel enrichment 燃料富集度
fuel fission product 燃料裂变产物
fuel handling accident 燃料操作事故
fuel handling area exhaust 燃料装卸区域排风
fuel handling machine 燃料装卸机
fuel handling operation 燃料装卸操作
fuel handling system 燃料装卸系统
fuel inspection 燃料验查
fuel inventory 燃料装载量
fuel loading and unloading 装料与卸料
fuel loading pattern 燃料装载布置方案
fuel management 燃料管理
fuel mis-positioning accident 错装料事故，燃料错位事故
fuel oil cooler 燃油冷却器
fuel oil day tank vault exhaust fan 燃

油日用油箱房间排风机
fuel oil day tanks　燃油日用油箱
fuel oil heater　燃油加热器
fuel oil storage tank　燃油储存罐，储油箱
fuel pellet　燃料芯块
fuel pellet stack　燃料芯块垒柱
fuel pool bridge　燃料池桥架
fuel replacement program　换料程序
fuel rod　燃料棒
fuel shipping container rig　燃料运输容器吊具，燃料运输容器成套器械
fuel storage　燃料贮存
fuel storage and handling　燃料贮存和操作
fuel swelling　燃料肿胀
fuel system　燃料系统
fuel tank　燃料罐
fuel transfer canal (FTC)　燃料运输通道
fuel transfer conveyor　燃料转移运输机
fuel transfer shielding　燃料转运屏蔽
fuel transfer system　燃料转运系统
fuel transfer tube　燃料运输管
fuel transfer tube blind flange　燃料转运管道盲板法兰
fuel transfer tube isolation valve　燃料转运管道隔离阀
fuel transfer tube penetration　燃料转运贯穿段
fuel transfer tube test connection　燃料转输管道试验接口
fuel transportation　燃料运输
fuel tube　燃料包壳管，燃料管
fulfill an order　履行订单
full lift safety valve　全启式安全阀
full load　全载荷，满负荷，满功率
full load steam flow　满负荷蒸汽流量
full penetration weld　全透焊
full power　满功率
full scope simulator (FSS)　全范围模拟机
full vacuum electron beam welder　高真空电子束焊机
full-fledged PRA model　成熟的 PRA 模型
full-load rejection　满功率甩负荷
full-load torque　满载转矩/扭矩
function allocation　功能分配
function key　功能键
function test　功能试验
functional assignment　职责分工，功能分配
functional capability　功能范围，实有功能
functional requirement analysis　功能需求分析
functional test　功能试验
funnel　漏斗，烟窗
furnace brazing　炉中钎焊
furnace soldering　炉中钎焊
fuse　熔断器，熔丝
fuse isolator　熔断器刀闸
fuse panel　熔丝屏
fuse switch　熔断器开关
fused flux　熔炼焊剂
fusible backing　保留垫板

fusion type plasma arc welding　熔透型等离子弧焊

fusion welding　熔焊

fusion zone　母材熔化区

G

gage　标准度量,计量器具

gallery　走廊,美术陈列室,画廊,图库

galvanic corrosion　电偶腐蚀

gamma radiation　γ辐照

gamma radiation intensity　γ辐射强度

gamma spectrophotometer　γ谱仪

gamma-ray inspection　γ射线探伤

ganged reopening　同轴再开

ganged support hanger　成组的支吊架

gantry　龙门架

gantry crane　龙门吊

gap conductance　间隙热导

gap heat transfer coefficient　间隙传热系数

gap release　间隙释放

gas accumulated flowmeter　气体累计流量计

gas bill　气费

gas cavity　气孔

gas chromatograph　气体色谱仪,气相色谱仪

gas cooler　气体冷却器

gas cutting　气割

gas cutting machine with magnetic wheels　磁轮式气割机

gas cylinder　气瓶

gas flux　气体钎剂

gas forming constituents　造气剂

gas heater　气体加热器

gas insulated switchgear (GIS)　气体绝缘开关装置

gas jet laser cutting　喷气激光切割

gas laser　气体激光器

gas plenum　气腔

gas pore　气孔

gas release　气体释放

gas shielded arc welding　气体保护电弧焊

gas shielded arc welding machine　气体保护弧焊机

gas stripper　脱气塔

gas stripping　脱气

gas stripping unit　脱气装置,脱气单元

gas waste management system　放射性废气管理系统

gas welder　气焊工

gas welding　气焊

gaseous　气相的,气体的

gaseous nitrogen storage tube　氮气瓶

gaseous purification　气体净化

gaseous radioactivity　气态放射性

gaseous sample　气体取样,气样

gaseous space　气空间,气腔

gaseous waste treatment system　气体废物处理系统

gasification speed　气体发生速度

gasket　(密封)垫片

gas-tight　气密性

gate　门,闸门

gate valve　闸阀

gateway　网关,门卫通道

gateways to common-Q equipment interface　common-Q设备的网关接口

应用软件
gauge pressure　表压
G-BOP weld metal crack test　G-BOP 焊缝金属裂纹试验
gear　齿轮,排挡,用具,设备
gear coupling　齿轮联轴器
gear pump　齿轮泵
gear reducer　齿轮减速器
Geiger-Müller counter　盖革-弥勒计数器
Geiger-Müller tube　盖革-弥勒计数管,盖弥计数管(G-M 管)
general arrangement drawing　总体布置图,总布置图
general corrosion　整体腐蚀
general scintillation counter　通用闪烁计数器
general standard　通用标准
general test acceptance criteria and methods　试验验收总则和方法
general valve　通用阀门
generating　发电的
generation　生成,代
generator & main transformer system　发电机和主变压器系统
generator circuit breaker (GCB)　发电机出口断路器
generator protection subsystem　发电机保护子系统
generic data　通用数据
generic refueling toolbox　通用换料工具箱
generic safety issues　共性安全问题
genetic effect　遗传效应
geography　地理
geology　地质
geometrical interference　几何干涉
geometry　几何
get down to brass tacks / get down to business　进入正题
get nowhere　不会有任何结果
get stuck　纠缠
get the message　领会用意
get to the bottom of　寻根问底
get to the heart of　找到问题的关键所在
girder　主梁,钢桁的支架,撑柱
girth weld　环焊缝
give a briefing on　就…作一简单介绍
give and take　讨价还价
give away　泄漏
give ground　退让
give in　让步
gland　密封压盖,密封套,填料箱
gland packing　压盖填料
gland packing plug valve　填料压盖式旋塞阀
gland seal condenser　填料密封凝汽器
glass dosimeter　(荧光)玻璃剂量计
glass pump　玻璃泵
glassware　玻璃器皿
glide plane fracture　滑移面断口
global analyses　全球分析,整体分析
global trip　完全停堆
globe diaphragm valve　隔膜截止阀
globe valve　截止阀,球阀
glove box　手套箱
go into details　进行细节的探究

go out training 外出培训
go pulse 启动脉冲
goal-means analysis 目标一手段分析
gouging blowpipe 表面割炬
governance 管理
governor valve 调节阀
grab sample unit 手动取样,间断取样装置
gracious 有礼貌的,得体的
grain boundary ferrite 晶界铁素体
grain-size 晶粒度
grant 给予,授予
grant permission 准许,批准
granular 晶粒,颗粒状的
granular bainite 粒状贝氏体
graphic display 图形显示
graphic display technologies 图形显示技术
graphic package 图形软件包
graphic printer 图形打印机
graphic screen 图形显示屏
graphite 石墨
graphite packing 石墨填料
graphite type electrode 石墨型电极
grating 隔栅,窗栅栏,光栅
gravelly 多碎石的,碎石构成的
gravelly sand 砾砂
gravity coefficient of coating 药皮重量系数
gravity damper 自垂式阀门,自垂式风门
gravity head 重力压头
gravity relief damper 重力式泄压阀,重力式泄压风门
gravity welding 重力焊
gray/Gy 戈瑞
gray code 葛莱码
gray code shaper 葛莱码整形器
gray rod 灰(控制)棒
gray rod cluster assembly 灰棒控制组件
gray rod drive mechanism 灰棒驱动机构
grid 电网,网格,栅栏
grid cell 栅格,栅元
grid frequency 电网频率
grid frequency response 电网频率响应
grid impact testing 格架影响试验,电网影响试验
grid spring 格架弹簧
grid telephone network 电网通话系统
grinding 磨光
gripper 抓具,抓手
groove (焊接)坡口
groove welding 坡口焊
gross 粗的,毛的,总的
gross breach 破裂
gross failure 总体失效
gross weight 毛重
ground (工作)有所进展
ground deposition 地面沉积
ground fault 接地故障
ground horizontal acceleration 地面水平加速度
ground water 地下水
grounded electrode 接地电极
grounding conductor 接地导体
grounding grid 接地网

grounding resistance 接地电阻
grounding system 接地系统
grounding wire 接地线
groundwater drainage pump 地下水排水泵
groundwater level 地下水位
group constant 群常数
group display 成组显示
group selector 组选择器
guarantee 保证
guaranteed performance 保证性能
guard 保护,保卫,防护装置
guide 导则,指南
guide cylinder 导流筒
guide rail 导向轨道
guide stud 导向螺栓
guide thimble 导向子套管
guide tube 导向管,导管
guide tube assembly (控制棒)导向管组件
guide tube support plate 导向管支撑板
guide vane 导向叶片
guideline 导则,方针,指标
guillotine 断裂,斩断,铡刀,切断
gust 强风,大风
gust response factors 阵风反应系数
gutter 返回槽,回流槽,沟槽
gutter isolation valve 排水隔离阀
gyration 回转,旋转

H

habitability area (可)居留区
half load pump 半负荷泵
hammering 锻造,锻打
hand and foot monitor 手足监测仪
hand hole 手孔
hand pump 手泵
handheld digital multimeter 便携式数字万用表
handheld multimeter 手持万用表,便携式万用表
handheld phones 手持式电话机,手提电话
handling 搬运,装卸
handling opening 吊装孔,操作孔
handrail 扶手,栏杆
handset 手持式话机,手提电话,手机
hands-free speaking telephone set 免提扬声电话
hands-free type portable headset 免提移动耳机
hands-on experience 实践经验
hanger 支吊架
hanger locking pin 吊架锁定销
hard copy printer 硬拷贝打印机
hard facing 表面硬化
hard facing material 硬贴面材料,表面硬化材料
hard rock 硬质岩层,硬岩
hard rubber lined pump 硬胶衬里泵
hard surfacing 表面硬化
hardened steel 淬火钢
hardened zone 硬化区
hardfacing 耐磨堆焊
hardfacing electrode 耐磨堆焊焊条
hardfacing material 硬质堆焊材料
hardness 硬度
hardware 硬件

hardwired　硬接线
harmful effect　有害影响
harsh　严酷的，苛刻的，粗糙的
harsh environment　严酷环境
hash out　深入讨论
hatch　舱门，闸门
have full authority　完全有权
have good connections with/in　有好的渠道，有好的关系
have in common　共同之处
have no alternative but…　只能
have no intention of　没有…意向
have not much choice　没有选择余地
hazardous atmosphere　危险环境
head　封头
head block assembly　定滑轮组
head closure system　堆顶封闭系统
head cooling nozzle　顶盖冷却管嘴
head dome　顶盖球冠
head dome insulation　顶盖球冠绝热层
head flange insulation　顶盖法兰绝热层
head insulation　顶盖绝热层
head lifting rig　（反应堆）顶盖吊具
head loss　水头损失，压头损失
head office　总部，总公司
head plate　封头板，顶板
head vent　顶盖放气
headed ties　端部拉筋
header　联箱，集管
heading　标题
headquarters　总部
headset　耳机
headset phone　头戴式电话机
healing effect　愈合效应
health physics　保健物理，保健物理学
health physics and hot machine shop　保健物理和热检修车间
health physics area　保健物理区，（辐射防护）控制区
health physics calibration equipment　保健物理校准设备
health physics program　保健物理大纲
health physics record keeping software　保健物理数据记录保存软件
heat analysis　熔炼分析，热力分析
heat balance　热平衡
heat bank control　电加热组控制
heat capacity　热容量
heat conductivity　热导率
heat diffusion　热扩散
heat exchanger　热交换器
heat flow rate　热流量
heat flux　热流密度
heat flux engineering hot channel factor　热流密度工程热通道(管)因子
heat flux hot channel factor(FQ)　热流密度热通道(管)因子
heat generation rate　产热率，发热率
heat load　热负荷
heat loss　热损失
heat resisting steel　耐热钢
heat sink　热阱，冷源
heat sink maintenance　热阱维持
heat tracing　热跟踪
heat tracing system　热跟踪系统
heat transfer coefficient　传热系数
heat treatment　热处理

heat treatment for dehydrogenation　去氢热处理
heat up　加热
heat-affected zone　热影响区
heat-affected zone crack　热影响区裂纹
heater　(电)加热器
heater actuation setpoint　电加热器动作整定值
heaters　加热器
heating coils　加热盘管
heat-treatded zone　热处理区
heatup and cooldown rate　升温和降温速率
heavy equipment mobile transporter　重型设备运输工具
heavy handling system　重物(吊装)操作系统
heavy lift crane　重型起重机,重型吊车
heavy lift rigging wire rope　重型设备起吊索具/钢丝绳
heavy oil　重油
height under door　门底缝
height-to-diameter ratio　(堆芯)高径比
helical rotor pump　单螺杆泵
helical-wrap drive cable　螺旋形驱动电缆
helium　氦
helium leak check　氦气检漏
helium-arc welding　氦弧焊
helmet　面罩
hemispherical　半球形封头
hereby　特此,兹
herein　此中,于此,据此看来
hereinafter　在下文
hereunder　据此
herewith　与此一起
hermetic　密封的,与外界隔绝的
hexadecimal　十六进制的
hexagon　六边形的,六角形的
hierarchical　等级的
hierarchical two-tiered scaling (HZST)　按谱系的双层次比例分析
hierarchy　等级制度
high-and moderate-energy pipe　高能和中能管道
high cellulose (type) electrode　高纤维素型焊条
high confidence low probability of failure (HCLPF)　高置信度低概率失效
high differential pressure alarm　高压差报警
high efficiency filter　高效过滤器
high efficiency particulate and air filter　高效空气微粒过滤器
high energy pipe　高能管道
high frequency spraying　高频感应喷涂
high ground water table　高地下水位
high inertia rotor　(主泵)大惯量转子
high iron oxide type electrode　氧化铁型焊条
high manganese steel surfacing electrode　高锰钢堆焊焊条
high neutron flux　高中子注量率
high neutron flux trip　中子注量率高紧急停堆

high point 高点
high point venting 高点放气
high pressure heater drains pump 高压加热器疏水泵
high pressure melt ejection 高压熔堆喷射
high pressure nitrogen storage tank 高压氮气贮罐
high pressure pump 高压泵
high pressure valve 高压阀门
high primary coolant system leakage rate 主冷却剂系统高泄漏率
high purity water units 高纯水制备装置
high range 高量程
high range gamma radiation monitor for accident 事故用高量程γ辐射监测仪
high range ionization chamber area gamma monitor 高量程电离室区域γ监测仪
high rate of neutron flux increase trip 中子注量率高增长率紧急停堆，短周期停堆
high solids fraction 高颗粒度，高固态份额
high speed centrifugal pump 高速离心泵
high speed pump 高速泵
high speed recorder (4CH) 4通道高速记录仪
high speed recorder (8CH) 8通道高速记录仪
high speed turbo pump 高速涡轮泵
high speed welding flux 高速焊剂
high speed, redundant communications network 高速冗余通信网络
high temperature pump 高温泵
high temperature valve 高温阀门
high titania (type) electrode 高钛型焊条
high toughness super low hydrogen electrode 高韧性超低氢焊条
high voltage power supply 高压电源
high-current plasma arc welding 大电流等离子弧焊
high-cycle fatigue 高周疲劳
high-energy line 高能管线
high-energy photon 高能光子
high-energy pipe 高能管道
high-energy piping system 高能管道系统
higher specific speed pump 高比转速泵
high-level noble gas gamma monitor 高放惰性气体γ监测仪
high-low pressure interface 高低压力界面
high-pressure cartridge filter (reactor coolant filter) 高压柱式过滤器(反应堆冷却剂过滤器)
high-pressure ionization chamber 高压电离室
high-pressure reactor trip setpoint 高压停堆整定值
high-purity germanium 高纯锗
high-speed cutting 高速切割
high-speed nozzle 高速割嘴

hinge　铰链,枢纽
hinge pin　铰链销轴
hinged panel　带铰链的台屏板
historical storage retrieval　历史数据调用
historical work order　历史工单
hoist　电动葫芦,起重机械,提升机构
hoist and monorail　单轨电动葫芦
hoist overload　提升超载
hoist overspeed　提升超速
hoist overtravel　提升超行程
hoist speed　提升速度
hoisting　起重,提升
hold /conduct(meeting)　举行(会议)
hold consultations with　与…进行协商/磋商
hold down spring　压紧弹簧
hold point　停工待检点
hold…covered　给…保险
hold-down assembly　压紧组件
hold-down spring　压紧弹簧
hold-down support assembly　压紧支撑结构
holding brake　制动器,制动闸
holding time of brazing　钎焊保温时间
holdup capacity　滞留能力,贮存能力
hold-up tank　滞留箱
holdup volume　滞留容积,贮存容积
homogeneous equilibrium model(HEM)　均匀化平衡模型
hood　通风柜
hook　吊钩
hoop　环箍
horizontal acceleration　水平加速度
horizontal bend test　横弯试验
horizontal guide wheel　水平导向轮
horizontal multistage centrifugal pump　卧式多级离心泵
horizontal pipe　水平管道,横管
horizontal plane oscillation　水平面(功率)振荡,径向振荡
horizontal position weld　水平焊
horizontal velocity　水平速度
horn loudspeaker　喇叭形扬声器,喇叭
hose　软管,水龙带
hose house　水龙带站,消防箱
hose rack station　消火栓
hose reel station　水龙带卷盘,消防卷盘
hose stream　水柱
hot cell　热室
hot channel　热通道
hot channel factor　热通道因子,热管因子
hot condensate capacity　凝热能力
hot crack　热裂纹
hot extrusion　热挤压
hot functional testing　热态功能试验
hot gas soldering　热风钎焊
hot laundry ventilation system　放射性洗衣房通风系统
hot leg　(主管道)热段
hot leg flow instrument root　热段流量仪表根部
hot leg insulation package　热段保温包
hot leg level instrument root　热段水位仪表根部
hot leg nozzle　热段接管

hot leg sample isolation 热段取样隔离

hot machine shop area 放射性设备检修区间

hot performance test 热态性能试验

hot plate 热板,加热片

hot point factor 热点因子

hot rod integrated power 热棒积分功率

hot short 热脆的,热脆性的

hot shower 热淋浴器

hot shutdown 热停堆

hot shutdown condition 热停堆状态

hot sink 热阱

hot spot 热点

hot spot analysis 热点分析

hot standby 热备用

hot startup 热启动

hot straining embrittlement 热应变脆化

hot test 热态试验

hot wall type vacuum brazing furnace 热壁真空钎焊炉

hot water production and distribution system 热水生产与分配系统

hot water system 热水系统

hot water unit heater 热水装置加热器

hot wire MIG welding 热丝 MIG 焊

hot wire plasma arc surfacing 热丝等离子弧堆焊

hot wire submerged-arc welding 热丝埋弧焊

hot wire TIG welding 热丝 TIG 焊

hot zero power(HZP) 热态零功率

hot-dip galvanize 热浸电镀

hot-ductility test 热塑性试验

hotwell 热井

hourly variation coefficient 单位时间(小时)变化系数

house load operation 带厂用电运行

housekeeping 场地管理

HP counting room equipment 高压蒸汽统计间设备

HP sampling cooler 高压蒸汽取样冷却器

HP-heater & LP heater drain system 高—低压加热器疏水系统

H-type cracking test H 形裂纹试验

hub 接口,网络集线器,网络中心

hull's equivalent factor 赫尔当量因子

human error 人员差错,人员失误

human factors activity 人因活动

human factors engineering 人因工程

human factors engineering program 人因工程大纲

human factors engineering program review model 人因工程大纲评审模式

human performance monitoring 人员效能监测

human reliability analysis(HRA) 人员可靠性分析

humidifier 湿度调节器,加湿器

hunting 抖动

hydrant 室外消火栓

hydration 水合,水合作用

hydraulic 水力的,水压的,液压的

hydraulic actuator 液动装置,液压执行器

hydraulic condition 水力条件

hydraulic drag　水力阻力
hydraulic element　液压元件
hydraulic equivalent diameter　水力当量直径
hydraulic expand　液压胀
hydraulic load　水力载荷
hydraulic press　水压机
hydraulic resistance　水力阻力
hydraulic test　水压试验
hydrazine　联氨
hydro laser with pipe cleaning nozzle　带管道清洗喷嘴的液压激光器
hydrodynamic journal bearings　水力润滑径向轴承
hydrodynamic load　水力载荷
hydrodynamic stability　水力稳定性
hydrogen　氢
hydrogen add containment isolation　加氢安全壳隔离
hydrogen add containment isolation test connection　加氢安全壳隔离试验接口
hydrogen add containment isolation valve　加氢安全壳隔离阀
hydrogen add line containment penetration　加氢管线安全壳贯穿件
hydrogen ambient air vaporizer　氢气常温气化器
hydrogen analyzer　氢分析仪
hydrogen concentration monitoring subsystem　氢气浓度监测子系统
hydrogen embrittlement　氢脆
hydrogen grab sample cylinder　氢气取样瓶
hydrogen igniter　氢点火器
hydrogen ignition subsystem　氢点火子系统
hydrogen ignition system　氢点火系统
hydrogen induced cracking　氢致开裂
hydrogen measuring device　氢气测量装置
hydrogen peroxide　过氧化氢
hydrogen pickup　吸氢
hydrogen recombination subsystem　氢复合子系统
hydrogen recombiner　氢气复合器
hydrogen system　氢气系统
hydrogen test　测氢试验
hydrogenate　氢化，使与氢化合
hydrogen-induced crack　氢致裂纹
hydrostatic　流体静力学的
hydrostatic pressure　静水压力
hydrostatic test　水压试验
hydrostatic testing　水压试验
hygrothermograph　温度湿度计，双球温度计
hypochlorite　次氯酸盐
hypothesis，assumption　假设
hypothetical event　假设事件，假想事件

I

I&C functional requirement　仪控功能要求
I/O module　输入/输出模块
ice point reference　冰点参考
ice point reference well　冰点参考池
icicle　毛刺
ID grit blasting　内部喷丸处理

identical 共同的,一致的
identification 标记,编码,标志
identification tagging 编码,标记
identified leakage 可识别泄漏
identify 弄清,识别,验明
identify common areas of interest and establish rapport 寻找共同之处,建立融洽关系
idle 空闲的,空转的,无用的
igniter (氢气)点火器
ignition 点燃,触发
incomplete fusion 未熔合
IIS guide tube 堆内测量系统导管
IIS processing cabinet 堆内测量系统处理机柜
IIWcracking test IIW 裂纹试验
ill-advised 失策的,愚蠢的
illuminated pushbutton 带灯按钮
illuminated switch 带灯开关
illumination 照明
illustrate,elaborate,elucidate,spell out 说明,阐明
ilmenite type electrode 钛铁矿型焊条
immediate reactor trip 反应堆紧急停堆
immersion dose 浸没(照射)剂量
immersion method 液浸法超声波探伤
impact handwheel 撞击手轮
impact test 冲击试验
impact test of haz 热影响区冲击试验
impact velocity 冲击速度
impact,affect,influence 影响
impedance 阻抗
impedance voltage (transformer) 短路电压,(变压器)阻抗电压
impeller 叶轮
impingement plate 冲击板
implant test 插入试验,植入试验
implement,put into effect 实施
implementation 执行,履行,贯彻,落实
imply 暗示
importance analysis 重要度分析
imposed 强加,施加影响
improvise 临时准备,临时凑成,即席创作
impulse pressure 脉冲压力
impulse steam trap 脉冲式凝汽阀
impulsed laser welding 脉冲激光焊
impurity 杂质
in accordance with 依照,与…一致
in advance 预先
in charge of 负责
in compliance with 遵照
in conformity to 遵照,和…相适应,和…一致
in consideration of 考虑到,由于
in current use 在使用中
in due course 在适当的时候,到一定的时候
in due time 按时,在通常的时间内
in duplicate 一式两份
in good agreement with 符合…,照…
in line with 与…相符
in mass production 投入批量生产中
in motion 正在运动
in one lot 一次性,全部
in proportion to 按…的比例,…成比例,与…相称

in regard to 关于
in series 连续地,串联
in terms of 依照,按照
in that case 既然是那样
in the context of 在…背景下
in the form of 以…形式
in the interest of 符合…利益
in the light of 根据
in triplicate 一式三份
in view of 鉴于
inaccessible equipment 不可接近的设备
inadvertent actuation 误触发,意外触发
inadvertent boron dilution 意外硼稀释
inadvertent opening 误开
inadvertent operation 误动作
inadvertent safeguards actuation 安保误触发
inadvertent 无意的,疏忽的,意外的
incandescent 白热的,遇热发光的
incandescent lamp 白热灯
incandescent light 白炽灯
incentive 刺激,鼓励
incident 事件
incident of moderate frequency 中等频率事故
incineration 焚烧,焚化
inclined position welding 倾斜焊
inclusion 夹杂
inclusion content 杂质含量
incoming breaker 进线断路器
incomplete penetration 未焊透
incompletely filled groove 未焊满的坡口
incompletely filled weld 未焊满
inconel 718 alloy 因科镍718合金
inconel alloy 因科镍合金
inconspicuously 不太显眼地
in-containment fuel storage rack 安全壳内燃料贮存格架
in-containment refueling water storage tank 安全壳内换料水箱返回槽
in-containment refueling water storage tank injection 安全壳内置换料水箱注射
in-core 堆芯内,堆内
in-core control component 堆内控制部件
in-core detector 堆内探测器,堆芯探测器
in-core flux measurement 堆内中子注量率测量
in-core flux thimbles 堆内中子注量率测量指套管
in-core instrument guide tubes 堆芯测量导管
in-core instrumentation 堆内测量
in-core instrumentation room 堆内测量室
in-core instrumentation thimbles 堆内测量指套管
in-core instruments 堆内测量仪表
in-core instruments support system (IISS) 堆内测量支持系统
in-core neutron detector 堆内中子探测器
in-core neutron flux monitoring system

堆芯中子注量率监测系统
in-core temperature monitoring system　堆芯温度监测系统
in-core thimble assemblies　堆芯指套管组件
in-core thimble tubes　堆芯测量指套管
incur　引起,遭到
indefinite period　无限期
indefinite period of time　无限期的
indemnification　赔偿
indemnify　赔偿
independence from cost and schedule　不受经费和进度的约束
independent review　独立审查
index　索引,指数
index file　索引文件
indicate　指示,表示,指出
indicating lamp　指示灯
indication　指示,指出,标识
indication of inspection status　检查状态标识
indicator　指示器,标识剂
indicator lamp　指示灯,标识灯
indisputable　无可争辩的
individual dosimetry and management device　个人剂量监测和管理装置
individual training plan (ITP)　个人培训计划
indoor switchgear　室内配电装置
induced　感生的
induced radioactivity　感生放射性
inductance　电感,感应
induction　感应
induction brazing　感应钎焊
inductively coupled plasma(ICP) spectrometer　电感耦合等离子谱仪
industrial standards　工业标准
industrial wastewater　工业废水
industry /economic sector　工业行业/经济部门
industry safety(IS)　工业安全
inelastic scattering　非弹性散射
inert aerosol species　惰性气溶胶核素
inert gas　惰性气体
inert-gas backing　惰性气体衬垫
inertia　惯性,惰性
inertia impulse pump　惯性脉冲泵
inertia load　惯性荷载
inevitable　不可避免的,不可逃避的;必然的
inferior to　低于,小于
infinite lattice　无限栅格
infinite medium multiplication factor　无限介质增殖因数
inflammable limit of the gaseous　混合气体可燃浓度限值
information management system　信息管理系统
information processing and display interface　信息处理和显示接口
infrared brazing (soldering)　红外线钎焊
infrequent incidents(condition Ⅲ)　稀有事故(Ⅲ类工况)
ingot　铸块,锭
inhalation dose　内照射剂量,吸入剂量
inherent stability　固有稳定性
inherent strain method　固有应变法,

本征应变法
inherent strain zone　固有应变区,本征应变区
inherent stress　固有应力,本征应力
inhibitor　抑制剂,缓解剂
initial(fuel)cycle　初始(燃料)循环
initial core　首炉堆芯,初始堆芯
initial criticality　首次临界,初次临界
initial diffusible hydrogen　初始扩散氢
initial excess reactivity　初始剩余反应性,初始过剩反应性
initial fuel load　首次装料
initial inspection　初始视察,首次检查
initial postulated event(IPE)　假想始发事件
initial power　初始功率
initial startup　首次启动
initial test conditions　初始试验条件
initial test program　初始试验大纲
initiating event　始发事件
injection　注入,注射
injection flow　注入流,注射流
injection line　注入管线,注射管
injection valve　喷射阀,注射阀
injector　喷射器
injector torch　射吸流焊炬
inlet header　入口集管
inlet isolation valve　入口隔离阀
inlet nozzle　进口接管
inlet nozzle insulation　进口接管保温层
inlet temperature　进口温度
inlet valve　进水阀,进口阀门
in-line check valve　直通单向阀
in-line piping temporary strainer　直管道临时过滤器
in-line piston pump　单列式柱塞泵
inner annulus　内环腔
inner cone　焰芯
inner strap　(格架内)条带
inoperable　不可操作的
inorganic material　无机材料
inorganic zinc coating　无机锌涂料
inorganic zinc primer　无机锌底漆
in-pile densification　堆内(燃料)密实化
in-plane　平面内
in-plate area radiation monitoring system　区域辐射监测系统
input device　输入设备
input floor acceleration　输入楼面加速度
input forcing function　输入强制函数
input isolator module　输入隔离组件
input signal conversion block　输入信号转换块
input/output　输入/输出
inquiry　询盘(买方),询价
insert　嵌条
insertion of reactivity　反应性引入
in-service aging mechanism　在役老化机理
in-service inspection　在役检查
in-service maintenance　在役维修
in-service test　在役试验
inset T joint　嵌入T形接头
inside reactor containment　安全壳内
inside screw nonrising stem type gate

valve 暗杆闸阀
insight 见解,看法,洞察力
in-site controlled area environment radiation 厂内控制区环境辐射监测系统
in-site test 现场试验,厂内试验
insofar as 在…的情况下,在…范围内
inspectability 可检查性
inspection 检查
inspection and enforcement bulletin 检查和强制执行通告
inspection at delivery point 交接检查,到货检查
inspection interval 检查的间隔时间
inspection of work 工作检查
inspection plan 检验计划
inspection program 检查大纲
inspection record 检查记录
inspection report 检查报告
inspection requirement 检查要求
inspector 检查员
installation 安装,装配
installation(erection)drawing 安装图
installation drawing 安装图
installation instruction 安装说明
installment 分期付款
instantaneous 瞬时的,瞬刻的,即时的
instantaneous concentration heat source 瞬时集中热源
instantaneous relay 瞬时继电器
instruction sheet 指令单,通知单
instructor 教员
instructor lesson guide 教案
in-structure response 构筑物响应
instrument 测量仪器、仪表
instrument air 仪用空气
instrument air supply outside containment isolation 仪用空气供气安全壳外隔离
instrument room 仪表室
instrument saturation 仪表满量程
instrument span 仪表的量程
instrumental joint 仪表接头
instrumentation 测量仪表
instrumentation (stop) valve 仪表(截止)阀
instrumentation and control architecture 仪控总体结构
instrumentation and control system 仪控系统
instrumentation application 仪表应用
instrumentation metering valve 仪表计量阀
instrumentation thimble 仪表管
instrumentation tube 仪表管
insufficient 不够的,不足的,不充分的,不适当的
insulation 绝缘,隔热
insulation level 绝缘等级,保温隔热等级
insulation panel 保温层面板
insulator 隔热,隔热片(块)
insulator bolt connector wrenches 绝缘螺栓连接扳手,隔热螺栓连接扳钳
insurance policy 保险单
in-surge 波动
intact 完整无缺的,未用过的
intact containment (IC) 完整的安

全壳

intact loop　完整的环路

intake　进口,入口,摄入

intake enclosure　吸风罩

intake structure　取水构筑物

intangible　不现实的

integral　积分,积分的,整体,一体化

integral flow restrictor　(取样和仪表阀组)整体限流器

integral system test(IST)　系统整体试验

integrated flux　积分注量

integrated head package shroud　一体化堆顶结构围筒

integrated layout　总体布置

integrated leak rate testing　整体泄漏率试验

integrated worth　(控制棒)积分价值

integrity　完整性,性质,特性

intelligent　智能

intensification　增强,激烈化,增强对比度

intensity of back washing　返洗强度,返冲洗强度

interaction　交互作用,相互作用

interactive mode　交互方式

interactivity　互动性

interbedding　(地址)交互层

intercept valve　(MSR)再热截止阀,中间截止阀

interchangeability　可换性

intercom network　内部通信网

intercom system　调度电话系统,内部通信系统

intercrystalline crack　晶间裂纹

interface　分界面,焊接连接面,接口

interface drawing　接口图

interfacial region　钎缝界面区

interfacing system loss-of-coolant accidents　界面系统失水事故

interference　干涉,介入

intergranular　晶间的,晶粒间的

intergranular attack　晶间腐蚀

intergranular attack test　晶间腐蚀试验

intergranular corrosion　晶间腐蚀

intergranular fracture　晶间断裂

intergranular stress corrosion cracking　晶间应力腐蚀开裂

interlock　联锁

interlock mechanism　联锁机理,联锁机制

interlock protection　联锁保护

interlock signal　联锁信号

interlock system　联锁系统

interlocks important to safety　安全重要的联锁

intermediate containment failure(CFI)　中期安全壳失效

intermediate floor　夹层

intermediate flow mixer(IFM)grid　中间(流动)搅混格架

intermediate grid　中间格架

intermediate range　中间量程

intermediate range channel　中间量程(测量)通道

intermediate range high neutron flux reactor trip　中间量程中子注量率高

紧急停堆
intermediate range neutron flux preamplifier panel 中间量程中子注量率前置放大器配电板
intermediate-current plasma arc welding 中电流等离子弧焊
intermittent 间断的,断断续续的
intermittent weld 间歇焊,间断焊缝
internal 内部
internal contamination 内污染
internal department procedure(IDP) 部(处)内管理程序
internal exposure 内照射
internal flame 内焰
internal flooding 内部水淹
internal interface 内部接口
internal pressure 内压
internal source 内部(放射)源
internal support ledge 内部支撑凸台
internal training 内部培训
internals 堆内构件,内部构件
internals lift rig 堆内构件吊具
internals lifting support stand 堆内构件起吊支撑架
international nuclear trade 国际核贸易
interpass temperature (焊件)层间的温度
interposing relay 中间继电器
interpreter 解释程序,口译,翻译
interrupt 中断
interrupting capacity 遮断容量
inter-system LOCA 界面失水事故
intervention 干预,介入,干涉
intervention level 干预水平
intrinsical 固有的,本质的
intrinsically safe equipment 固有安全设备
introduce 引进,引入
inventory 存量
inventory balance 装载平衡,存量平衡
inverted 反向的,倒转的
inverted bucket 钟形罩
inverter 逆变装置
inverter welding power source 逆变式焊接电源
inverter 逆变器
in-vessel release 压力容器内释放
in-vessel retention(IVR) (堆芯熔融物)压力容器内滞留
investigating level 调查水平
investigation 调查
investment 投资
invoice 发票
invoice processing 发票处理手续
involuntary 不知不觉的,无意识的
iodine 碘
iodine-131 monitor 碘-131 监测仪
iodine cartridge 碘吸附盒
iodine detector 碘探测器
iodine removal efficiency 除碘效率
iodine spiking 碘释放峰,碘尖峰释放
iodine well 碘坑
ion chamber 电离室
ion chromatograph (IC) 离子色谱
ion chromatograph(IC) for anion analysis 阴离子色谱分析仪
ion chromatograph(IC)for canion analy-

sis 阳离子色谱分析仪
ion exchange resin 离子交换树脂
ion exchanger 离子交换器
ion exchanger after-filter 离子交换器后置过滤器
ion exchanger media replacement 更换离子交换树脂
ion exchanger pre-filter 离子交换器前置过滤器
ionic purification 离子净化
ionization chamber 电离室
ionization vacuum pump 电离真空泵
ionizing 电离
iron atom displacement 铁原子置换
iron out 解决,处理
iron oxide type electrode 氧化铁型焊条
iron soldering 烙铁钎焊
irradiated fuel 已辐照燃料
irradiated fuel assembly 辐照过的燃料组件,已辐照燃料组件
irradiated sample handling tool 辐照样本装卸工具
irradiation 辐照
irradiation creep 辐照蠕变
irradiation damage 辐照损伤
irradiation embrittlement 辐照脆化
irradiation growth 辐照生长
irradiation induced growth 辐照感生生长
irradiation sample handling tool 辐照样品操作工具
irradiation stability 辐照稳定性
irradiation surveillance capsule 辐照监督管
irradiation surveillance program 辐照监督大纲
irradiation tube plug seat jack 辐照监督管塞堵插孔
irradiation-assisted stress corrosion cracking 辐照促进应力腐蚀开裂
irrational arguments 非理性的争论
irrespective of 不管,不顾,不论,不考虑
irrigation pump 灌溉泵
IRWST bottom temperature 安全壳内置换料水箱底部温度
IRWST gutter 安全壳内置换料水箱返回槽
IRWST gutter bypass isolation valve 安全壳内置换料水箱返回槽旁路隔离阀
IRWST heat up test 安全壳内置换料水箱升温试验
IRWST injection 安全壳内置换料水箱注射
IRWST injection check 安全壳内置换料水箱注射止回阀
IRWST injection check test 安全壳内置换料水箱注射止回阀试验
IRWST injection isolation valve 安全壳内置换料水箱注射隔离阀
IRWST level transmitter isolation 安全壳内置换料水箱液位变送器隔离
IRWST screen 安全壳内置换料水箱滤网
IRWST surface temperature 安全壳内置换料水箱水的表面温度

IRWST to containment sump　安全壳内置换料水箱至安全壳地坑
IRWST water level　安全壳内置换料水箱液位
ISO 9000 family　ISO 9000 系列标准
isochoric　等容的
isokinetic sampling nozzle assembly　等速取样接管装置
isokinetic sampling nozzle for stack　烟囱等速取样接管
isolating amplifier (IA)　隔离放大器
isolating switch　隔离开关
isolating valve　隔离阀
isolation　隔离,隔绝
isolation damper　隔离风阀
isolation valve angle configuration　角形隔离阀
isolator　绝缘子,绝缘体,隔离器
isometric　等轴的,等轴图
isophase bus　离相(封闭)母线
isophase generator bus　发电机离相母线
isoseisms　等震线
isothermal temperature coefficient　等温温度系数
isotope　同位素
isotopic abundance　同位素丰度
isotopic material　同位素材料
isotopic stripping fraction　同位素萃取份额
issue　发给,发行,放出,(争论)问题,议题
issue management　问题管理,事务管理
it should be noted here that…　应该注意…
it will be appreciated if…　如果…,本人将不胜感激
item　物项,条款,项目
item important to safety　安全重要物项
itemize　开列

J

jet　喷射
jet impingement　喷射冲击,喷射撞击
jet pump with long Venturi tube　配有长文丘利管喷射泵
jet pump with short Venturi tube　配有短文丘利管喷射泵
jib　(起重机)臂
jib crane　悬臂式起重机
j-integral approach　j-积分法
job　岗位工作
job analysis　岗位工作分析
job and task analysis(JTA)　岗位工作和任务分析
job batching　分批加工
job competencies analysis(JCA)　岗位全面工作能力分析
job shadowing training　岗位影子培训
jockey pump　稳压泵
joggled lap joint　平面搭接头
Johnson screen　约翰逊滤网
joint　接头,接合处
joint design　接头设计
journal　轴颈
journal bearing　径向轴承

jumper　跨接线，搭接线
junction　接管，接头，接点
jurisdiction　管辖范围，权限
justification for conciliation　妥协的理由

K

keeper　保管员
kerf　切口
kerf angle　切口角
kerf width　切口宽度
key　键，销
key plan　索引图
keyboard　键盘
keyhole effect　小孔效应
keyhole-mode welding　小孔型等离子弧焊
key-locked switch　带钥匙开关
killed steel　脱氧钢，镇静钢
kind of　些须，有些
kinematic viscosity meter　动黏度计，动黏滞率计
kinetic characteristics　动特性
kinetic energy released in material　比释动能
Kinzel test　肯泽尔(Kinzel)弯曲试验
kirk-key interlock switch　kirk 键联锁开关
knife corrosion　刀口腐蚀
knife line attack　刀状腐蚀
knob　旋钮
knob-and-tube wiring　穿墙布线
knot　节(航速单位)
know-how　专门技能，知识，技巧
know-how training (KH)　专业技能培训课程
knowledge, skill and attitude (KSA)　知识、技能和态度
knowledge-intensive　知识密集型的

L

label　标号，标签，标记
labeling　编码，标记
labyrinth　迷宫，迷津
lack of fusion　未熔合
lack of penetration　未焊透
ladle　钢水包
ladle analysis　熔炼分析，桶样分析，炉前分析
lag time　滞后时间，延迟时间
lamellar segregation　层状偏析
lamellar tearing　层状撕裂
laminated　薄板状的，层积的，层压的
laminated part　轧件
lamp holder　灯座
lancing system　(蒸汽发生器管板清洗用)冲洗系统
lantern based pump　灯笼架支承泵
lantern mounted pump　灯笼架固定泵
lanthanide group　镧系元素
lap joint　搭接接头
lap welding　搭接焊
lapped T-joint/flanged T-joint　搭接 T 形接头
large break　大破口
large release frequency(LRF)　大量放射性释放概(频)率
large screen display in main control room　主控制室大屏幕显示器
large volume air sampler (LVAS)　大

容积空气取样器
large-bore three-way 大口径三通阀
laser beam cutting 激光切割
laser beam welding 激光焊
laser brazing (soldering) 激光钎焊
laser cutting(LC) 激光切割
laser printer 激光打印机
laser welding 激光焊
laser welding equipment 激光焊机
latch 钩爪,插销,碰锁,门栓
latch arm 钩爪,销爪
latch assembly 闭锁组件,插销组件
latch housing 插销盒,闭锁盒体
latching device 闭锁装置
latching relay 闭锁继电器
latch-locking plunger 钩爪锁紧柱塞,插销锁紧柱塞
lateral earth pressure 侧向土压力
lateral expansion 侧向膨胀
lateral load 侧向载荷
lateral movement 侧向运动
lath ferrite 条状铁素体
lath martensite 板条马氏体
lattice pitch 栅格间距
launch(a new product) 推出(一个新产品)
laundry system 洗衣房系统
lavatory 卫生间
lay on 安排
layout drawing 布置图
layup 接合处
leachable concentration 可浸出浓度
lead 铅,领导,引线
lead acid battery 铅酸蓄电池
lead auditor 主监查员
lead burnup rod 燃耗最深的燃料棒
lead container 铅室
lead test assembly 先导试验组件
lead time 产品从设计到交货的时间
lead-lag compensation 超前—滞后补偿
lead-lag compensation module 超前—滞后补偿模块
lead-lag compensator 超前—滞后补偿器
leak 泄漏
leak before break 先漏后破,破前漏
leak detector 泄漏探测器
leak proof 密封的,防漏的
leak test 密封性检验,泄漏试验
leakage flaw size 泄漏裂纹尺寸
leakage isolation 泄漏隔离
leakage rate 泄漏率
leak-off 引漏
leak-off connection 引漏接头
leak-off line 引漏管线
leak-test 检漏试验
leak-tight barrier 密封屏障
leak-tight integrity 密封完整性
least developed countries 欠发达国家
least square adjustment 最小二乘法
ledge 壁架,架状突出物,支撑凸台
leg 肢,段
legal force 法律效力,法律约束力
legal metrological unit 法定计量单位
legibility 字迹清楚
lehigh restraint cracking test 里海抑制裂纹试验

lesson plan　课程计划
let down　下泄
letdown containment isolation irc　下泄安全壳内隔离
letdown heat exchanger　下泄热交换器
letdown line containment penetration　下泄管安全壳贯穿件
letdown orifice　下泄节流孔板
letdown path　下泄通道
letdown pressure instrument root　下泄压力表根管
letdown purification line　下泄净化管道
letter of credit(L/C)　信用证
level detection instrumentation　液位检测仪
level differential pressure transmitter　差压式液位变送器
level reference leg　水位参考段
level switch　液位开关
level tap connection　液位引出接头
level transmitter root isolation valve　液位变送器根管隔离阀
levelized cost of electricity (LCOE)　平准化发电成本
levelized fuel cycle cost　平准化燃料循环成本
lever　杠杆
lever and weight loaded safety valve　杠杆式安全阀
lever reducing valve　杠杆式减压阀，杠杆式卸压阀
liability　责任，义务
liable to　易于…的，有…倾向的
liaison　联络
license　准许，发许可证给
licensee　接受授权者(接受方)
licensing　颁发执照
licensing document　许可证申请文件
licensing procedure　许可证申领程序
licensor　授权者(转让方)
lick side seam joint　平面锁缝接头
lift　提升，举起，开启
life (time) test　寿命试验
lifetime　寿期
lift armature　提升衔铁
lift check valve　提升式止回阀
lift coil　提升线圈
lift columns　提升支柱
lift height　起重高度，提升高度
lift legs　提升杆
lift off　提升脱开
lift pole　提升磁极
lift station　提升泵站
lifting and lowering speed　升降速率
lifting capacity　提升能力
lifting lug　提升吊耳
lifting of reservation　解除预留，解除预约
lifting rig　吊具，提升装置
lifting system　提升系统
light fixture　灯具，照明装置
light oil　轻油
light soldering (brazing)　光束钎焊
light source　光源
lighting panel　照明配电盘
lightning arrester　避雷针，避雷器
lightning protection　避雷保护

likelihood　可能性
likewise　同样
lime titania type electrode　钛钙型焊条
lime type covered electrode　碱性包覆焊条
limit approval　有限批准
limit load approach　极限载荷法
limit switch　行程开关,限位开关
limited access area　监控区,限制出入区
limited condition　极限工况
limited fault　极限故障,极限事故
limiting conditions for operation　运行极限工况,运行限制工况
limiting failure　极限失效,极限故障
limiting faults (condition Ⅳ)　极限事故(Ⅳ类工况)
limiting fuel rod　极限(燃耗)燃料棒
limiting power distribution (shape)　极限功率分布(形状)
limiting value　限值
line　行,线,电线,管线
line balance assembly　线平衡组件
line break　管线破裂,管道破裂
line of business　业务范围
line of communication　联络渠道,通信线
line routing　管道布置
linear analyses　线性分析
linear elastic fracture mechanics　线性弹性断裂力学
linear heat rate　线功率密度,线发热率
linear heat source　线热源
linear porosity　线性气孔率
linear power density　线功率密度
linear variety　线性变化
linearity　线性度
line-by-line basis analysis　(用于安全壳隔离阀再次开启的)逐根管线为基础的分析
line-mounted　管线连接,管线安装
link　(隔离)闸刀
linkage close　附带条件的成交
linkage editor　连接编辑
linking loader　链接加载程序
linking strategy　附加条件的战略
linters pump　剥绒机泵,短棉绒泵
liquation crack　液化裂纹
liquation of filler metal　钎料熔析
liquefaction potential　液化势,液化作用势
liquid and gaseous effluent monitor　液体和气体排出流监测仪
liquid carbon dioxide storage tank　液态二氧化碳贮存箱
liquid crystal test　液晶试验
liquid filter　液体过滤器
liquid hydrogen storage tank　液氢贮存箱
liquid level transmitter　液位变送器
liquid nitrogen pump　液氮泵
liquid nitrogen storage tank　液氮贮罐
liquid penetrant examination　液体渗透探伤
liquid phase　液相
liquid sample check valve　液体取样止回阀
liquid sample line containment isolation

IRC　液体取样管安全壳内隔离
liquid scintillation analyzer　液体闪烁分析仪
liquid space　液体空间
liquid waste management system　放射性废液管理系统
liquid waste receiver tank　废液接受箱
liquid waste treatment system　废液处理系统
lithium hydroxide monohydrate　氢氧化锂水化物
litigation　打官司,诉讼,起诉
live load　活(动)载荷
live part　带电部件
load　装,加载
load attachment point　(吊车吊钩)加载点
load block assembly　动滑轮组
load break switch　负载断路开关
load cell　负荷传感器
load cell (LC) with linkage　相互联系的负荷单元
load change　负荷变化
load combination　荷载组合
load cycling　负载周期,负荷循环
load factor　负荷因子
load follow　负荷跟踪
load follow capability　负荷跟踪能力
load follow operation　负荷跟踪运行
load group　负载组
load regulation　负载调节
load rejection　甩负荷
load rejection steam dump controller　甩负荷蒸汽排放控制器
load sensor　载荷传感器
load shedding　甩负荷
load steam loss　带载漏汽量
load surge　负荷冲击,负荷波动
load test fixture inside containment　安全壳内负荷试验装置
load test fixture outside containment　安全壳外负荷试验装置
load-deformation curve　应力—应变曲线,载荷—变形曲线
loader　加载程序,加料器
loading　装料,加载
lobe　圆形突出部位
local and remote indication (alarm)　就地和远程显示(报警)
local boiling　局部沸腾
local cabinet　就地机柜
local control room system　就地控制室系统
local control station　就地控制站
local deformation　局部变形
local drain valve in radwaste building　放射性废物厂房的就地疏水阀
local dry underwater welding　局部干式水下焊
local fire alarm panel　就地消防报警盘
local heat flux　局部热流密度
local leak rate testing　局部泄漏率试验,就地泄漏率试验
local missile damage　局部飞射物破坏
local panel　就地控制盘
local radiation processor　就地辐射处理器
local sampling　就地取样

local shielding 局部屏蔽
local stress 局部应力
local stress index 局部应力指数
local stress relieving 局部消除应力，局部应力释放
localing tube 定位(距)管
localization 本地化
localize 本土化，本地化
localized corrosion 局部腐蚀
locate 使定位，位于销
locating pin 定位销
lock corner joint 锁缝角接头
lock nut 锁紧螺母
lock out 闭锁
locked rotor 被卡转子
locked rotor accident (尤指主泵)转子被卡事故
locked-closed isolation valve 卡在“关”位的隔离阀
locking button 锁紧钮
locking cup 锁紧管，锁紧杯
locking device 加锁装置，闭锁装置
lockup pressure 锁定压力
lodge a claim 提出索赔
logbook (运行)日志
logic and interlock test 逻辑和联锁试验
logic bus 逻辑总线
logic error monitor 逻辑错误监测电路
logic isolator module 逻辑隔离组件
logic operation 逻辑运算
logic variable 逻辑变量
lone of cut 切割线
long ionization chamber 长电离室
long slag 长渣
longitudinal 纵向的
longitudinal crack 纵向裂纹
longitudinal magnetization 纵向磁粉检验
longitudinal seam 纵向焊缝
longitudinal section 纵剖面
longitudinal-bend test 焊道纵向弯曲试验
long-lived fission product 长寿命裂变产物
long-term cooling 长期冷却
long-term core burnup 长期燃耗
long-term loan 长期贷款
long-term monitoring function 长期监视功能
long-term safe shutdown 长期安全停堆
longtime rating 长期额定值，长期等级
loop 环路，回路
loop compartment 环路隔间
loop inaccuracy 环路误差
loop seal 环形水封(存水弯)
loop temperature sensor 回路温度传感器
loose part 松动部件，松脱零件
lorry 运货汽车，卡车
loss of coolant accident 冷却剂丧失事故，失水事故
loss of external load 丧失外负荷
loss of heat sink trip 丧失热阱停堆
loss of load 负荷丧失
loss of motive power position 失电位置，失去动力位置

loss of offsite power　丧失厂外电源
loss of power　失电
loss of reactor coolant flow　反应堆冷却剂流量丧失
loss-of-voltage　失(电)压
lot　批次,批量
lot number　批号
loudspeaker　扬声器
louver　百叶窗
louver damper　百叶窗式挡板
low activity α,γ detector　低放 α,γ 探测器
low alloy ferrite material　低合金铁素体材料
low alloy steel　低合金钢
low differential pressure opening check valve　低压差开启止回阀
low hydrogen type electrode　低氢型焊条
low hydrogen type flux　低氢型焊剂
low leakage　低泄漏
low lift safety valve　微启式安全阀
low load factor　低负荷因子
low poison flux　低毒焊剂
low population zone　规划限制区,低人口密度区
low pressure acetylene generator　低压乙炔发生器
low pressure injection　低压安注,低压安全注射
low pressure melt ejection　低压熔融物喷射
low pressure pump　低压泵
low pressure valve　低压阀门
low pressurizer pressure engineered safety features　稳压器低压专设安全设施触发信号
low profile debris filter bottom nozzle (LP DFBN)　短的带防异物滤网的燃料组件下管座
low radiation background work area　低本底工作区
low range　低量程
low range ionization chamber area γ monitor　低量程电离室区域 γ 监测器
low sound level pump　低噪音泵
low specific speed pump　低比转速泵
low temperature overpressure protection　低温超压保护
low temperature steel electrode　低温钢焊条
low temperature stress relieving　低温(拉伸)消除应力
low voltage distribution panel　低压配电盘
low voltage load centers　低压负荷中心
low-background β detector　低本底 β 探测器
low-current plasma arc welding　小电流等离子弧焊
low-cycle fatigue　低周疲劳
low-cycle strain fatigue　低周应变疲劳
lower compartment　下部隔间
lower core plate　堆芯下板
lower core support assembly　堆芯下部支承组件
lower core support plate　堆芯下部支

承板
lower core support structure　堆芯下部支承结构
lower flammability limit　较低的可燃限值
lower internals　下部堆内构件
lower internals storage stand　吊篮部件存放架,下部堆内构件存放架
lower limit　下限
lower plenum　(燃料棒)下气腔
lower shell course　(压力容器)下筒体段
lower support plate assembly　下栅格板组件
lower support plate fuel alignment pin　下部支承板燃料定位销
lower-tier contractor　低层次承包商
low-fume and low-toxic electrode　低烟尘低毒焊条
low-level noble gas β monitor　低放惰性气体β监测仪
low-level waste liquid monitor　低放废液监测仪
low-melting-temperature alloy　低熔点合金
low-pressure cartridge filter　低压过滤器
low-pressure cartridge pre-filter　低压前置过滤器
low-pressure engineered safety features actuation setpoint signal　低压专设安全设施触发设定信号
low-speed pump　低速泵
low-temperature physics testing　低温物理试验
LP sampling cooler　低压取样冷却器
L-pattern three way　L形三通
lube oil system　润滑油系统
lubricant　润滑剂
lubricated plug valve　润滑式旋塞阀
lubricator　润滑器,加油器
lug　耳状物,吊耳,拖,拉
luminous flux　光通量
lumped mass stick model　集中质量杆状模型
lumped parameter model　(热工)集中参数模型

M

mach shop DIST XFMR　机械车间配电变压器
machine tool　机械工具
machined part　机加工件
macro assembly　宏观汇编程序
macrofractography　断口宏观分析
macrograph　目视检查
macroscopic cross section　宏观截面
magnetic disk unit　磁盘机
magnetic fluid actuating pump　磁流泵
magnetic flux　磁性焊剂
magnetic force　磁力
magnetic inspection　磁粉检查
magnetic particle examination　磁粉检验
magnetic particle inspection　磁粉探伤
magnetic particle test　磁粉探伤
magnetic pump　磁泵,磁感应驱动泵
magnetic tape　磁带
magnetic tape unit　磁带机

magnetic trip breaker　磁脱扣断路器
magnetic valve　电磁阀
magnetically coupled gear pump　磁力联轴节齿轮泵
magnetization　磁化
magnitude of release　释放量
mail transfer(M/T)　信汇
main AC power system　主交流电源系统
main air flow path isolation valve　主空气流道隔离阀
main arc　主弧
main condensate pump　主凝结水泵
main condenser　主凝汽器
main contractor　总承包商,主承包商
main control room (MCR)　主控制室,中央控制室
main control room alarm processing system　主控制室报警处理系统
main control room emergency habitability system　主控制室应急可居留系统
main control room habitability ventilation system　主控制室可居留通风系统
main control room multiplexers　主控制室多路传输器
main control room supply air duct　主控制室送风管道
main control room system　主控制室系统
main coolant　主冷却剂
main distribution frame　总配线架
main equipment hatch　主设备闸门
main equipment hatch test connection　主设备闸门测试接口
main feedwater　主给水
main feedwater check valve　主给水逆止阀
main feedwater control valve　主给水控制阀
main feedwater isolation　主给水隔离
main feedwater isolation valve　主给水隔离阀
main feedwater piping penetration　主给水管道贯穿件
main feedwater pump　主给水泵
main feedwater pump lube oil cooler　主给水泵润滑油冷却器
main feedwater system　主给水系统
main fire protection/alarm panel　主消防/报警盘
main generator system　主发电机系统
main hoist　主吊,主起升机构
main hook　主钩
main steam and turbine bypass system　主蒸汽及汽轮机旁路系统
main steam dump control system　主蒸汽排放控制系统
main steam isolation valve　主蒸汽隔离阀
main steam line break　主蒸汽管道破裂
main steam line drain isolation valve　主蒸汽管疏水隔离阀
main steam line isolation bypass valve　主蒸汽管隔离旁通阀
main steam line isolation valve　主蒸汽

管隔离阀
main steam line penetration 主蒸汽管贯穿件
main steam line radiation monitor 主蒸汽管道辐射监测仪
main steam line vent isolation 主蒸汽管线排汽隔离
main steam pipe line break accident 主蒸汽管道破裂事故
main steam piping penetration 主蒸汽管道贯穿件
main steam power-operated relief valve 主蒸汽电动卸压阀,大气释放阀
main steam safety valve 主蒸汽安全阀
main steam supply system 主蒸汽供应系统
main stepup transformer 主升压变压器
main stop valve 主汽阀
main transformer 主变压器
main turbine control and diagnostic system 主汽轮机控制和诊断系统
main turbine lube oil cooler 主汽轮机润滑油冷却器
main turbine system 主汽轮机系统
main valve 主阀
mains outage 主电路故障
maintain composure 保持冷静
maintainability 可维修性
maintenance 维修
maintenance AC power source 检修交流电源
maintenance engineer 维修工程师
maintenance hatch 维修闸门
maintenance hatch hoist 维修闸门提升装置
maintenance tool 维修工具
major pipe break 大口径管道破裂
make(take) a note of 记录(下)
make a concession 做出让步
make a corresponding move 开始相应的行动
make a deal 成交
make a point of 强调,重视
make a presentation on /give a brief account of 介绍,阐述
make a prompt move 立即行动
make a sacrifice 做出牺牲
make an offer 报价
make concession 做出让步
make enquire about 咨询,询问,查询
make it 做到,办成功
make some headway 取得了一些进展
make up water 补水
make up water system 补水系统
makeup air stop valve 补气断流阀,补气截止阀
makeup filter 补水过滤器
makeup line containment isolation 补水管安全壳隔离
makeup line containment penetration 补水管安全壳贯穿件
makeup miniflow heat exchanger 小流量补水热交换器
makeup pump 补水泵
makeup stop valve 补水断流阀,补水截止阀
make-up water pump 补水泵

make-up water tank　补水箱
malfunction　故障
malleable　可锻的,有延展性的
malleable (cast) iron　(可)锻(铸)铁
management　管理部门,经理部
management review　管理评审
management training(MTT)　管理培训课程
managerial expertise　管理经验
managerial personnel　管理人员
managing director　总经理,总裁
mandatory standard　强制性标准,法定标准
manganese　锰
manhole　人孔,检查井,检修孔
manifold　多方面的,多样的
man-in-the-loop testing　(人机回路)工程试验
manipulating　搬运,装卸
manipulator crane　装卸料机
man-machine interface　人一机接口
manometer　压力表
manual　手册,细则
manual arc welder　手工电焊工
manual arc welding machine　手工弧焊机
manual block　人工闭锁,手动闭锁
manual bypass switch　手动旁路开关
manual charging handle　(开关箱)人工起动手柄
manual control　手动控制,人工控制
manual control mode　手动控制模式
manual electroslag welding　手工电渣焊
manual emergency drive　应急手动操作机构
manual fire suppression system　人工灭火系统
manual ground and test device　手动接地和试验装置
manual head vent　手动顶盖排气
manual initiation　手动触发,手工启动
manual oil pump valve　手动油泵阀
manual operation　手动操作
manual over-ride　手动干预,人工纠正
manual oxygen cutting　手工气割
manual valve　手动阀门
manual welding　手工焊
manual-electric changeover switch　手动一电动转换开关
manufacturer　制造商
manufacturing supervision　监造
manufacturing tolerance　制造公差
manway　人孔
manway cover　人孔盖
margin　裕量
margin of safety　安全裕量
marine pump　船用泵
mark　标记
mark the occasion　庆祝这个重大的盛会
marking　标记,编码,标志
marshalling box　编线盒,编组盒,信号编集盒
martensite-austenite constituent　m-a组元
martensitic　马氏体的
masonry structure　石工砌体结构
mass and energy release　质能释放
mass velocity　质量速度

master clock 母钟
master contract 主合同
master cycler 主循环器
master cycler output module 主循环器输出组件
mat foundation 底筏基础,筏基
material coefficient for penetration 材料穿透系数
material coupon 材料试块
material handling load cells unit 材料操作负荷传感器单元
materialize 实现,使具体化
mating surface 配合面
matrix 基体,母体,矩阵,模型
matrix material 基体材料
maximum air concentration 空气中最大浓度
maximum allowable cooldown rate 最大容许冷却速率
maximum allowable pressure 最高允许压力
maximum allowable temperature 最高允许温度
maximum angle of incidence 最大入射角
maximum credible accident (MCA) 最大可信事故
maximum daily output 最高日供应量,最大日输出
maximum differential pressure 最大压差
maximum flow rate 最大流量
maximum hardness test 最高硬度试验
maximum letdown flow 最大下泄流
maximum operating back pressure 最高工作背压
maximum operating pressure 最高工作压力
maximum operating temperature 最高工作温度
maximum potential vibratory ground motion 潜在的最大地面震动
maximum power rod 最高功率燃料棒
maximum precipitation runoff 最大降雨径流
maximum radius of rotational speed 最大转速半径
maximum rate of back pressure 最高背压率
maximum rate of pressure drop 最大压降速率
maximum reactivity feedback 最大反应性反馈
maximum rotational speed 最大转速
maximum shear theory 最大理论剪应力
maximum subcooled temperature 最大过冷度,最大欠热度
maximum temperature at discharging 最高排水温度
maximum translational speed 最大平移风速
maximum wind speed 最大风速
MCR air delivery flowrate 主控制室送风流量
MCR air delivery isolation valve 主控制室送风隔离阀
MCR air storage bottle pressure 主控

制室储气瓶压力
MCR isolation test valve 主控制室隔离试验阀
MCR penetration test valve 主控制室贯穿件试验阀
MCR return air isolation damper 主控制室回风隔离风阀
MCR RO WS A desk 主控制室反应堆操纵员工作站 A 桌
MCR RO WS B desk 主控制室反应堆操纵员工作站 B 桌
MCR SRO/SHIFT foreman desk 主控制室反应堆高级操纵员/值长办公桌
MCR supply air isolation damper 主控制室送风隔离风阀
MCR toilet exhaust isolation damper 主控制室卫生间排风隔离风阀
MCR wall panel displays 主控制室墙面大屏幕
MCR/CSA supplemental air filtration units MCR/CSA 增补空气过滤装置
MCR/ECR transfer switch 主控制室/应急控制室切换开关
MCR/LL transfer switch 主控制室/就地切换开关
MCR/RSW transfer panel 主控制室/远程停堆工作站切换台屏
MCR/RSW transfer switch panel 主控制室/远程停堆工作站转换开关盘
mean 平均值
mean dose 平均剂量
mean solidification rate 平均结晶率
measured rod position 测量的(控制)棒位
measurement box 测量箱
measurement channel for particle, noble gas, and iodine activity monitor 用于粒子、惰性气体和碘放射性检测仪测量通道
measurement error 测量误差
measurement pump 计量泵
measurement uncertainty 测量不确定性
measuring tank 计量槽,测量箱
measuring value 测量值
mechanic 机械的
mechanical clutch 机械离合器
mechanical connection 机械连接
mechanical damage 机械损伤
mechanical design flow 机械设计流量
mechanical draft cooling tower 机械通风冷却塔
mechanical fittings 机械装配
mechanical interlock 机械联锁
mechanical load 机械载荷
mechanical property 机械性能,力学性能
mechanical seal 机械密封
mechanical shim 机械补偿
mechanical stop 机械止挡块
mechanical strength 机械强度,力学强度
mechanical stress relieving 机械消除应力
mechanical test 机械试验,力学试验
mechanism 机制,机理,机构,机械装置
mechanism coil 机构线圈

mechanism of removal　去除机制，清除机理

mechanism seismic support structure　机械抗震支承结构

mechanistic pipe break　机械管道破裂

mechanistic pipe break evaluation　机械管道破裂评估

media　介质，媒介，媒体

media converter　介质转换器

median　中间的，中等的，中值，介质

median shell　中部筒体段

mediate　协调，调停

medical surveillance　医学监督

medicine protection　医学保护

medium sand　中砂

medium voltage　中间电压

medium-level noble gas γ monitor　中放惰性气体 γ 监测仪

medium-term loan　中期贷款

megawatt-days per metric ton of uranium（MW·d/tU）　兆瓦日/吨铀（燃料燃耗的单位）

melting characteristics of an electrode　焊条熔化性能

melting temperature　熔点

membrane　膜，隔膜

membrane force　薄膜力

memorandum　备忘录

memorandum of understanding　谅解备忘录

memory module　记忆组件

mercurad 3D simulation software for dose rate & shielding calculations　用于剂量率和屏蔽计算用的 mercurad 三维模拟软件

mercury air pump　汞（蒸汽）泵

mercury diffusion pump　汞扩散泵

mercury ejector pump　汞喷射泵

mercury vapor jet pump　汞蒸汽喷射泵

mesh　网格，筛孔

messenger tray　（用于 CRDM 电力/堆内仪表/棒位指示的）电缆托架

metal active-gas arc welding　活性气体保护金属极电弧焊

metal arc welding　金属极电弧焊

metal clad　金属包壳

metal impact　金属撞击，金属碰撞

metal impact monitoring system　金属撞击监测系统

metal inert-gas welding machine　金属极惰性气体保护弧焊机

metal reflective equipment insulation　金属反射型设备隔热层

metal reflective insulation　金属反射隔热层

metal reflective type (insulation)　金属反射型（隔热）

metal spraying　金属喷涂

metallic debris　金属碎片

metallographic examination　金相检验

metallurgical structure　金相结构

metallurgical weldability　冶金焊接性

metallurgy　冶炼，冶金

meteorological data　气象资料，气象数据

meteorological observatory　气象观测台

meteorological station　气象站

meteorological system interface application software　气象系统接口应用软件
meter calibrator　测量装置校准器，计量设备标定仪
meter needle type globe valve　表针式球芯阀，表针式截止阀
methodically　有条不紊地，有组织地，有次序地
metrological　计量的，度量衡学的
microalloying　微量合金化
microbiological　微生物的
microcomputer　微机
micro-crack/micro-fissure　微裂纹
microfissuring　微裂隙，微裂纹
micrograph　显微检查
micro-plasma arc welding　微束等离子弧焊
micro-plasma welding equipment　微束等离子弧焊机
microscopic cross-section　微观截面
microsoft windows platform　微软视窗平台
mid-arm　分支管
middle of life (MOL)　寿期中
middle pressure valve　中压阀门
mid-loop operation　半充水运行，半水位运行
milestone　里程碑
MIM reactor coolant pump instrument　金属碰撞检测设置在反应堆冷却剂泵的仪表
MIM reactor vessel upper head instrument　金属碰撞检测设置在反应堆压力容器上封头的仪表
MIM steam generator instrument　金属碰撞检测设置在蒸汽发生器的仪表
mimic　模拟的，模仿的
mimic diagram　模拟(流程)图
mimic panel　模拟盘
miniature　小型的，缩小的，缩图，缩微模型
miniature fission chamber　微裂变室，小型裂变室
mini-flow circulation mode　小流量循环方式
miniflow orifice　小流量孔板
minimize　使减到最少
minimum burnup rod　燃耗最浅(燃料)棒
minimum differential pressure　最小压差
minimum flow recirculation　最小流量再循环
minimum inventory　最低库存
minimum measured flow　最小被测流量，最小可测流量
minimum operating pressure　最低工作压力
minimum reactivity feedback　最小反应性反馈
minimum service head　最小使用压头
minimum subcooled temperature　最小欠热度，最小过冷度
minimum yield strength　最小屈服强度
minor pipe break　管道小破口
minute　备忘录，记录
misalignment　(控制棒)未对中，错对

中,错边
misc AHUs 其他空气操作单元
misc I&C orifice plate 其他仪控节流孔板
miscellaneous 各种混在一起,混杂的
miscellaneous BOP building ventilation system 各种 BOP 厂房通风系统
mismatching deformation 错边变形,失配变形
missile barrier 飞射物屏障
missile load 飞射物荷载
missile shape factor 飞射物形状因子
misunderstanding 误会,误解,曲解
mitigate 减轻,减缓
mitigate the consequences of postulated accident 缓解假想事故后果
mitigation 缓解,减小
mixed bed 混床
mixed bed demineralizer 除盐混合床
mixed flow pump with blades adjustable when stationary 半可调式混流泵
mixed gas arc welding 混合气体保护电弧焊
mixed resin 混合树脂
mixer 搅拌器
mixing 混合
mixing chamber 混合室
mixing coefficient 搅混系数
mixing ratio 混合比
mixing vane 搅混翼,搅混叶片
mobile (processing) equipment 移动式(处理)设备
mobile (processing) system 移动式(处理)系统
mobile system facility 移动式系统设施
mock-up 模拟体
mode of operation 运行方式
mode of transport 运输方式
model 模型
model for quality assurance 质量保证模式
model test 模型试验
modem 调制解调器
moderate energy piping 中能管道
moderator 慢化剂
moderator defect 慢化剂亏损
moderator density coefficient 慢化剂密度系数
moderator pressure coefficient 慢化剂压力系数
moderator reactivity coefficient 慢化剂反应性系数
moderator temperature coefficient (MTC) 慢化剂温度系数
moderator temperature defect 慢化剂温度亏损
moderator void coefficient 慢化剂空泡系数
moderator-to-fuel ratio 慢化剂—燃料比,氢铀比
modification observations 修改观测报告
modified checkerboard pattern 修正的棋盘式交叉换料方案
modified mercalli intensity 修正的麦加利烈度(地震术语)
modify,revise,amend 修改,修正

modulation　调制
module　组件,模块
module boundary　模块边界
module calibration adjustment　模块标定调整
module interaction　模块交互作用
module plan　单元计划
modules of water glass　水玻璃模数
modulus　模量,模数,系数
modulus of elasticity　弹性模量
moisture content　含水量,湿度
moisture separator　汽水分离器
moisture separator reheater　汽水分离器再热器
mold for thermit weld　热剂铸模,热熔模铸
molecular　分子
molecular air pump　分子空气泵
molecular drag pump　分子拖曳泵
molecular pump　分子泵
molten core concrete interaction　熔融堆芯—混凝土相互作用
molten debris　堆芯熔化碎片
molten metal bath dip brazing　金属浴钎焊
molybdenum　钼
moment　力矩
momentary　瞬时的,瞬态的
momentary cooling rate　瞬时冷却速度
momentary rating　瞬时额定值
momentum　动量
momentum exchange　动量交换
monel electrode　蒙乃尔焊条
monitor　监督,监视,监控
monitor tank　检测槽
monitor tube　检测管
monitoring　监测
monitoring instrumentation　监测仪器(仪表)
monitoring system　监测系统
monitoring unit　检测装置,监测单元
mono energetic　单能的
mono pump　单泵
mono stable module　单稳态组件
monoblock scoop pump　单体斗式泵
monorail　单轨
monorail hoist　单轨吊车
Monte Carlo theory　蒙特卡罗方法,蒙特卡罗理论
monthly training plan　月度培训计划
most damaging direction　最不利的方向,最有害的方向
most stringent loading condition　荷载最严酷的工况/条件
motion　(议会中的)提议
motivating mechanism　激励机制
motivation　动机
motor　电机,马达
motor control center　电机控制中心
motor driven gear positive displacement type　电动齿轮正位移型
motor driven pump　电动泵
motor operated valve　电动阀
motor operator　电动执行机构
motor stand　电机支座
motor terminal box　电机端子箱
motor-driven single beam suspending crane　电动单梁悬挂吊车

motor-generator set 电动发电机装置
motorized valve 电动阀
mounting 安装,装配
mounting plate 安装盘
movable armature 活动衔铁,保持衔铁
movable detector 可移动探测器
movable gripper coil 移动钩爪线圈
movable gripper latch 移动钩爪闭锁
movable miniature fission chamber 移动式微型裂变室
movable neutron detector 可移动中子探测器
movable part 可动部分
movable pole 移动磁极
move on to 继续
move on, proceed with 进行
MP sampling cooler 中压取样冷却器
MSHIM(mechanical shim) 机械补偿
MSIV bypass control isolation 主蒸汽隔离阀旁路控制隔离
MSIV bypass isolation 主蒸汽隔离阀旁路隔离
MSIV monorails hoist 主蒸汽隔离阀单轨吊车
mud 泥浆
mud drum 泥浆桶
mud scraper 刮泥机
mud settler 泥浆澄清槽
mudmat 垫层
multi cell pump 泵组
multi channel impeller pump 多流道叶轮泵
multi cylinder(reciprocating) pump 多缸往复泵
multi discharge pump 多出口泵
multi flow pump 多流量泵
multi lobe pump 多凸轮泵
multi nozzle jet pump 多喷嘴喷射泵
multi plunger pump 多柱塞泵
multi screw pump 多螺杆泵
multi throw(reciprocating) pump 多缸往复泵
multi throw crank pump 多缸曲柄泵
multichannel spectrum analyzer 多道—能谱分析仪
multi-disc swing foot valve 旋启多瓣式止回阀
multi-layer welding 多层焊
multimeter 万用表
multi-pass weld 多道焊缝
multi-pin socket 多芯插座
multiple failures 多重失效
multiple gear pump 多齿轮泵
multiple package contract 零散分包合同,多分包合同
multiple setpoint 多重整定值
multiple transfer switch 多路切换开关
multiple wire submerged arc welding 多丝埋弧焊
multiplexers 多路传输器
multiplication factor 倍增因子
multi-purpose 多用途的
multipurpose pump 多用泵
multistage jet pump 多级喷射泵
multistage pump 多级泵
multistage pump with single impeller

单叶轮多级泵
multistage segmental type pump 分段式多级泵
multistage vertical turbine pump 长轴深井泵,多级垂直涡轮泵
multizone core 多区堆芯,分区堆芯
murex hot cracking test 缪雷克丝热裂纹试验
mushroom valve 盘阀,菌形阀,蘑菇阀
mutual benefit 互利

N

N-16 radiation monitor 氮-16 辐射监测仪
name plate 铭牌
narrow gap welding 窄缝焊
narrow groove welding system 窄坡口焊接系统
narrow range 窄量程
narrow range low level setpoint 窄量程低水位整定值
narrow-gap electroslag welding 窄间隙电渣焊
narrow-gap submerged arc welding 窄间隙埋弧焊
national plumbing code valve 符合(美国)给排水规范的阀门
national standard 国家标准
national standard body 国家标准机构
natural background 天然本底
natural circulation 自然循环
natural circulation test 自然循环试验
natural draft 自然通风
natural draft cooling tower 自然通风冷却塔
natural frequency 固有频率
natural isotopic abundance 天然同位素丰度
natural phenomena 自然现象
natural processes 自然过程
natural radioactivity 天然放射性
natural removal mechanism 自然去除机理
natural uranium dioxide 天然二氧化铀
navigation 航海,航空,导航,领航
navigation displays for nuclear island 核岛的导航标志
NC cutting machine 数控切割机
near-term operation license 近期运行许可证,短期运行许可
neatness 清洁
needle valve 针形阀
needless to say 不用说
negative moderator temperature coefficient 慢化剂负温度系数
negative power coefficient 负功率系数
negative pressure 负压
negative rate 负率
negative reactivity 负反应性
negative reactivity feedback 负反应性反馈
negative reactivity margin 负反应性裕量
negotiable 可协商的,可兑现的
negotiation 协商,谈判,交涉,协定
negotiation consciousness 谈判意识
nephelometer 浊度计
net free volume (安全壳)净自由容积

net weight　净重
network　电网,网络
network system　电网系统
neutral flame　中性焰
neutral point　中性点,中和点
neutralizer tank　中和槽
neutron absorber　中子吸收剂
neutron absorber material　中子吸收材料
neutron absorber rod　中子吸收棒
neutron absorption　中子吸收
neutron absorption cross section　中子吸收截面
neutron activated corrosion product　中子活化腐蚀产物
neutron activation　中子活化
neutron albedo　中子反射率,中子反照率
neutron counting rate　中子计数率
neutron density　中子密度
neutron detector　中子探测器
neutron diffusion theory　中子扩散理论
neutron dose rate-equivalent monitor　中子剂量当量率监测仪
neutron energy group　中子能群
neutron exposure　中子辐照,中子照射
neutron flux　中子注量率
neutron flux detector　中子注量率探测器
neutron flux doubling　中子注量率倍增
neutron flux range　中子注量率量程
neutron generation time　产生每代中子的平均时间
neutron hardening　中子谱硬化
neutron irradiation　中子辐照
neutron leakage　中子泄漏
neutron lifetime　中子寿命
neutron multiplication　中子增殖
neutron pad　中子衬垫
neutron perturbation theory　中子微扰理论
neutron poison　中子毒物
neutron reflector　中子反射层
neutron source　中子源
neutron source assembly　中子源组件
neutron source rod　中子源棒
neutron spectrum　中子能谱
neutron streaming　中子束,中子流
neutron temperature　中子温度
neutron thermalization　中子热化
neutron transport　中子输运
neutron transport equation　中子输运方程
new fuel assembly handling tool　新燃料组件装卸工具
new fuel assembly inspection operating　新燃料组件检查操作
new fuel assembly measuring handling tool　新燃料组件测量操作工具
new fuel ELEV & hoist　新燃料电梯/吊车
new fuel elevator　新燃料升降机
new fuel elevator and hoist　新燃料升降和起吊机构
new fuel handling tool　新燃料操作工具
new fuel JIB crane　新燃料悬臂吊车

new fuel pit　新燃料贮存间
new fuel rack　新燃料格架
new fuel shipping container　新燃料运输容器
new fuel shipping container console　新燃料运输容器控制台
new fuel storage rack　新燃料贮存格架
new fuel storage rack system　新燃料贮存格架系统
new RCC handling tool　新控制棒束装卸工具
NFPA code valve　符合美国国家防火协会法规的阀门
NI fire protection system　核岛消防水系统
nickel　镍
nickel base alloy covered electrode　镍基合金焊条
nickel equivalent　镍当量
nickel-based alloy　镍基合金
nickel-chromium-iron alloy　镍-铬-铁合金
nickel-chromium-iron alloy 718　inconel-718 合金，镍-铬-铁 718 合金
nil-ductility transition　无延性转变
nil-ductility transition reference temperature　无延性转变参考温度
nil-ductility transition temperature (NDTT)　无延性转变温度
nitriding　氮化，渗氮(处理)
nitrogen　氮
nitrogen blanket　氮气覆盖层
nitrogen low-pressure ambient air vaporizer　氮气低压常温气化器
nitrogen makeup containment penetration　氮气补给管安全壳贯穿件
nitrogen skid　氮气成套设备，氮气装置
nitrogen system　氮气系统
nitrogen-arc welding　氮弧焊
no unique design basis accident　非单一设计基准事故
no xenon　无氙
noble gas　惰性气体
noble gas detector　惰性气体探测器
nodal theory　节块理论，节块法
no-load　零负荷，空载荷
no-load condition (zero load condition)　零负荷工况
no-load loss　空载损失
no-load operation　空载运行
no-load RCS pressure and temperature conditions　反应堆冷却剂系统零负荷压力、温度条件
no-load steam loss　无负荷漏汽量，空载漏汽量
nominal (rated) pressure　额定压力
nominal diameter　公称通径
nominal operating loads　额定运行负荷
nominal power　名义功率，额定功率
nominal range　额定量程，额定范围
nominate　指定，任命
non uniformity　不均匀性
nonbinding　非约束力的
nonbrittle　不易破裂的，结实的
non-clog pump　不堵塞泵
noncombustible material　不燃物

noncondensable gas 不凝气体
non-conformance 不符合项
non-conformance control 不符合项管理,不符合项控制
non-conformance review 不符合项审查
nonconforming item 不符合项
non-conforming processing 不符合项处理
non-conforming processing,tracking and resolution 不符合项处理,跟踪和解决
non-consumable electrode 不熔化电极
noncorrosive 无腐蚀性的,不锈的
non-destructive examination（NDE） 无损检验
non-destructive test(NDT) 无损检测
non-essential chilled water cooling water system 非重要冷冻水冷却水系统
non-essential chilled water system 非重要冷冻水系统
non-fission capture （中子）非裂变俘获
non-hermetic screw pump 非气密螺杆泵
non-isolable leak 不可隔离性的泄漏
non-linear analyses 非线性分析
non-metallic inclusion 非金属夹杂物
non-negotiable 不可兑现的
non-operational load 非运行载荷
non-polluted industrial wastewater 未被放射性污染的工业废水
non-proprietary 非专利,未申请专利
non-radiologically controlled area（non-RCA） 非放射性控制区,非RCA区
non-return valve 逆止阀,止回阀
non-safety related 跟安全不相关的,非安全级的
non-safety related components 非安全相关部件
non-safety related display 非安全相关显示
non-safety related functions 非安全相关功能
non-safety related instrumentation and control function 非安全相关仪控功能
non-safety related plant control system 非安全相关电厂控制系统
non-safety related plant system 非安全相关电厂系统
non-safety related system 非安全相关系统
non-scale 无氧化皮的,不起皮的
non-segregated bus load center tie 整体式母线负载中心结,非隔离母线负载中心结
non-segregated phase bus 非隔离相母线,整体式相母线
non-seismic 非抗震的
non-seismically-supported 非抗震支承的
nonsoluble 不溶解的,不可溶的
non-stochastic effect 非随机效应
non-transferred arc 非转移弧
non-vacuum electron beam welder 非真空电子束焊机
nonvented structure 无通风设施的构

筑物
norm, standard　标准，规范
normal　正常，正常的
normal and abnormal operating procedures　正常运行和异常运行规程
normal and upset plant conditions　电厂的正常与扰动工况
normal auxiliary power system　正常厂用电源系统
normal condition　正常工况
normal fashion　正常工况，正常方式
normal idle condition　正常备用状况
normal incidence　（飞射物）正入射
normal operation　正常运行
normal operation and operational transients (condition Ⅰ)　正常运行和运行瞬态(Ⅰ类工况)
normal power supply　正常电源
normal range　正常测量范围，正常量程
normal residual heat removal discharge line penetration　正常余热导出出口管线贯穿件
normal residual heat removal suction line penetration　正常余热导出入口管线贯穿件
normal residual heat removal system isolation　正常余热排出系统隔离
normal telephone system　正常电话系统
normally close (NC) contact　常闭接点
normally closed type　常闭式
normally open (NO) contact　常开接点
normally open type　常开式
notch　缺口
notch bend test　缺口弯曲试验
notch toughness　缺口韧性
note that…　注意…
notification　通知
notify　通知
notwithstanding　虽然，尽管…仍…
nozzle　喷嘴
nozzle belt　（反应堆容器）接管段
nozzle dam　盲板
NRC Branch Technical Position　美国核管会技术立场
nuclear auxiliary building lighting system　核辅助厂房照明系统
nuclear boiling　泡核沸腾
nuclear control and protection system　核控制和保护系统
nuclear data link　核数据通信线，核数据链接
nuclear design　核设计
nuclear enthalpy rise factor　核焓升因子
nuclear enthalpy rise hot channel factor　核焓升热管因子
nuclear heat flux hot channel factor　核热流密度热管因子
nuclear instrumentation system　核测仪表系统
nuclear island 220V AC instrumentation PSS　核岛 220 V 交流仪表电源系统
nuclear island 220V DC power supply system　核岛 220 V 直流电源系统

nuclear island 24V DC power supply system　核岛 24 V 直流电源系统
nuclear island 380V AC power supply system　核岛 380 V 交流电源系统
nuclear island and conventional island fire protection　核岛常规岛消防水分配系统
nuclear island building stair positive pressure supply　核岛厂房楼梯间正压送风系统
nuclear overpower trips　核超功率停堆
nuclear power　核功率，核电
nuclear power plants　核电厂
nuclear safety (NS)　核安全
nuclear safety enforcement　核安全法规，核安全强制执行
nuclear safety inspection　核安全检查
nuclear uncertainty factor　核不确定性因子
nucleate boiling　泡核沸腾
nuclide　核素
numbering　编号，标记
numerical-control cutting (NC)　数控切割
nuts　螺母

O

object code　目标程序，目标代码
object code for embedded third-party technology　内含第三方技术的目标程序
object program　目标程序
objective　目的，目标，任务
objective evidence　客观证据
observe　遵守
obtain, secure　获得
occasional　偶然的，不经常的
occupancy factor　占用率，居留因子
occupant　居留人员，居住者
occupation standard　行业标准
occupational exposure　职业照射
occupational health　职业健康
occupational radiation worker　辐照从业人员
occurrence　(故障、事件的)发生，出现
occurrences of moderate frequency　中等频率事件
offense　冒犯，得罪
offer　报价
offer firm　报实盘，报实价
offer subject to our final confirmation　报价以最后我方确认为准
offeree　受盘人(买方)
offeror　发盘人(卖方)
off-line　离线
off-nominal conditions　异常工况
offset　偏移，失调
offsite　厂(区)外
offsite AC power sources　厂外交流电源
offsite dose　厂外剂量
offsite emergency plan　厂外应急计划
offsite power　厂外电
offsite power source　厂外电源
offsite release　厂外释放
off-the-shelf hardware and software　现货供应的硬软件
oil ejector　注油箱
oil injection　注油

oil pit　油槽，油坑
oil reservoir　储油箱，油库
oil tank　油箱
oily waste drain　含油废水疏水
omega seal　Ω 密封
on/off input　开关量输入
on balance　总的来说，总而言之
on line　在线，在役
on site standby diesel generator　厂内备用柴油发电机组
on the subject of　关于
on the whole　总的说来，总而言之
once for all　一次性地
once-through fuel cycle　开始燃料循环，一次通过式燃料循环
once-through stop valve　直流式截止阀
once-through ventilation system　直流式通风系统
one breaker and a half type　一个半断路器型
one way slab　混凝土单行道
one-time alignment　一次对中
one-way composite concrete slab and steel beam　混凝土和钢梁组合单行道
online core monitoring system　在线堆芯监测系统
online learning　在线学习
online monitoring　在线监控
online power distribution monitoring system　在线功率分布监测系统
online test ability　在线可测性能
online testing　在线试验
on-position　处于工作状态
on-site　厂内，现场
on-site AC power sources　厂内交流电源
on-site diesel generator A package　厂内柴油发电机 A 包
on-site diesel generator package　厂内柴油发电机机组
on-site operational support center　厂内运行支持中心
on-site source　厂内电源
on-site technical support center　厂内技术支持中心
onsite weld　现场焊接
on-the-job training(OJT)　在岗培训
open　断开
open a saving account　开设活期存款户头
open arc welding　明弧焊
open bucket steam trap　浮桶式疏水阀
open circuit　开路
open circuit voltage (no load voltage)　空载电压
open ditch　明沟
open funnel　敞开式烟窗
open sprinkler　开式喷头
open umbrella-type transition piece　开伞形过渡件
open wiring　明配线，布明线
open-cycle system　开式循环系统
opening　孔洞，开启，开度
opening bank　开户行
opening offer　首次报价(开盘)
opening valve temperature　开阀温度

operability 可使用性,可操作性
operability test 操作性能测试
operable 可操作的
operable louver damper 可操作百叶窗式挡板
operate on CIF terms 按到岸价格成交
operating (service, working) pressure 工作压力
operating and testing experience 运行和试验经验
operating back pressure 工作背压
operating basis earthquake (OBE) horizontal acceleration 运行基准地震水平加速度
operating basis earthquake (OBE) vertical acceleration 运行基准地震垂直加速度
operating capacity factor 运行能力因子
operating characteristics of an electrode 焊条使用性
operating condition 运行工况,操作条件,运行状态
operating deck 操作平台
operating differential pressure 工作压差
operating flexibility 运行灵活性
operating floor 工作平台,操作层
operating license 运行许可证,运行执照
operating organization 营运部门,运营组织
operating overvoltage 操作过电压
operating parameter 运行参数
operating personnel 运行人员
operating platform 操作平台
operating pressure 工作压力
operating procedure 运行规程
operating process 运行过程,操作过程
operating staff 操作员,运行人员
operating temperature 工作温度,运行温度
operation 运行,运营
operation and control centers system 运行控制中心系统
operation basic ground acceleration 运行基准地面加速度
operation coils 工作线圈
operation deck 操作平台
operation engineer 运行工程师
operation floor 操作平台,操作层
operational experience 运行经验
operational limits and conditions 运行限值和条件
operational records 运行记录
operational safety assessment review team(OSART) 运行安全评审团
operational sequence analysis 操作顺序分析
operational states 运行状态
operational support 运行支持
operational test 运行试验
operator 运行人员,操纵员
operator console 操纵员控制台
operator control 操纵员控制
operator decision-making mode 操纵员决策模型
operator desk set 操纵员办公设施

operator station　操作员站，操作站
operator training simulator　操纵员培训模拟机
operator workload　操纵员工作负荷，操作员工作量
operator workstation　操作员工作站
opponent　对手，竞争者
opposed blade damper　对开片式风阀
opposed piston pump　对置活塞泵
opposed-blade type　对开片式
optical disk　光盘
optimization of radiation protection　辐射防护最优化
optimum moderation　最佳慢化
orbital welding　轨道式焊
order sheet　订单
orderly shutdown　顺序停堆
organic　有机的
organic iodine　有机碘（碘的有机化合物）
organic material　有机材料
organization freedom　组织独立性
organization structure　组织结构
oriental　东方的
orifice　孔板
orifice diameter　节流孔板直径
orifice throat length　节流孔板孔长
orifice throat ratio　节流孔板节流比
O-ring　O 形环
orthographic drawing　正投影图
oscillation　振荡，振动
oscillator　频率振荡器
oscilloscope　示波器
other BOP building water supply and drainage system 其他 BOP 厂房给排水系统
other specific training（OST）　其他专项培训课程
out of line with　与…偏离，与…不相符
outage time　停运时间
outage，shutdown　停运，停机，停堆
outcome，result　结果
outdoor design mean daily temperature for summer air conditioning　夏季空调的室外日平均设计温度
outdoor design relative humidity for summer ventilation　夏季通风的室外设计相对湿度
outdoor design relative humidity for winter air conditioning　冬季空调的室外设计相对湿度
outdoor design temperature for heating 采暖的室外设计温度
outdoor design temperature for summer air conditioning　夏季空调的室外设计温度
outdoor design temperature for summer ventilation　夏季通风的室外设计温度
outdoor design temperature for winter air conditioning　冬季空调的室外设计温度
outdoor design temperature for winter ventilation　冬季通风的室外设计温度
outdoor design wet-bulb temperature for summer air conditioning　夏季空调的室外设计湿球温度

outer annulus 外环腔，外环廊
outgoing breaker 出线断路器
outlet 批发商，销售公司
outlet nozzle 出口管嘴，出口接管
outlet nozzle insulation 出口接管保温层
outlet temperature 出口温度
outline 叙述(要点)，概括地论述
outline drawing 外形图，轮廓图
out-of-plane 平面外，脱离平面的
output 输出，产量，供水量
outside reactor containment (反应堆)安全壳外
outside strap 围筋，围带
outstanding 未解决的，未完成的
out-surge 负波动
Ovation base software package Ovation 基础软件包
Ovation power tool database Ovation 电力工具数据库
Ovation real time data network electronic module Ovation 实时数据网络电气模块
overall 全面的，总的
overall containment leakage rate 安全壳总泄漏率
overall response time 整体响应时间
overall responsibility 全面责任，总责任
overcurrent 过电流
overcurrent relay 过电流继电器
overdue 过期的，过期未付的
overestimate 估计过高，过分评价
overfill (蒸汽发生器)满溢，充水过量
overflow 溢出，过流
overflow valve 溢流阀
overhaul 大修
overhead and gantry cranes 桥式和门式起重机
overhead crane 桥式起重机，行车
overhead position welding 仰焊
overheated structure 过热组结构
overheated zone 过热区
overlap 焊瘤
overlaying 堆焊
overload 过负荷，过载
overlying structures 上部结构
overpower ΔT 超功率 ΔT
overpower ΔT trip 超功率 ΔT 停堆
overpowerΔT trip setpoint 超功率 ΔT 停堆保护整定值
overpower limit 超功率限值
overpower transient 超功率瞬态
overpressure protection 超压保护
overpressure transient 超压瞬态
override 超越，优先
overspeed 超速
overspeed protecting switch 超速保护开关
overspeed safety brake 超速安全制动器
overtemperatureΔT (OTDT) 超温 ΔT
overtemperatureΔT reactor trip 超温 ΔT 反应堆停堆
overtemperatureΔT turbine runback 超温 ΔT 汽轮机降功率
overturning moment 倾覆力矩
overview 全貌

overvoltage relay 过压继电器
owner 业主
oxidation scale 氧化皮
oxide melting slag 氧化物型熔渣
oxidizing flame 氧化焰
oxy-acetylene flame 氧乙炔焰
oxy-acetylene torch 氧乙炔焊炬
oxy-acetylene welding 氧乙炔焊
oxy-arc cutting 氧气电弧切割
oxy-coal gas flame 氧煤气焰
oxygen 氧
oxygen analyzer 氧分析仪
oxygen cutting 氧气切割
oxygen evaporator 液氧气化器
oxygen free waste gas 无氧废气
oxygen lancing/oxygen lance cutting 氧气切割
oxygen plasma arc cutting 氧等离子切割
oxygen scavenger 除氧剂
oxygen-free flux 无氧焊剂
oxy-hydrogen flame 氢氧焰
oxy-hydrogen welding 氢氧焊

P

package (blanket contract) 成套设备(总承包)
package unit 柜式机组,包式单元
packages bids 投标承揽项目的各个部分,打包投标
packages-radiation monitoring 包一放射性监测
packaging 包装法,打包
packaging box friction 填料盒摩擦
packaging withdrawal tool 衬垫取出工具,填料取出工具
packing 填料
packing seat 填料垫
packless globe (metal diaphram) valve 无填料(金属膜片)球阀
packless valve 无填料阀
pad 衬垫、垫(块)
page-party system 呼叫通话系统
paint 油漆,涂料
palladium 钯
panel 台屏,盘
pan-tilt control facility (摄像机)姿态控制器
parallel blade damper 平行叶片风阀
parallel gate valve 平行式闸阀
parallel single disk gate valve 单碟双面平行闸阀
parallel slide valve 平行式滑阀
parallel-blade type 平行叶片式(风阀)
parametric sensitivity analysis 参数敏感性分析仪
paramount 首要的,最高的,至上的
parapet 胸墙,低墙,栏杆,扶手
parasitic absorption (中子)寄生吸收,寄生俘获
parent metal 母材
Paris formula 帕瑞斯(Paris)公式
partial load pump 部分负荷泵
partial melting region 部分熔化区
partial penetration weld 离子渗透焊
partial trip signal 局部停堆信号
partial vacuum electron beam welder 低真空电子束焊机
participant 参加者,与会代表

particle　粒子,颗粒
particle counter meter　粒子计数器
particular average　单独海损(水渍险)
particulate　微尘,微粒
particulate detector　微尘探测器,微粒探测器
particulate filter　微粒过滤器,微尘过滤器
particulate iodine　粒子碘(碘无机化合物)
particulate-iodine sampler　微粒—碘取样器
partition　分配,划分,隔离
partition coefficient　分配系数
partition factor　分配因子
partition plate　隔板,分配版
partition wall　隔墙
partitioning　(蒸汽发生器)汽水分离
partitioning and transmutation　分离与嬗变
part-length rod　短(控制)棒
partner　伙伴,配手,搭档
parts retainers　(吊车)夹轨装置,固定器
part-time instructor　兼职教员
pass sequence　焊道工序
pass test　合格性试验
passivation　钝化
passive　非能动的,被动的
passive autocatalytic recombiner　非能动自催化(氢)复合器
passive components　非能动部件
passive containment cooling　非能动安全壳冷却
passive containment cooling actuation　非能动安全壳冷却触发
passive containment cooling ancillary water storage tank　非能动安全壳冷却辅助水贮存箱
passive containment cooling system water storage tank　非能动安全壳冷却系统贮存水箱
passive containment cooling water flow　非能动安全壳冷却水流(量)
Passive containment cooling water storage tank　非能动安全壳冷却水贮存水箱
passive containment spray　非能动安全壳喷淋
passive earth pressure　被动土压力
passive heat sink　非能动热阱
passive residual heat removal heat exchanger alignment　非能动余热排出热交换器调整
passive residual heat removal test　非能动余热排出试验
passive RHR heat exchanger　非能动余热排出热交换器
passward　口令,密码
patching panel　配电板,接线板
path repeatability　轨迹可重复性
path selector (rotary)　(旋转)通路选择器
pattern recognition for welding　焊接图像识别
pay on dispatch of the goods　发货付款
PCCWST discharge line cross-connect isolation valve　非能动安全壳冷却水

贮存箱出口桥管隔离阀
PCCWST isolation valve leakage detection cross-connect valve 非能动安全壳冷却水贮存箱隔离阀泄漏探测桥管阀门
PCCWST isolation valve leakage detection drain 非能动安全壳冷却水贮存箱隔离阀泄漏探测疏水
PCCWST long-term makeup check valve 非能动安全壳冷却水贮存箱长期补水止回阀
PCCWST long-term makeup isolation drain valve 非能动安全壳冷却水贮存箱长期补水隔离疏水阀
PCCWST supply to FPS isolation valve 非能动安全壳冷却水贮存箱至消防水系统供水隔离阀
PCCWST to recirculation pump suction isolation valve 非能动安全壳冷却水贮存箱至再循环泵入口隔离阀
PCCWST water level instrument 非能动设备冷却水贮存箱水位仪表
PCS cooling flow 非能动安全壳系统冷却流量
PCS flow instrument 主冷却剂系统流量仪表
PCS storage tank water level 非能动安全壳系统冷却水贮存箱液位
PCS water storage tank isolation valve 非能动安全壳系统冷却水箱隔离阀
peak 峰值,尖峰,顶点
peak clad temperature (燃料棒)包壳峰值温度
peak ground acceleration 地面峰值加速度
peak linear heat rate(PLHR) 峰值线功率密度,峰值线性发热率
peak load pump 高峰负荷泵
peak temperature 峰值温度
peak value 峰值
peak xenon 最大氙,峰值氙
peaking factor limits 峰值因子限值
pearlitic heat resistant steel electrode 珠光体耐热钢焊条
peening 喷丸硬化(焊缝)
pellet (燃料)芯块
pellet conductivity (燃料)芯块热导率
pellet stack (燃料)芯块柱
pellet-cladding interaction (PCI) (燃料)芯块与包壳相互作用
penalty 亏损,惩罚
penalty factor for rod bow 燃料棒弯曲亏损因子
pendant pushbutton control box 悬挂式按钮控制箱
pending 悬而未决的,未定的
penetrating radiation 贯穿辐射
penetration 贯穿件,贯穿
penetration capability 穿透能力,贯穿能力
penetration depth 穿透深度
penetration inspection 渗透探伤
penetration resistance 耐穿透性
penetration sleeve 穿墙套管
percent by weight 重量百分比
percent of contact area 吻合度,接触面份额
percentage of moisture for covering 焊

条药皮含水量
perfect,improve 改善,改良
perforated plate (下管座的)下格板,多孔板
performance 业绩
performance assessment 业绩评价
performance criteria 运行准则,性能标准
performance curve 性能曲线
performance requirements 性能要求
performance test 性能试验
performance-based testing program 基于性能的试验大纲
peril 危险
periodic inspection 定期检查
periodic safety review 定期安全评审
periodic test 定期试验
peripheral 外围的,外围设备
peripheral support ring 外部支撑环
permanent backing 保留垫板
permanent deformation 永久变形
permanent nonsafety load 永久性非安全负载
permanent storage 永久储存
permanently installed equipment 固定安装设备
permeability 渗透性
permissive circuit 允许电路
permissive logic module 容许逻辑组件
permit 许可,许可证,执照
perpendicular acceleration 垂直加速度
personal dosimeter 个人剂量计
personal radiation dosimetry 个人辐射剂量仪
personal respirators 个人呼吸器
personnel and population exposure 个人和集体照射
personnel exposures 工作人员照射,人员受照
personnel hatch 人员闸门
personnel hatch test connection 人员闸门试验接口
personnel radiation exposure 人员辐照,人员受照
persuade,talk into 劝说,说服
pertinent 相关的,有关的
pertinent information 相关信息
perveance 导流系数
pH adjuster pH 调节器
pH adjustment basket pH 调节范围
pH adjustment chemical pH 调节剂,pH 调节化学药品
pH meter (bench top) (台式)pH 计
pH meter (portable) (便携式)pH 计
pH value pH(值)
phase control and firing circuit 移相触发电路,相位控制和启动电路
phase difference 相位差
phase displacement 相位移
phase transition 相变
phenomena identification and ranking table (pirt) 现象识别与归类表
phoresis 电泳现象
phosphate treatment 磷酸盐处理
phosphorous 磷的
phosphorus 磷
photo-electric tracing cutting 光电跟

踪切割机
photomultiplier　光电倍增管
photon run-time software(flat panel display operating system)　光子传输时间软件(平屏显示操作系统)
physical barriers　实体屏障
physical mechanism　物理机理
physical plant boundary　电厂实体边界
physical separation　实体隔离
physically independent　实体独立
physically-based source term　基于物理规律的源项
physics laboratory　物理实验室
physics startup test　物理启动试验
physics test　物理试验
physiography　地理学,地文学
pickling　酸洗,腐蚀
pictorial　有插图的,用图表示的
pier foundation　墩式基础
pigtail connection　软辫线连接
pillow test　枕形气密检验
pilot arc　维弧
pilot arc current　维弧电流
pilot device　先导装置
pilot operated pressure relief valve　先导式卸压阀
pilot operated safety valve　先导式安全阀
pilot valve　先导阀
pilot-operated reducing valve　先导式减压阀
pinch valve　夹片阀
pinhole　针尖状气孔
pinion　小齿轮
pipe break　管道破裂,管道裂口
pipe chase　管沟
pipe flame cutting machine　火焰切管机
pipe hanger　管道支吊架
pipe leakage event　管道泄漏事件
pipe monitoring displacement transducer　管道位移监测变送器
pipe rupture condition　管道破裂工况
pipe section　管道节,管段
pipe whip　管道甩击,甩管
pipe whip restraint　管道防甩件
pipeline canned motor pump　管道屏蔽泵
pipeline mounted pump　管道泵
piping　管道,水管,布管
piping assembly　管道组件
piping penetration　管道贯穿件
piping safety valve　管道安全阀
piping support　管道支承件
piping system　管道系统
piping vibration effects　管道振动效应,管道振动作用
piston pump　活塞泵
piston pump with rotary gate　带回转闸阀的活塞泵
piston reducing valve　活塞式减压阀
piston valve　活塞阀
piston valve line　配汽活塞阀缸套
piston with shakeable cylinder　摆动缸活塞泵
pit　凹坑
pitot pump　斗式泵,皮托泵

pitting corrosion　点蚀
place an order　提交订单
plain glass　防护镜片
plan　计划,设计图,平面图
plane heat source　面热源
planeness of cutting surface　切割面平面度
planner　规划员
plant alarm processing system　电厂报警处理系统
plant alarm system　电厂报警系统
plant auxiliary electrical system (CI)　厂用电系统
plant availability　电厂的可利用率
plant cleanup system　电厂净化系统
plant computer system　电厂计算机系统
plant condition　电厂工况
plant cool down　电厂冷却
plant critical safety function　电厂关键安全功能
plant date logging　电厂数据日志
plant department manager　电厂部门经理
plant design objective　电厂设计目标
plant design transient　电厂设计瞬态
plant fire detection and alarm system　电厂火灾探测及报警系统
plant gas system　电厂气体系统
plant layout drawing　电厂总平面布置图
plant life　电厂寿期
plant maintenance software server　电厂维修(护)软件服务器
plant management　电厂管理
plant manager　电厂经理
plant operating life　电厂运行寿期
plant operating, emergency, and surveillance procedure　电厂运行、应急和监督规程
plant personnel　电厂从业人员
plant protection system　电厂保护系统
plant protection, control, and monitoring function　电厂的保护、控制和监测功能
plant safety-related function　电厂安全有关功能
plant startup　电厂启动
plant training manager　电厂培训经理
plant trip　电厂紧急停堆,电厂停堆
plant trip controller　电厂停堆控制器
plant vent　电厂通风管(排气道)
plasma (welding) torch　等离子弧焊枪
plasma arc　等离子弧
plasma arc cutting　等离子弧切割
plasma arc powder surfacing　粉末等离子弧堆焊
plasma arc surfacing　等离子弧堆焊
plasma arc welding　等离子弧焊
plasma arc welding machine　等离子弧焊机
plasma gas/orifice gas　等离子气
plasma jet　等离子流
plasma MIG welding　等离子—混合气体保护焊
plasma spraying　等离子喷涂
plastic deformation　塑性变形
plastic lined pump　塑料衬里泵

plastic spraying 塑料喷涂
plastic tensile strain 塑性拉伸应变
plate 板,板材
plate tectonic (地质)板块构造
plate type heat exchanger 板式热交换器
plate valve 平板阀
plating 电镀
platinum 铂
platinum resistance thermometer 铂电阻温度计
please rest assured 请放心
plenum 空腔,腔室
plug 插头,塞子
plug resistant orifice 管塞阻力孔板,塞阻孔板
plug valve 旋塞阀
plug weld 管塞焊
plug welding 塞焊
plugging 堵管
plug-in breaker 插入式断路器
plug-in locking type disconnects 插入闭锁式断开装置
plumbing 给排水管道工程
plumbing material and equipment 铅管材料及设备
plume 烟羽,汽团
plume exposure pathway 烟羽照射路径
plunger 柱塞,活塞
plunger globe valve 柱塞截止阀,柱塞球阀
plunger pump 柱塞泵
plunger valve 活塞阀,柱塞阀
PMS cabinets 保护与安全监督系统机柜
PMS QDPS MCR qualified display units 保护与安全监督系统、合格1E级数据处理系统、主控制室经鉴定的显示器装置
pneumatic (-operated) valve 气动阀
pneumatic actuator 气动装置
pneumatic pump 风动泵,气动泵
pneumatic test 气动试验
pneumatically controlled variable inlet vane 气动控制可调进口导向叶片
pneumatic-hydraulic 气—水力的
pneumatic-hydraulic actuator 气—液动装置
pocket dosimeter 袖珍式剂量计
point heat source 点热源
point out 指出
polar crane 环吊,环形吊车
polarity 极性
policy statement 政策声明
polluted industrial wastewater 被(放射性)污染的工业废水
polluted item 被污染的物项
polygonization boundary 多边化边界
polygonization crack 多边化裂纹
polypropylene culture flasks 聚丙烯培养瓶
ponding 积水(库),蓄水,人工池塘
population dose 公众剂量,集体剂量
population equivalent 公众当量,人口当量
pore 气孔,微孔
porosity 密集气孔,多孔性

porous slag　多孔焊渣
portable　便携式,手提式
portable air sampler　便携式空气取样器
portable clamp on ultrasonic flow meter　超声流量计上的便携式夹具
portable exhaust air filtration unit　便携式排风过滤单元
portable extinguisher　手提式灭火机
portable extinguisher system　移动式灭火器系统
portable fire extinguisher　便携式灭火器
portable handset　便携式手机
portable high pressure abrasive decontamination unit　便携式高压研磨去污装置
portable high pressure water decontamination unit　便携式高压水去污装置
portable leak rate test panel　便携式泄漏率试验屏
portable multi channel analyzer　便携式多道分析仪
portable neutron survey instrument　便携式中子探测仪
portable personal dose rate and survey meter　便携式个人剂量仪(表)
portable survey instrument for alpha/beta/gamma　α/β/γ便携式测定仪
portable transceiver　便携式无线收发机
portable underwater lights　便携式水下照明灯
portable vacuum integrity tester　便携式真空试验装置
portable workstation　移动式工作站
portable γ exposure rate meter　便携式γ照射率仪
portal frame　门式构架,龙门架
portal γ monitor　门框式γ监测仪
portfolio analysis　组合分析
PORV block valve　动力卸压阀闭锁阀
PORV discharge condensate drain isolation　动力卸压阀泄出凝结水疏水隔离
position　位置
position indicating system　位置指示系统
position indication　位置指示
position indicator　位置指示器
position limit switch　限位开关
position of welding　焊接位置
position switch　行程开关
positioner　定位器
positioning part　定位部件
positioning system　定位系统
positive displacement pump　容积式泵,正位移泵,往复泵
positive pressure　正压
positive reactivity　正反应性
positive reactivity addition　正反应性引入
positive shutoff　正关闭
post accident inadequate core cooling monitor　事故后堆芯不充分冷却监测,事故后堆芯冷却监测
post accident monitoring　事故后监测

post accident sampling　事故后取样
post indicator　(事故)后用的指示器
post-processor　(事故)后用的信息处理机
post weld heat treatment　焊接后热处理
post-72-hour action　事故72小时后的行动
post-72-hour temporary power supply unit　事故72小时后临时供电机组
post-72-hour ventilation　事故72小时后通风
post-accident monitor　事故后监测仪
post-accident monitoring system　事故后监测系统
post-audit meeting　(质保)监查后会议
post-design basis accident　设计基准事故后
post-filter　(事故)后用的过滤器
post-fuel load flow　装料后流量
post-heat　(事故)后的释热,后热
post-hydrotest ultrasonic map of welds　水压试验后焊缝超声检查图像
post-irradiation examination　辐照后检查
post-processing　后处理工艺
post-trip review　脱扣过程追溯,停堆过程复审
postulated accident　假想事故
postulated accident condition　假想事故工况
postulated design basis event　假想设计基准事件
postulated event　假想事故,假想事件
postulated failure　假想失效
pot　罐,壶
potential　潜在的
potential energy release　潜能释放
potential transformer　电位互感器
potentiometer　电位计
powder　粉末
powder cutting　氧熔剂切割,粉末切割
powder dispenser　粉剂罐
powder lancing　熔剂氧切割
powder washing　氧熔剂表面修整
powdered filler metal　粉状钎料
power ascension　功率上升
power cable　动力电缆
power capacity　电厂发电能力
power channel comparison facility　功率通道比较装置
power coefficient　功率系数
power control rod　功率控制棒
power coolant mismatch　功率与流量失配
power cord　电源线,动力线
power defect　功率亏损
power density　功率密度
power distribution　功率分布
power distribution limit　功率分布限制
power effect　功率效应
power factor　功率因子
power generation　发电
power generation design basis　发电设计基准
power level　功率水平
power line carrier　电力线塔架
power loss　失电

power modes of operation　功率运行模式
power operated relief valve(PORV)　动力卸压阀,动力释放阀
power operation　功率运行
power oscillation　功率振荡
power overshoot　功率超调
power peaking factor　功率峰值因子
power plant operation log　电厂运行日志
power pump　动力泵
power range　功率量程
power range channel　功率量程测量通道
power range high neutron flux reactor trip　功率量程高中子注量率停堆
power range high voltage distribution boxes　功率量程高压配电箱
power shape　功率分布形态,功率形状
power spectra density　功率普密度
power spike factor　功率峰因子
power station pump　电厂泵
power supply　电源,供电,电力供应
power transmission　输电
power transmission line　输电线
PRC　公司级管理程序
preaction sprinkler system　先行启动喷洒系统
preaction valve　(火警探测系统)先动阀
preamble　序言
pre-audit meeting　(质保)监查前会议
precast　预制,预浇制
precast concrete　预制混凝土
precaution　预防措施
precaution sign　预防,防范,防止
precedent　惯例,先例,范例
precedent desoxydation　先期脱氧
precipitation　沉淀
precipitation desoxydation　沉淀脱氧
precise-pulse generator　精密脉冲发生器
precision　精(密)度
precision aneroid barometer　精密无液气压计/真空计
precision decade capacitor　精密十进制电容器
precision thermometer　精密温度计
pre-configured ovation report　预先设定好的 ovation 报告
predetermined setpoint　预置设定值
pre-existing iodine spike　先期存在的碘峰
prefabrication(piping)　预制(管道)
preferential　优惠的
preferential terms　优惠条件
preferential treatment　优惠待遇
preferred power supply　优先电源,优先供电
pre-filter　预过滤器
preformed filler metal　成形钎料
pre-fuel load flow　装料前流量
preheat　预热
preheat flame　预热火焰
preheat oxygen　预热氧
pre-heater　预热器
pre-irradiation　辐照前
preliminary　初步的

preliminary exercising　预先演练，预先练习
premise　厂房，前提，所在地
premium　保险金
preoperational　预运行的，运行前的
preoperational and/or startup testing　预运行和/或启动试验
preoperational test　调试试验，预运行试验
preoperational test program　预运行试验大纲
preparation of procedures　规程编制，规程准备
pre-pressurization　预加压，预打压
prerequisite　先决条件，前提
prescribed limit　规定限值
prescriptive-based testing approach　试验方法基本描述
pre-service inspection (PSI)　役前检查
preset　事先调整，边框形的
preset bias　预设偏差，偏差范围
preset gain　预设增益
preset pressure　预设压力
president　院长，总统
press　锻压，压机
pressure balance line　压力平衡管线
pressure boundary　压力边界
pressure boundary integrity　压力边界完整性
pressure boundary weld　压力边界焊缝
pressure characteristics　压力特性
pressure characteristics derivation　压力特性推导
pressure coefficient　压力系数
pressure compensation diaphragm　压力补偿膜片
pressure control operator　压力控制器
pressure controlled valve　压力调节阀，压力控制阀
pressure decay method　压力衰减法
pressure drop　压降
pressure gauge　压力表
pressure housing　承压壳体
pressure housing assembly　耐压壳组件
pressure increasing ratio　增压比，压力增长比
pressure loss coefficient　压降系数，失压系数
pressure reducing ratio　减压比
pressure reducing valve　减压阀
pressure reduction　降压，减压
pressure regulating valve　压力调节阀
pressure regulator/gas regulator　减压器
pressure release valve　卸压阀
pressure relief damper　卸压风阀
pressure relief isolation valve　卸压隔离阀
pressure relief valve　卸压阀
pressure response　压力响应
pressure retaining　承压，压力维持
pressure seat　内压自封
pressure sensors　压力传感器
pressure spike　压力峰
pressure surge　压力波动
pressure tap　测压阀

pressure test 耐压检验,压力试验
pressure transmitter 压力变送器
pressure transmitter root isolation valve 压力变送器根部隔离阀(根阀)
pressure vacuum variator 真空调节器
pressure valve 压力(增压)阀
pressure vessel(assembly) 压力容器(组件)
pressure-containing material 承压材料
pressure-retaining boundary 承压边界
pressure-temperature limit 压力—温度限值
pressurized 加压的
pressurized gas bottle 加压气瓶
pressurized thermal shock 承压热冲击
pressurizer 稳压器
pressurizer auxiliary spray 稳压器辅助喷雾
pressurizer compartment wall 稳压器隔间墙壁
pressurizer heater 稳压器(电)加热器
pressurizer heater block 稳压器电加热器隔离
pressurizer heater controller 稳压器加热器控制器
pressurizer heaters modulating control 稳压器电加热器调制控制
pressurizer heaters trip 稳压器电加热器跳闸
pressurizer insulation 稳压器保温
pressurizer level and pressure control 稳压器液位和压力控制
pressurizer level liquid space instrument root 稳压器水位计水腔根管
pressurizer level steam space instrument root 稳压器水位计汽腔根管
pressurizer liquid isolation 稳压器液相隔离
pressurizer liquid phase 稳压器液相
pressurizer manual vent 稳压器手动排气
pressurizer pressure 稳压器压力
pressurizer pressure control system 稳压器压力控制系统
pressurizer reference leg temperature 稳压器参考管温度
pressurizer relief tank 稳压器卸压箱
pressurizer safety and relief valve (PSARV) 稳压器安全阀和卸压阀
pressurizer safety and relief valve (PSARV) module 稳压器安全阀和卸压阀(PSARV)模块
pressurizer safety valve 稳压器安全阀
pressurizer spray 稳压器喷雾
pressurizer spray block valve 稳压器喷雾隔离阀
pressurizer spray bypass 稳压器喷雾旁路
pressurizer spray valve 稳压器喷雾阀
pressurizer spray valve setpoint 稳压器喷雾阀设定值
pressurizer steam phase 稳压器汽相
pressurizer steam space sample isolation 稳压器汽空间取样隔离
pressurizer surge line 稳压器波动管线
pressurizer surge line stratification evaluation 稳压器波动管热分层评估

pressurizer water inventory　稳压器水装量
pressurizer water level　稳压器水位
pressurizer water level control　稳压器水位控制
pressurizer water level control system　稳压器水位控制系统
pressurizer water level trend　稳压器液位(变化)趋势
pre-stressing　预应力
presume　假设
prevention　预防
preventive action　预防措施
preventive maintenance　预防性维修
previously　预先,在前
PRHR HX control valve　非能动余热排出热交换器调节阀
PRHR HX flow　非能动余热排出热交换器流量
PRHR HX flow transmitter isolation　非能动余热排出热交换器流量变送器隔离
PRHR HX inlet isolation valve　非能动余热排出热交换器进口隔离阀
PRHR HX inlet temperature　非能动余热排出热交换器入口温度
PRHR HX outlet temperature　非能动余热排出热交换器出口温度
PRHR HX pressure transmitter isolation　非能动余热排出热交换器压力变送器隔离
primary auxiliary building ventilation system　一回路辅助厂房通风系统
primary combustion　一次燃烧
primary containment　一次安全壳,主安全壳
primary fire water tank　一次消防水箱,主消防水箱
primary membrane stress　一次隔膜应力
primary neutron source　初级中子源
primary neutron source assembly　一次(初级)中子源组件
primary neutron source rod　一次(初级)中子源棒
primary overpressure trip　一回路超压停堆
primary plant performance application software　主要的电厂性能应用软件
primary radiation　初级辐照
primary shield　一次屏蔽
primary shield wall　一次屏蔽墙
primary side　一次侧
primary solidification structure　一次结晶组织
primary system leak rate application software　主系统泄漏率应用软件
primary treatment　初步处理
primary water chemistry　一回路水化学
primary winding　(变压器)初级绕组
primary-side leakage test　一次侧泄漏试验
prime out-form dimensions　主要外形尺寸
primer　底漆,雷管
priming tundish valve　漏斗形启动注液阀

primordial　理所当然的
principal　委托人,委托方
principal stress　主应力
printer　打印机
printer station　打印机站
prioritize　分清重点,明确轻重缓急
priority　优先(权)
priority call　优先传呼,优先呼叫
private automatic branch exchange　专用自动电话小型交换机
private automatic branch exchange (PABX)system　专用自动电话小型交换机系统
probabilistic　概率的,或然的
probabilistic risk analysis (PRA)　概率风险分析
probabilistic risk criteria　概率风险准则
probabilistic safety assessment (PSA)　概率安全评价
probable maximum storm surge　可能最大风暴潮
probe　探头
procedure　程序,手续,方法
procedure development　规程开发,程序开发
procedure qualification test　程序评定试验,规程评定试验
procedure qualification test report　程序评定试验报告,规程评定试验报告
proceed to,take up　动手,开始,着手
process　过程,工艺
process and decay heat removal system　工艺和衰变热排出系统
process and effluent radiological monitoring and sampling subsystem　工艺和流出物放射性监测和取样子系统
process control　工艺控制,过程控制
process drain　工艺疏水
process fluid monitoring　工艺流体(放射性)监测
process monitoring system　过程监测系统,工艺监测系统
process qualification　工艺评定
process radiation monitoring system　工艺过程辐射监测系统
process signals　工艺信号
process specification　工艺规格书
process water　工艺用水,生产用水
processed material　流程性材料
procrastinate　拖延,耽搁
procurement　采购
procurement payment　采购支付
procurement specification　采购规格(书)
prod method　(磁粉检验)触头法,刺法
product analysis　产品分析
product liability　产品责任
product pump　产品泵,物料泵
product receipt and inspection　产品接收和检验
production drainage and storm water yard pipe　生产排水及雨水厂区管网系统
production permit　生产许可
profile　简况,简介
profit margin　利润,利润率
program　大纲,规划,方案,程序

program manager　项目经理
programmable logic controller　可编程的逻辑控制器
programme　大纲
programmed water level　程控水位
progress,come along　进展
progressive　推进的,渐进的,累进的
progressive cavity positive displacement pump　旋进挖洞正排泵
progressively　渐进地,逐渐地
project control　项目控制,项目管理
project coordination　工程协调
project engineering manager　项目工程经理
project expenses　工程费用
project management　项目管理
project management office (PMO)　项目管理办公室
project manager　项目经理
project on a turnkey basis　交钥匙工程
project proposal　项目建议书
project supervision　项目监督
promote　促进
prompt critical　瞬发临界
prompt gamma radiation　瞬发γ辐射
prompt neutron　瞬发中子
prompt neutron fraction　瞬发中子份额
proof　证据,试验,考验,防…的
propagation　蔓延,扩展
propeller type fan　旋桨式风机
property　特性,性能
proportion　比例
proportional (group electrical) heater　比例组电加热器
proportional boron lined counter　正比涂硼计数管
proportional counter　正比计数管
proportional plus integral (PI)-compensated　正比加积分(PI)补偿的
proportional spray valve　比例喷雾阀
proportioning pump　比例泵,配料泵
proposal　提议,建议,提案
propose claims　提出索赔
propose,move　提议,动议
proprietary　专利
prospect　前途,希望,远见,潜在的客户
prospective　预期的,将来的
protection and safety monitoring system　保护和安全监测系统
protection mechanism　防护机理
protection relay　保护继电器
protection system　保护系统
protection system hardware　保护系统硬件
protection system pressure set point　保护系统压力整定值
protective barrier　防护屏障
protective clothing　防护衣具
protective clothing pickup and suitup room　防护服取换室
protective coating　防护涂层
protective cover　防护罩
protective function　防护功能
protective glove　防护手套
protective grid　保护格架
protective shoe cover　防护鞋套

protective structure　防护结构
protocol　草案
prototype　原型,样机
prototype test　原型试验,样机试验
protracted　拖延的
provention　预防
provision　(预防)措施,防备,储备
provisional　临时的
provisional acceptance report　临时验收报告
PTP repeatability　点位可重复性
public　公众
public address system　公众广播系统,(车内)播音系统
public health　公众健康
public telephone network　公用电话网
pull out　失步,离开,拔出
pull out torque　失步转矩
pull-out unit　失步单元,失步件
pulsation　脉动
pulse　脉冲,脉动
pulse distributor module　脉冲分配组件
pulse fission chamber　脉冲裂变室
pulse input　脉冲输入
pulse monitor module　脉冲监测组件
pulse signal generator module　脉冲信号发生器组件
pulse-code modulation　脉冲编码调制
pulsed arc exciter　脉冲激弧器
pulsed arc stabilizer　脉冲稳弧器
pulsed arc starter　脉冲引弧器
pulsed arc welding machine　脉冲弧焊机
pulsed arc welding power source　脉冲弧焊电源
pulsed arc welding rectifier　脉冲弧焊整流器
pulsed electron beam welding　脉冲电子束焊
pulsed plasma arc welding　脉冲等离子弧焊
pulse-height analyzer　脉冲幅度分析器
pulser　脉冲发生器,脉冲器
pump　泵
pump characteristic curve　泵运行特性曲线
pump flow coastdown curve　泵流量惰走曲线
pump for acid washing of boiler　锅炉酸洗泵
pump for pipe system　管网泵
pump for water-borne solids　泥浆泵
pump speed　泵转速
pump with cyclic electromagnetic drive　电磁驱动泵
pump with enclosed impellers　闭式叶轮泵
pump with nonrotating cylinder　固定缸体泵
punching　冲孔
punctual　准时的
punctually　准时地
purchase order format and review　采购订单格式和审核
purchased items　已购物项
purchaser　买方
purchasing department　采购部

pure iron electrode 纯铁焊条
pure nickel cast iron electrode 纯镍铸铁焊条
purge 净化,清除,吹扫
purge gas 吹扫气体
purification 净化
purification of coolant 冷却剂净化
purification piping 净化管线
purification process effluent 净化工艺流出物
purification section 净化段
purification stop value 净化截止阀,净化断流阀
purifier 净化器
push-button 按钮
push-button dial 按钮拨号
put forward 提出
put into operation 投入运行
PWR-GALE code 压水堆GALE程序
PXS accumulator sample isolation 非能动堆芯冷却系统蓄压安注箱取样隔离
PXS leak test line isolation 非能动堆芯冷却系统泄漏试验管路隔离
PZR HTRS control GRP MCCs 稳压器加热器控制GRP MCCs

Q

QDPS thermocouple reference junction panel 合格1E级数据处理系统热电偶参比端盘
quadrant 象限
quadrant power tilt 象限功率倾斜
quadrant tilt difference 象限(功率)倾斜偏差
qualification 鉴定,赋予资格
qualification certificate 资格评定证书
qualification criteria 鉴定准则
qualification process 鉴定过程
qualification program 人员资格评审培训大纲
qualification requirement 鉴定要求
qualification specification 鉴定规格书
qualification test 鉴定试验
qualified 鉴定合格
qualified data processing system 合格的数据处理系统
qualified isolation device 合格的隔离装置
qualified person 合格人员
qualified quality assurance program 获批准的质保大纲
qualify 合格鉴定
quality assurance(QA) 质量保证
quality assurance audit 质量保证监查,质保监查
quality assurance program 质量保证大纲,质保大纲
quality assurance record 质量保证记录,质保记录
quality audit 质量监查
quality audit observation 质量监查与观察结果
quality control(QC) 质量控制
quality control program 质量控制大纲
quality evaluation 质量评价
quality factor 质量因子,质量因素
quality group classification 质量分级
quality inspection 质量检查

quality inspector　质量检查员
quality loop　质量环
quality management　质量管理
quality observation report(QOR)　质量观察报告
quality of cut face　切割面质量
quality planning　质量策划
quality policy　质量方针,质量政策
quality qualification　质量鉴定,质量评定
quality release　质量释放
quality system　质量体系
quality treads analysis　质量趋势分析
quality-related costs　质量相关成本
quantity　量,数量
quartz powder cutting　氧—石英砂切割
quasi-cleavage fracture　准解理断口
quasi-stationary temperature field　准稳定温度场
quasi-steady condition　准稳态
quench　淬火
quick disconnect　快速断开,快断
quick draining valve　快速排泄阀
quick-acting　快动作
quotation　报价,报盘
quote　报价(虚盘)
quoted standard　引用标准

R

raceway　电缆敷设道,电缆沟,布线通道
rack　机架,架,台架
rack and pinion type elevator　齿轮和齿条型提升机构
radial and axial temperature profile　径向和轴向温度分布形态
radial arm　径向臂
radial buckling　径向曲率
radial grounding system　辐射状接地系统
radial peaking hot spot factor　径向峰值热点因子
radial power distribution　径向功率分布
radial power peaking factor　径向功率峰因子
radial power tilt　径向功率倾斜
radial support　径向支承件
radial thermal expansion　径向热膨胀
radial vane　径向翼
radially split pump　径向剖分泵
radiation accident　辐射事故
radiation area　辐射区域
radiation background　辐射本底
radiation chemistry　辐射化学
radiation corrosion　辐射腐蚀
radiation detector　辐射探测器
radiation dose rate　辐射剂量率
radiation effect　辐射效应
radiation exposure　射线照射,辐射照射,辐照
radiation field　辐射场
radiation map　辐射分区图
radiation monitor　辐射监测仪
radiation monitoring　辐射监测
radiation monitoring channel　辐射监测通道
radiation monitoring system interface

application 辐射监测系统接口应用软件
radiation precaution sign 辐射危险标志，辐射警告标志
radiation protection 辐射防护
radiation quantity 辐射量
radiation shield wall penetration 放射性屏蔽墙贯穿件
radiation shielding 辐射屏蔽
radiation streaming 辐射流束
radiation supervisor 辐射监督员
radiation surveillance specimen 辐照监督样品
radiation survey 辐射巡检
radiation thermal luminescent dosimeter reader 辐射热荧光剂量仪读数装置
radiation thermal luminescent dosimeter 辐射热释光剂量仪
radiation work permit 辐射工作许可证
radiation zone 辐射区
radiation zone map 辐射分区图
radical zone 放射区
radio frequency 射频
radio graphic testing examination 射线探伤
radio transmission 无线寻呼系统
radioactive 放射性的
radioactive concentration 放射性浓度
radioactive decay 放射性衰变
radioactive fallout 放射性尘降，放射性尘降物
radioactive hydrogenated effluent 含氢化物的放射性流出物
radioactive material 放射性物质，放射性材料
radioactive particulate 放射性颗粒，放射性微粒
radioactive release 放射性释放
radioactive waste management system 放射性废物管理系统
radioactivity 放射性
radioactivity contamination 放射性污染
radiochemical analysis 放射化学分析，放化分析
radiochemistry 放射化学
radiochemistry data management software 放射化学数据管理软件
radiographic examination 射线检查
radiographic inspection 射线探伤，射线检查
radiographic inspection/ radiography 射线探伤
radiographic test 射线试验
radiography 射线照相
radioisotope 放射性同位素
radiological consequence 放射性后果
radiological habitability 放射性可居留性
radiological risk profile 放射性风险分布状况
radiologically controlled first-aid 可控的放射性急救
radiologically restricted area 放射性限制区
radiolysis 辐照分解
radionuclide 放射性核素

radionuclide retention 放射性核素存留(量)
radio-paging system 无线移动通信系统
radius of gyration 回转半径
radwaste 放射性废物
radwaste BLDG MCC 放射性废物厂房电机控制中心
radwaste building 放射性废物厂房
radwaste building power transformer 放射性废物厂房电力变压器
radwaste solidification building ventilation system 放射性废物固化厂房通风系统
rail car bay 有轨车车场
rainfall 降雨,降雨量
rainfall intensity 降雨强度,暴雨强度
rainwater 雨水
raise,collect,make up 筹集,自筹
ramp 斜坡,坡道,使有斜面
ramp change 线性变化
ramp insertion of reactivity 反应性线性引入
ramp load change 线性负荷变化,负荷线性变化
range of welding current regulation 焊接电流调节范围
rapid load reduction 快速降功率,快速降负荷
rapid power reduction 快速降功率
rapid power reduction (RPR) system 快速降功率系统
rapid response 快速响应
rapid shutdown 快速停堆
rapidly propagating fracture 快速扩展的断裂
rate 率,比率,等级,评价
rate of back pressure 背压率,背压
rate of change application software 变化率应用软件
rate of dilution 稀释(速)率
rate of flow 流量
rate of load steam loss 有负荷漏汽率
rate of no-load steam loss 无负荷漏汽率
rate of pressure change 压力变化率
rate of reactivity insertion 反应性引入速率
rate-of-rise 上升速率
rated 额定的
rated capacities 额定容量
rated duty cycle/standard service 额定负载持续率
rated HP 额定马力
rated NSSS power 核蒸汽供应系统额定功率
rated power 额定功率
rated welding current 额定焊接电流
rather than 而不
rating 等级,种类
raw 未加工的,生的,原始的
raw liquor pump 原液泵
raw water 原水
raw water pump 原水泵
raw water purification system 原水净化系统
RCDT gas outlet containment isolation IRC 反应堆冷却剂疏排水箱气体出

口安全壳隔离(壳内)
RCP bearing water temperature instrument 反应堆冷却剂泵轴承水温计
RCP flange leak off 反应堆冷却剂泵法兰泄漏
RCP flush 反应堆冷却剂泵冲洗
RCP flywheel assembly 反应堆冷却剂泵飞轮组件
RCP pump speed instrument 反应堆冷却剂泵泵速仪表
RCP stator temperature instrument 反应堆冷却剂泵定子温度计
RCP switchgear 反应堆冷却剂泵开关装置
RCP variable frequency drive unit 反应堆冷却剂泵变频驱动机组
RCP variable speed drive unit 反应堆冷却剂泵变速驱动装置
RCP vibration instrument 反应堆冷却剂泵振动测量仪表
RCS cold overpressure 反应堆冷却剂系统冷态超压
RCS cool down rate 反应堆冷却剂系统冷却速率
RCS heat up/cool down monitoring application software 反应堆冷却剂系统加热/冷却监测应用软件
RCS pressure 反应堆冷却剂系统压力
RCS purification return line check valve 反应堆冷却剂系统净化返回管止回阀
RCS purification stop 反应堆冷却剂系统净化中断
RCS subcooling 反应堆冷却剂系统欠热度
reach an agreement 达成协议
reach rods extending 延伸杆,穿墙杆
reaction flux 反应钎剂
reaction rate 反应速率
reactive power 无功功率
reactivity 反应性
reactivity accident 反应性事故
reactivity and power distribution anomalies 反应性和功率分布异常
reactivity change 反应性变化
reactivity change rate 反应性变化率
reactivity coefficient 反应性系数
reactivity compensation 反应性补偿
reactivity control 反应性控制
reactivity control component 反应性控制部件
reactivity control system 反应性控制系统
reactivity feedback 反应性反馈
reactivity feedback weighting factor 反应性反馈权重因子
reactivity importance 反应性价值
reactivity insertion rate 反应性引入(速)率
reactivity power 无功功率
reactivity shutdown margin 反应性停堆裕量
reactivity worth 反应性价值
reactivity worth of control rod 控制棒反应性价值
reactor 反应堆
reactor building 反应堆厂房
reactor building lighting system 反应

厂房照明系统

reactor building water supply and drainage system 反应堆厂房给排水系统

reactor cavity 反应堆堆腔,堆腔

reactor cavity cooling system 堆腔冷却系统

reactor cavity discharge flow element 堆腔排风流量元件

reactor cavity seal ring 堆腔密封圈,堆腔密封环

reactor closure head stud 反应堆顶盖螺栓

reactor containment 反应堆安全壳

reactor coolant 反应堆冷却剂

reactor coolant average temperature 反应堆冷却剂平均温度

reactor coolant boundary 反应堆冷却剂边界

reactor coolant depressurization sparger 反应堆冷却剂降压喷头

reactor coolant drain pump 反应堆冷却剂疏排水泵

reactor coolant drain tank 反应堆冷却剂疏排水箱

reactor coolant drain tank WLS connection penetration 反应堆冷却剂疏排水箱与废液系统连接贯穿件

reactor coolant filter 反应堆冷却剂过滤器

reactor coolant inlet temperature (T_{cold}) 反应堆冷却剂进口温度

reactor coolant loop 反应堆冷却剂环路

reactor coolant loop piping 反应堆冷却剂回路管道

reactor coolant outlet temperature (T_{hot}) 反应堆冷却剂出口温度

reactor coolant pressure boundary 反应堆冷却剂压力边界

reactor coolant pressure boundary material 反应堆冷却剂压力边界材料

reactor coolant pump (RCP) 反应堆冷却剂泵,堆冷却剂泵,主泵

reactor coolant pump (RCP) bearing water temperature 反应堆冷却剂泵轴承水温度

reactor coolant pump insulation package 反应堆冷却剂泵保温包

reactor coolant pump locked rotor 反应堆冷却剂泵被卡死的转子

reactor coolant pump seal water system 反应堆冷却剂泵轴封水系统

reactor coolant pump trip 反应堆冷却剂泵脱扣

reactor coolant pump underspeed 反应堆冷却剂泵低转速

reactor coolant system(RCS) 反应堆冷却剂系统

reactor coolant system(RCS) effluent 反应堆冷却剂系统排出流

reactor coolant system hot log 反应堆冷却剂系统热段

reactor coolant system inventory monitoring 反应堆冷却剂系统装量监测

reactor coolant system makeup 反应堆冷却剂系统补水

reactor emergency cooling system 反应堆应急冷却系统

reactor head vent system 反应堆顶盖放气系统
reactor inherent protection 反应堆固有保护
reactor integrated head package 反应堆一体化顶盖包
reactor internals 反应堆内部构件,堆内构件
reactor internals lifting rig 堆内构件吊具
reactor kinetics 反应堆动态学
reactor lower internals 反应堆下部堆内构件
reactor operating deck 反应堆操作台
reactor operator (RO) 操纵员
reactor operator console in main control room 主控制室反应堆操作员控制台
reactor power burst 反应堆功率猝增
reactor power control 反应堆功率控制
reactor power control system 反应堆功率控制系统
reactor power excursion 反应堆功率偏差
reactor power versus turbine load mismatch signal 反应堆功率和汽轮机负荷失配信号
reactor pressure vessel support 反应堆压力容器支座
reactor pressure vessel vent system 反应堆压力容器放气系统
reactor protect system 反应堆保护系统
reactor pump 反应堆泵,主泵
reactor shutdown 反应堆停堆,停堆
reactor startup 反应堆启动,启堆
reactor trip 反应堆紧急停堆,停堆
reactor trip breaker 反应堆停堆断路器
reactor trip circuit breaker cabinet 反应堆停堆电路断路器机柜
reactor trip function 反应堆紧急停堆功能
reactor trip switchgear 反应堆停堆开关装置
reactor upper internals 反应堆上部堆内构件
reactor upper internals assembly 反应堆上部堆内构件组件
reactor vessel 反应堆压力容器
reactor vessel and reflective insulation 反应堆压力容器和反射保温层
reactor vessel cavity 反应堆堆腔
reactor vessel core region plate 反应堆压力容器堆芯段
reactor vessel flange 反应堆压力容器法兰
reactor vessel flange leakoff 反应堆压力容器法兰引漏
reactor vessel head 反应堆压力容器顶盖,压力容器上封头
reactor vessel head and flange area reflective insulation 压力容器顶盖和法兰区反射型保温层
reactor vessel head lifting device 反应堆容器顶盖吊具
reactor vessel head penetration 反应堆

压力容器顶盖贯穿件
reactor vessel head vent 反应堆压力容器顶盖放气
reactor vessel head vent valve 反应堆压力容器顶盖放气阀
reactor vessel insulation 反应堆压力容器保温层
reactor vessel integrated head package 反应堆压力容器一体化堆顶组合件
reactor vessel internals vibration testing 反应堆压力容器堆内构件振动试验
reactor vessel missile shield 反应堆压力容器飞射物屏蔽
reactor vessel reflective insulation (all other) 反应堆压力容器反射绝热层(其他部分)
reactor vessel reflective insulation in reactor cavity (IVR) 在反应堆内(用于堆芯熔融物压力容器内滞留)的反应堆压力容器反射绝热层
reactor vessel stud tensioner system 反应堆压力容器螺栓拉伸器系统
reactor vessel support 反应堆压力容器支承
reactor vessel temporary cover 反应堆压力容器临时盖板(假盖)
reactor vessel-hot leg water level 反应堆压力容器热段液位
reactor vessel-hot leg water level readout 反应堆压力容器热段液位读取
reactor water chemistry 反应堆水化学
readout 读取,指示
readout and control cabinet 读取和控制柜
real practice 实践性
real time 实时
real time operating system 实时操作系统
real time spectroscopy detector 实时能谱探测器
realignment 重新校正,再对中
realistic source terms 现实源项
real-time data network 实时数据网,实时数据网络
real-time data network cables and connector 实时数据网络电缆和连接器
real-time data network electronic module 实时数据网络电子模块
real-time portable gamma-ray imaging system 便携式实时γ射线影像系统
rear 后方,背后
rear bracket 后方支撑托架
receiving tank 接收箱
reciprocating 往复的,交替的,摆动的
reciprocating charging pump 往复式上充泵
reciprocating pump with counter running piston 对向活塞往复泵
reciprocating pump with pistons in V-type arrangement V型活塞往复泵
reciprocating pump with unidirectional piston 单向活塞往复泵
recirculation 再循环
recirculating cooling water 再循环冷却水
recirculation fuel oil return line 再循环燃油回油管线

recirculation heater　再循环加热器
recirculation mode　再循环模式
recirculation phase　再循环阶段
recirculation pump discharge check valve　再循环泵出口止回阀
recirculation pump fire suction isolation valve　再循环泵消防入口隔离阀
recirculation pump suction from long term makeup　再循环泵长期补水入口隔离阀
recirculation pump suction isolation valve　再循环泵入口隔离阀
recirculation pump throttle valve　再循环泵节流阀
recirculation valve　再循环阀
recognize　承认
recommended standard　推荐标准
reconciliation　调和，和解，顺从
reconditioning　修理，修补
reconstitutable　可复装的，可拆式的
record　记录
recorder　记录仪
recording level　记录水平
recovery　回收，恢复，防御
recovery plan　赶工计划，补救计划
recovery safety instruction　恢复安全措施
rectifier　整流器
recurrence　再发生
recurrence interval　重现期
recycle　再循环
redistribution effect　再分配（分布）效应
reduced-voltage motor controller　降压电机控制器
reducer　缩减，减压器，还原器
reducing valve　液压阀
reduction　减少
redundancy　冗余，冗余度，多重
redundant　冗余的，备份的，多重的
redundant division　冗余序列，多重序列
redundant equipment　多重设备，冗余设备
redundant GPS time server system　多重 GPS 时间服务器系统
redundant measurement　多重测量
redundant measurements program application software　多重测量程序应用软件
reel　线轴，卷绕，旋转
reentrant program　重入程序
reeving　（吊索）卷筒
reeving system　（环吊）卷绕系统
refer to　参考，参照，叫某人参看，叫某人查询
reference　编号，证明，提及，参考
reference block　基准试块
reference leg temperature　参考段温度
reference level　参考水平，基准液位
reference power distribution　参考功率分布
reference temperature　参考温度
reference value　参考值，基准值
reference voltage　基准电压
reference water level　参考液位
references　参考文献
refill　再充水

refill check valve　再充水止回阀
refill header manual vent valve　再充水集管手动排气阀
refilling stage　再灌水阶段
reflective stainless steel-type　不锈钢反射型
reflector (radial/axial)　反射层(径向/轴向)
reflector saving　反射层节省
reflood　再淹没
reflux　回流,逆流
refresh　刷新
refuel outage work order　换料大修工单
refueling　换料
refueling canal　换料通道
refueling cavity isolation　换料水池隔离
refueling cavity overflow to SG compartment　换料水池溢流至蒸汽发生器隔间
refueling cavity tank　换料水池
refueling machine　换料机
refueling operation　换料操作
refueling outage　换料停堆
refueling pool　换料水池
refueling pool underwater filter system　换料水池水下过滤器系统
refueling system　装换料系统
refueling water storage tank (RWST)　换料水贮存箱
refurbishment　整修
regardless of　不管,不顾,不拘,不注意
regenerated boric acid solution　再生硼酸溶液
regenerative heat exchanger　再生式热交换器
registered capital　注册资金
registration　注册,登记
regular test　定期试验
regularly　经常地
regulating transformers　调节变压器
regulating valve　调节阀
regulation spring　调节弹簧
regulatory guide　核安全法规导则
regulatory requirement　核安全法规导则要求
regulatory treatment of non-safety system　非安全系统的法规规定处理方法
reheat crack　再热裂纹
reheater　再加热器
reimburse　报销,补偿
reinforcement　加强
reinforcement envelope　配筋包络值
reinforcement provided　实际配筋
reinforcement required　需要配筋,计算配筋
reinforcing plate　补强板
reiterate　重申
rejection　报废,拒收
related to　与…有关
relative biological effectiveness (RBE)　相对生物效应
relative flow characteristics derivation　相对流量特性演绎
relative humidity　相对湿度
relative pressure characteristics deriva-

tion　相对压力特性演绎
relative static characteristics derivation　相对静演绎
relaxation　松弛,放宽,缓和,减轻
relaxed axial offset control (RAOC)　放宽轴向偏移控制
relay　继电器
relay actuator module　继电器驱动器组建
release　交付,产品的放行
release category　释放类型,释放种类
release for shipment　释放运输许可,允许运输
release location　释放点,释放位置
release pathway　释放途径
reliability　可靠性
reliable　可靠的
relief　免除,减轻,释放,调剂,安慰
relief damper　卸压风阀
relief valve　排放阀,卸压阀,安全阀
relieving capacity　排放量
relieving pressure　排放压力
reload　换料
reload core (cycle)　换料堆芯(循环)
reload design　换料设计
reload enrichment　换料富集度
remain composure　保持冷静
remain in force　保持有效
remain in good terms with　与…保持良好关系
remarks　备注,意见
remedial action　纠正行动,矫正措施
remit　汇寄,付(款),开发(支票)
remittance　汇款,汇款额,支付(金额)
remote　遥远的,偏僻的,细微的
remote control　遥控
remote shutdown station console　远程停堆站控制台
remotely located sensor　远程定位传感器
remotely operated　远程操作的,遥操的
remotely operated containment isolation valve　远程操作安全壳隔离阀
remotely operated valve　远程操作阀门
removable insulation modules　可拆卸保温组件
removal　去除,拆下,排出
removal tank　排放箱
render　提出,提供
render advisory service　提供咨询服务
renew　续签合同
renew an offer　更新报盘
re-occurrence　(事件)再次发生
repair　返修,修补
repair and replacement parts　修复和更换部件
repair welding　补焊
repayment　还款
repeater　转发器
repeater converter boxes　循环转换箱
replacement　替代,替换
replacement power　替代电源
report　报表,报告
report to　向…汇报
representative　代表,代理(人)
representative sample　代表性样品
requirement society　要求

requisite 必要的,必不可少的
reseating 回座
reseating pressure of a safety valve 安全阀回座压力
resetting 复位,重设,重新安排
residual 剩余的,残留的
residual chlorine 残余氯
residual heat
余热,(反应堆)剩余释热
residual heat power 剩余功率
residual heat removal (RHR) exchanger 余热排出热交换器
residual heat removal (RHR) pump 余热排出泵
residual heat removal pump A 余热排出泵 A
residual heat removal pump B 余热排出泵 B
residual heat removal system relief valve 余热排出系统卸压阀
residual hydrogen 残余氢
residual sample 剩余样品
residual stress analysis 残余应力测定
resilient 弹回的,有回弹力的
resilient seal 弹性密封,有回弹力的密封
resin 树脂
resin addition tank 树脂添加箱
resin bed 树脂床
resin fine 树脂碎片,树脂细粒
resin flush IRC isolation 树脂冲排安全壳内隔离
resin flush pump 树脂冲排泵
resin flush water 树脂冲排水
resin regeneration 树脂再生
resin retention screen 树脂截留滤网
resin sampler package 树脂取样装置
resin sluice line relief valve 树脂冲排管线泄放阀
resin transfer pump 树脂输送泵
resin trap 树脂捕集器,树脂陷阱
resistance 电阻,阻力,抵抗
resistance box 电阻箱
resistance brazing 电阻钎焊
resistance curve 阻力曲线
resistance temperature detector 电阻温度探测器,热电阻温度计
resistance to earth 对地电阻
resistance to irradiation damage 抗辐照损伤性
resistance to sliding 抗滑移性
resistivity 电阻率
resistor 抗蚀剂,保护层
resolution 分辨,分解,分辨率,清晰度
resolution time 分辨时间
resolution-annealing heat treatment 重新固熔退火热处理
resolve 解决,处理
resolve conciliation 调停
resonance 共振
resonance absorption 共振吸收
resonance capture 共振俘获
resonance escape probability 逃脱共振概率
resonance integral 共振积分
resonance neutron 共振中子
resonance region 共振区
resonance width 共振宽度

resort to arbitration　诉诸仲裁
resource manager　资源经理
respecting　关于
respirator　呼吸面具，口罩，呼吸器
response　反应，响应
response spectra　反应谱，响应谱
response time　反应时间，响应时间
responsible organization　责任单位
rest with　在于，归于
restate　重申，再声明，重新陈述
restraint　约束，限制，阻尼器
restraint coefficient　约束系数
restraint deflections　约束变形
restraint intensity　约束强度
restricted area boundary　限制区边界
restriction　限制
restrictor valve　限流阀，节流阀
result from　由…引起(产生，造成)
result in　(引起)导致
retaining screen　截留滤网，滞留筛
retention　滞留，保存
retention area　保存区
retention categories　保存分类
retention of records　记录保存
retention period　保存期
retrieval　取回，恢复
retrieval and accessibility　检索和可达性
return chilled water temperature　冷冻水回水温度
return fan　回风机
return line　回流管线
return on investment(ROI)　投资回报
returning spring　复位弹簧
reusing water　复用水
reusing water pool　复用水池
reusing water pump　复用水泵
reusing water tank　复用水箱
reveal　揭露
reverse acting reducing valve　反向作用式减压阀
reverse voltage　反向电压
reverse-power　逆功率
reviewer　审查者
revised pipe-break criteria (leak-before-break)　修改的管道破裂准则(先漏后破)
revised thermal design procedure　修正的热工设计程序
revision　修订版
rewarding　有益的，值得做的
rewelding　返修焊，重焊
rewetting temperature　再湿温度
rework　返工
rib　加强肋
ridge　脊，脊梁，脊状延伸
ridiculously　不合理地，可笑地
rig　索具装备，吊具，成套器械
rigging　搬运，装卸
rigid　刚性的，坚硬的，无弹性的
rigid dimple　(格架的)刚性搅混翼
rigid frame　刚架
rigid pipe　刚性管
rigid restraint cracking test　刚性约束裂纹试验(RRC 试验)
rigid support　刚性支撑
ring girder　环梁
ring header　环状集管

ring support 环形支架
ringing bell 响铃
ripple 结晶层状线
rising characteristic arc welding power source 具有上升特性的弧焊电源
risk analysis assumptions 风险分析假设
risk importance 风险重要度
risk important tasks 风险重要作业
risk informed 风险告知，风险引导
risk profile 风险概貌
risk reduction worth (RRW) 风险降低因子，风险降低值
risk-oriented accident analysis methodology 风险导向事故分析方法
risk-significant 重大风险影响
river flooding 河流洪水
RNS discharge containment isolation valve-IRC 正常余热排出系统排放安全壳隔离阀—壳内
RNS discharge containment isolation valve-ORC 正常余热排出系统排放安全壳隔离阀—壳外
RNS discharge containment isolation valve test 正常余热排出系统排放安全壳隔离阀试验
RNS discharge containment isolation valve test connection 正常余热排出系统排放安全壳隔离阀试验接头
RNS discharge containment penetration isolation valve test 正常余热排出系统排放安全壳贯穿件隔离阀试验
RNS discharge to IRWST isolation 正常余热排出系统排放至内置换料水箱的隔离
RNS HL suction and containment isolation-inner 正常余热排出系统热段入口和安全壳的壳内隔离
RNS HL suction and containment isolation-outer 正常余热排出系统热段入口和安全壳的壳外隔离
RNS HL suction pressure relief 正常余热排出系统热段入口压力释放
RNS hot leg suction isolation valve 正常余热排出系统热段入口隔离阀
RNS HX bypass flow instrument isolation 正常余热排出系统热交换器旁路流量仪表隔离
RNS HX channel head drain isolation 正常余热排出系统热交换器管侧封头疏水隔离
RNS HX outlet flow control 正常余热排出系统热交换器出口流量控制
RNS HX shell drain 正常余热排出系统热交换器壳侧疏水
RNS pump casing drain isolation 正常余热排出系统泵壳疏水隔离
RNS pump discharge isolation 正常余热排出系统泵出口隔离
RNS pump discharge pressure instrument isolation 正常余热排出系统泵出口压力表隔离
RNS pump discharge relief 正常余热排出系统泵出口释放
RNS pump discharge to spent fuel pool isolation 正常余热排出系统泵出口至乏燃料池的隔离
RNS pump miniflow isolation 正常余

热排出系统泵小流量隔离

RNS pump suction containment isolation test connection　正常余热排出系统泵入口安全壳隔离试验接口

RNS pump suction from spent fuel pool isolation　从乏燃料池到正常余热排出系统泵入口的隔离

RNS pump suction isolation　正常余热排出系统泵入口隔离

RNS pump suction piping drain isolation　正常余热排出系统泵入口管疏水隔离

RNS pump suction pressure instrument isolation　正常余热排出系统泵入口压力表隔离

RNS pump suction to cask loading pit isolation　正常余热排出系统泵入口至乏燃料装运池的隔离

RNS return from CVS containment isolation　从化学和容积控制系统至正常余热排出系统返回管的安全壳隔离

RNS suction from IRWST containment isolation　从内置换料水箱至正常余热排出系统入口的安全壳隔离

RNS suction header containment isolation　正常余热排出系统入口集管安全壳隔离

RNS train A discharge flow instrument isolation　正常余热排出系统 A 系列出口流量表隔离

robertson crack arrest test　罗伯逊止裂试验

robot working space　机器人工作空间

rocker arm pump　变行程往复泵，摇臂泵

rocking plunger valve　柱塞型摆动阀

rod bow/bowing　（燃料）棒弯曲

rod cluster　棒束

rod cluster cabinet　控制棒控制柜

rod cluster control assembly（RCCA）and thimble　棒控系统（控制棒与阻力塞组件）

rod cluster control assembly and thimble plug　控制棒与阻力塞抽插机

rod cluster control assembly out of bank measurements 控制棒失步，控制棒组件棒位偏离测量值

rod cluster dead band　棒控死区

rod control cabinet　控制棒控制柜，棒控机柜

rod control cluster assembly change fixture　控制棒组件更换装置

rod control cluster handling tool　棒控束装卸工具

rod control cluster storage station　控制棒束贮存站

rod control system　控制棒控制系统，棒控系统

rod drive M/G set　控制棒驱动电动发电机组

rod drive mechanism　控制棒驱动机构

rod drive M-G control cabinet　控制棒驱动电动发电机组控制机柜

rod drive motor set　控制棒驱动马达

rod drive shaft　控制棒驱动杆/轴

rod driven M-G set　棒电源机组

rod drop　落棒

rod ejection （控制棒）弹棒
rod ejection accident （控制棒）弹棒事故
rod internal gas pressure （燃料）棒内气体压力
rod let 棒，特指单根控制棒或灰棒
rod misalignment （控制）棒失步/失调
rod motion 棒运动
rod operating/power history （燃料）棒运行/做功史
rod pitch （燃料）棒中心距，（燃料）棒间距
rod position detector 棒位探测器
rod position detector assembly 棒位探测器组件
rod position digital display 棒位数字显示
rod position indication 棒位指示
rod position indication cabinet 棒位指示机柜
rod position indication system 棒位指示系统
rod position indicator 棒位指示器
rod position LED indicator 棒位光点指示器
rod travel housing （控制）棒驱动机构耐压壳
rod wise power distribution （燃料）棒功率分布
rod withdrawal （控制）棒提升，控制棒抽提
rod withdrawal accident 提棒事故
rod worth 控制棒价值
rod-at-bottom signal 控制棒插到底信号
rodlet 棒，特指单根控制棒或灰棒
roller pump 滚轮泵
rolling 辊轧
roof drain 屋顶排水沟
roof ventilator 屋顶通风装置
room 房间，室
room module 房间模块
room temperature 室温，常温
room to maneuver 回旋余地
root 根部(焊接)
root bend test 根部弯试验
root concavity 根部凹陷
root crack 根部裂纹
root face 钝边
root gap 根部间隙
root isolation valve 根部隔离阀
root mean square value 均方根值
root valve 根阀
rope 钢丝绳
rosin flux 松香钎剂
rotary piston pump with slide gate 滑门式回转活塞泵
rotary-block radial pump 旋转缸体径向活塞泵
rotating arc welding 旋转电弧焊
rotating inertia 转动惯量
rotating seal ring 动密封环
rotational speed 转动速度，旋转速度
rotor 转子
rotor seizure 转子卡死
rough machining 粗加工
roughness 粗糙度
round bar cracking test 圆棒裂纹试验

route PM Action　例行预防性维修活动
routine inspection　例行检查
routine monitoring　日常监测
routine shutdown　例行停堆
routine test　例行试验,常规试验
royalty　(专利权)使用费
RSR WS desk　远程停堆工作站办公桌
rubber bumper　橡胶缓冲器
rubber expansion joint　橡胶膨胀节
rubber lined diaphragm valve　衬胶隔膜阀
rubber lined globe valve　衬胶截止阀,衬胶球阀
rubber lined pump　衬胶泵
rudimentary　基本的
rudimentary idea　基本想法
run through　浏览,从头到尾过一遍
running characteristics of an electrode　焊条使用性
runoff　径流
run-off tab　引出板
run-on tab　引弧板
runway track　轨道
rupture　断裂,破裂
rupture disk　爆破盘
rupture strength　断裂强度
rupture test　断裂试验
RX coolant pump　RX 冷却剂泵
RXS non-threaded fasteners　RXS 非螺纹紧固件
RXS threaded structural fasteners　RXS 螺纹结构紧固件

S

saddle support　鞍式支座
safe end　安全端
safe operating region　安全运行/范围
safe shutdown　安全停堆
safe shutdown condition　安全停堆工况
safe shutdown earthquake　安全停堆地震
safe shutdown earthquake (SSE) horizontal acceleration　安全停堆地震水平加速度
safe shutdown earthquake (SSE) vertical acceleration　安全停堆地震垂直加速度
safe shutdown system　安全停堆系统
safeguards　安全保障
safeguards actuation　触发安全保障措施
safeguards actuation signal　安全保障触发信号
safekeeping　保卫,维安
safety　安全性
safety actuator　安全驱动器
safety analysis　安全分析
safety analysis limit　安全分析限值
safety analysis report　安全分析报告
safety bank　安全组(控制棒)
safety benefit　安全效益,安全收益
safety brake　安全制动器
safety class　安全等级
safety class interface　安全等级界面
safety classification　安全分级
safety code　安全法规

safety criteria 安全准则
safety culture 安全文化
safety design basis 安全设计基准
safety end (焊接)安全端
safety evaluation 安全评价,安全评估
safety factor 安全系数,安全因子
safety injection 安全注射,安注
safety injection deflector 安全注射导流板
safety injection nozzle (反应堆压力容器)安全注射接管
safety injection phase 安全注射阶段
safety injection pump (SIP) 安全注射泵
safety injection signal 安全注射信号
safety injection system 安全注射系统
safety injection tank 安全注射箱
safety instruction 安全细则
safety interlocks 安全联锁
safety latch 安全栓,安全插销
safety lighting 安全照明
safety limit 安全限值
safety monitoring system (SMS) 安全监测系统
safety parameter display system (non-safety) 安全参数显示系统(非安全级)
safety parameter display system (SPDS) 安全参数显示系统
safety related component 安全相关设备
safety related decay heat removal capability 安全相关衰变热导出能力
safety relief valve 安全释放阀,安全卸压阀
safety rod 安全棒
safety system 安全系统
safety system instrumentation 安全系统仪表
safety valve 安全阀
safety valve discharge chamber rupture disk 安全阀出口腔爆破盘
safety valve set point 安全阀整定值
safety-related 安全相关的,安全有关的
safety-related display 安全相关显示
safety-related display information 安全相关显示信息
safety-related display instrumentation 安全相关显示仪表
safety-related function 安全相关功能
safety-related parameter 安全相关参数
safety-related system 安全相关系统
salt bath dip brazing (soldering) 盐浴钎焊
salt melting slag 盐型熔渣
salt-oxide melting slag 盐—氧化物型熔渣
samarium concentration 钐浓度
samarium poisoning 钐中毒
sample cooler 取样冷却器,样品冷却器
sample cooler rack 样品冷却器机架
sample line 取样管
sample station 取样站
sample survey 抽样调查
sample vessel 取样容器

sampler　取样器
sampling and detecting assembly　取样和探测装置
sampling bottle　取样瓶
sampling cabinet　取样箱,取样间
sampling device　取样装置
sampling of document　文件抽样
sampling resister　样品抗蚀剂,样品保护膜
sampling subsystem　取样子系统
sampling system　取样系统
sand blast　喷砂
sand blasting　喷砂
sand filter　砂滤器
sandstorm　砂暴
sandwich filler metal　夹层状钎料
sandy silt　砂质粉土,砂粉
sanitary　卫生的
sanitary sewer treatment system　生活污水处理系统
sanitary sewer yard pipe network system　生活污水厂区管网系统
SAT analysis phase　SAT 的分析阶段
SAT design phase　SAT 的设计阶段
SAT development phase　SAT 的编制阶段
SAT evaluation phase　SAT 的评价阶段
SAT implementation phase　SAT 的实施阶段
saturated boiling　饱和沸腾
saturated steam　饱和蒸汽
saturated vapor　饱和蒸汽
saturation　饱和(状态)
saturation index　饱和指数
saturation temperature　饱和温度
scaffolding　脚手架
scale　比例,刻度,等级,范围
scaler　定标器,比例尺
scan　扫描,浏览,审视
scan rate　扫描速率
scarce　稀少的
scarfing　火焰表面清理
scattering　散射
scattering cross section　散射截面
scattering law　散射定律
scattering matrix　散射矩阵
scavenge　扫除,去除,提取
scavenge oxygen　除氧,去氧
scavenger　清洗剂,去除剂
scenario　情景,状态描述,剧本
Schaeffler's diagram　舍夫勒组织图
schedule　进度
schedule development, integration and management　进度编制,综合和管理
scheduler　进度计划员
scheduling　编制进度
scintillation　闪烁,闪光
scintillation detector　闪烁探测器
scintillator　闪烁体
scoop pump　铲斗泵
scope　范围,余地
SCR arc welding rectifier　晶闸管弧焊整流器
scram　紧急停堆
scrap　废弃,报废
screen　隔栅,滤网
screen cartridges　滤芯

screen flow area 滤网流通面积，滤网流通区
screening criteria 审批标准
screw 螺丝，螺钉，螺杆
screw pump with square thread 方形螺纹螺杆泵
screwed cap 螺纹帽
scrolling 滚动
scrubbing 擦洗，刷洗，去除
scrubbing decontamination factor 擦洗去污因子
scrubbing device for bolt 螺栓清刷装置
scum gate valve 浮渣闸阀，泡沫闸阀
scupper 排水孔，排水口，排水管
seal cover 密封盖
seal housing 密封壳
seal in 自保持
seal ledge 密封凸台，密封壁架
seal off 密封，封闭
seal ring 密封环
seal test 密封试验
seal test pressure 密封试验压力
seal type 密封式，封闭式
seal up 密封，封闭
seal water injecting filter 轴封注水过滤器
seal water reflux filter 轴封回流水过滤器
seal water reflux heat exchanger 轴封回流水热交换器
seal weld 密封焊
sealant 封孔剂
sealant compound 密封剂混合物
sealed plug and socket 密封接插件
sealed source 密封源
sealing face 密封面
sealing strip for valve seat 阀座密封嵌条
sealing structure on bottom of the refueling pool 换料水池底部密封结构
sealing 封孔处理
seam 缝，焊缝，接合处，层
seamless part 无缝件
seat ring 阀座环
seawater corrosion resistant steel electrode 耐海水腐蚀钢焊条
secondary auxiliary building ventilation system 二回路辅助厂房通风系统
secondary clock 子钟
secondary combustion 二次燃烧
secondary containment 二次安全壳
secondary core support 二次堆芯支撑
secondary ductwork distribution system (HVAC)分配支系统，二回路管网分配系统
secondary fire water tank 二次消防水箱
secondary heat removal system 二回路排热系统
secondary manway 二次侧人孔
secondary missile 二回路飞射物
secondary neutron source 次级中子源，二次中子源
secondary neutron source assembly 次级中子源组件
secondary radiation 次级辐射
secondary shield 二次屏蔽

secondary shield wall 二次屏蔽墙
secondary side 二次侧
secondary side leakage test 二次侧检漏试验
secondary solidification structure 二次结晶组织
secondary steam 二回路蒸汽
secondary stress 二次应力
secondary stress index 二次应力指数
secondary support assembly (堆芯)防断组件
secondary support base plate (堆芯)防断底板
secondary support plate (堆芯)防断支撑板
secondary system 二回路系统
secondary treatment 第二级处理,二次处理
section modulus of the pipe 管道截面模量
sectional view 剖面图
security central alarm station 保卫控制中心
security communication system 保安通信系统
security damper 安全阀
sediment pump 沉积物泵
sedimentation 沉淀
see to that 注意做到
seed (ing) sludge pump 腐泥引进泵,接种污泥泵
segmental 部分的
segmental type pump 分装式多级泵
segmented circular groove cracking test 分块圆周形槽热裂纹试验
segregation (冶金)偏析,熔析,分离
seiche 湖震,湖面波动
seism 地震
seism tectonic province 地震构造区
seismic 地震的
seismic acceleration channel 地震加速度通道
seismic and environmental qualification 抗震和环境评定
seismic basic intensity 地震(基本)烈度
seismic category 抗震分类
seismic category Ⅰ 抗震Ⅰ类
seismic category Ⅰ structure 抗震Ⅰ类构筑物
seismic classification 抗震分级
seismic design 抗震设计
seismic design basis dynamic loads 抗震设计基本动态载荷
seismic design basis loads 抗震设计基准载荷
seismic excitation 地震激励
seismic floor response spectra 地震楼面反应谱
seismic geology 地震地质
seismic instrumentation 地震仪表
seismic instrumentation system 地震仪表系统
seismic loads 地震载荷
seismic margin analysis (SMA) 抗震裕量分析
seismic monitoring system interface application software 抗震监测系统接

口应用软件
seismic qualification testing 抗震鉴定试验
seismic restraint 抗震阻尼器
seismic restraints for control rod drive mechanisms 控制棒驱动机构抗震阻尼器
seismic risk 地震风险
seismic stabilizer bracket 抗震加固支架
seismic standpipe 抗震立管,抗震管体式水塔
seismic standpipe system 抗震立管系统,抗震消防系统
seismically-supported 抗震加固的
seismotectonic province 地震构造区
selection of supplier 供方选择
selector 选择器,选择者
selector switch 选择开关
selector valve 换向阀
self-actuating containment isolation valve 自触发的安全壳隔离阀
self-adjusting arc welding 自调节电弧焊
self-contained 成套的,自持的,整体式的
self-contained breathing apparatus refill station 自储式呼吸器灌充站
self-fluxing alloy powder 自熔剂合金粉末
self-fluxing brazing alloy/ self-fluxing filler metal 自钎剂钎料
self-inspection 自检
self-leveling 自校平,水准自调整,自平整
self-leveling epoxy 自流校平环氧树脂
self-releasing slag 自动脱落焊渣
self-reliance 自主化
self-shielding 自屏蔽,自屏
self-shielding factor 自屏因子
seller 供方
semaphore 信号灯,信号,旗语
semi-automatic welding 半自动焊
semiconductor detector 半导体探测器
semienclosed 半封闭式
semi-graphic display 半图形显示
semi-infinite cloud 空间无限扩散的烟云
semi-rotary (wing) pump 摆动泵
semi-volatile covering 气渣联合保护型药皮
senior reactor operator (SRO) 高级操纵员
senior reactor operator console in main control room 主控制室反应堆高级操纵员控制台
sensible heat 显热
sensing 传感,敏感
sensitive volume (探测器的)灵敏体积
sensitivity analysis 敏感性分析
sensitization 敏化
sensitization temperature 敏化温度
sensitized stainless steel 敏化不锈钢
sensor 传感器,变送器
sensor rack 传感器格架
sentimentality 感情用事
sentinel valve 报警阀,信号阀,脉冲阀
separate PM action 单项预防性维修

活动

separate system　分流系统，单个系统

separate valve box type piston pump　分置阀室式活塞泵

separation　隔离

separator　分离器

septic tank　化粪池

sequence　次序，顺序，序列

sequence events　相继事件

sequence of events function　事件作用序列，事件影响序列

series submerged arc welding　横列多丝埋弧焊

service　服务

service air　厂用空气，服务用气

service air supply outside containment isolation　服务用气安全壳外隔离

service condition　工作工况，运行工况，使用工况

service delivery　服务提供

service modules　服务模块

service power operation　带厂用电运行

service pressure　工作压力

service program　服务程序

service water blowdown radiation monitor　厂用水排污辐射监测仪

service water cooling tower　厂用水冷却塔

service water cooling tower fan　厂用水冷却塔风机

service water pump　厂用水泵

service water strainer　厂用水过滤器

service water yard pipe network system　厂用水厂区管网系统

service weldability　使用可焊性

services　服务

servo valve　伺服阀

servomotor　伺服马达

session　会议，会议的一次（一届）

set out to do something　（带着某种目的的）开始做某事

set out to prepare　着手准备

set point　整定点，整定值

set pressure　整定压力

setpoint adjustment　整定值调整

setpoint signal　整定值信号

settle a claim　理赔

settle a dispute　解决纠纷

settle an account　结账

settlement　沉降，移居，沉积

settlement joint　沉降接合缝，沉降接合

settling　沉降

severance　断裂，切断，分离，断绝

severe accident　严重事故

severe accident management guidance　严重事故管理导则

severe accident mitigation design alternatives（SAMDA）　缓解严重事故后果的设计替代方案

severe accident sequence　严重事故序列

sewage　下水道，污水

sewage wastewater engineering　下水道排污工程

sewer　污水渠，下水管，阴沟

SFS cask loading pit drain isolation　乏燃料系统装运池排水隔离

SFS cask loading pit drain to WLS isolation 乏燃料系统装运池向低放废液系统的排水隔离
SFS containment isolation valve test 乏燃料系统安全壳隔离阀试验
SFS containment penetration test connection isolation 乏燃料系统安全壳贯穿件试验接口隔离
SFS CVS makeup reverse flow prevention 防止化学和容积控制系统向乏燃料系统补水倒流
SFS discharge to cask loading pit isolation 乏燃料系统至装运池的出口隔离
SFS fuel transfer canal drain isolation 乏燃料系统燃料传输通道排水隔离
SFS refueling cavity drain to containment sump 乏燃料系统换料水池向安全壳地坑排水的隔离
SFS refueling cavity drain to SGS compartment 乏燃料系统换料水池疏水至蒸汽发生器系统隔间的隔离
SFS refueling cavity suction isolation 乏燃料系统换料水池入口隔离
SFS suction line from IRWST isolation 乏燃料系统来自内置换料水箱补水管线入口的隔离
SG blowdown isolation 蒸汽发生器排污隔离
SG channel head divider plat 蒸汽发生器管侧封头隔板,蒸汽发生器下封头隔板
SG condensate pipe drain valve 蒸汽发生器凝结水管排水阀
SG feedwater distribution ring supports 蒸汽发生器给水分配环支撑
SG feedwater flow instrument 蒸汽发生器给水流量仪表
SG main steam line flow instrument 蒸汽发生器主蒸汽管线流量仪表
SG nitrogen sparging isolation 蒸汽发生器氮气鼓泡隔离
SG startup feedwater flow instrument 蒸汽发生器启动给水流量仪表
SG tube bundle support assembly 蒸汽发生器传热管束支撑组件
shaft (泵)轴
shaft seal assembly 轴封组件
shakeable 可摇动的,可摆动的
shape coefficient (风荷载)形状系数
shape cutting 仿形切割
share 共享
shared file 共享文件
shared resource 共享资源
shear 剪切
shear force 剪切力
shear lip zone 剪切唇区
shear modulus 切变膜量
shear stress 切变应力,剪应力
shear test 剪切试验
shear wave velocity 剪切波波速
shearing strength 剪切强度
sheathed cable 铠装电缆
sheathing and shielding cable 铠装屏蔽电缆
sheet 薄片
shell 壳体,筒体
shell insulation 筒身保温层

shell stream　壳程，沿壳壁流
shell test　壳体试验
shell-and-tube heat exchanger　管壳式热交换器
shell-side　壳侧
shield　屏蔽
shield building　屏蔽厂房
shielded compartment　屏蔽隔间
shielded gas plasma arc cutting　保护气体等离子弧切割
shielded hatch　屏蔽闸门
shielded metal arc welding　保护金属氩弧焊
shielded metal arc welding (SMAW)　保护电弧焊
shielded pipe chase　屏蔽管沟
shielding calculations　屏蔽计算
shift (operation personnel)　(运行人员)倒班，轮值
shift supervisor console　值班主任台
shipment/shipping/dispatch　装货，装运
shipper　托运人
shipping cask　(乏燃料)运输容器
shipping cask loading pool　(乏燃料)运输容器装料池
ships pump　船用泵
shock load　冲击载荷
shock resistance　抗冲击能力
shock strength　抗冲击强度
shock test　冲击试验
short circuit　短路
short circuit thermal capacity　短路热容量
short circuiting arc welding　短路过渡电弧焊
short life isotope　短寿命同位素
short slag　短渣
shortage　短缺
short-lived fission product　短寿期裂变产物
short-term safe shutdown　短期安全停堆
shot-blasting　喷丸处理，喷沙处理，喷净法
shoulder meltability/ melting degree of shoulder　上缘熔化度
shower　淋浴器
shrinkage　收缩
shrinkage cavity　(铸件的)缩孔
shrink-fit　热配合，冷缩配合
shroud　风罩，遮盖物，封闭，隐藏
shuffling　(反应堆)倒料操作，倒料
shunt　分流器
shunt trip　并联跳闸
shunt trip interposing relays　励磁停堆中间继电器
shut down　停堆
shutdown bank　停堆棒组
shutdown bank insertion limit　停堆棒插入限制
shutdown boron concentration　停堆硼浓度
shutdown condition　停堆工况
shutdown group　停堆(控制棒)组
shutdown rods　停堆棒，安全棒
shutoff damper　阻断风阀
shutoff head　关闭扬程

shut-off valve　截止阀，关断阀
shutter　百叶窗，闸板
shuttle valve　换向阀，梭形滑阀，往复阀
side bend test　侧弯试验
side channel pump　侧流道泵
side pot type piston pump　分置阀室式活塞泵
side suction pump　侧吸泵
side-by-side　并列，一个接着一个
sievert (Sv)　希沃特，希(剂量当量单位)
sight glass　观察窗
sign a contract　签订合同
sign, take on　签订
signal　信号
signal conditioning　信号处理，信号调制
signal selection algorithms　信号选择运算法则
signal selector algorithm　信号选择器运算法则
significant operating event report (SOER)　重大运行事件报告
signify　表示
silencer　消音器
silicon　硅
silicon arc welding rectifier　硅弧焊整流器
silicon-controlled rectifier　可控硅整流器
silty clay　粉质泥土
silty sand　粉砂
silver-indium-cadmium alloy　银-铟-镉合金
silver-plating　镀银
simplified emergency planning　简化应急计划
simplified logbook　简化的日志
simply put　简而言之
simulated　模拟的
simulated program　模拟程序，仿真程序
simulated signal　模拟信号，仿真信号
simulated test　模拟试验，仿真试验
simulation　模拟
simulator training　模拟机培训
simultaneously　同时地
single(power)pump　单杠曲柄泵
single(steam)pump　单杠蒸汽直动泵
single action vane pump　单作用滑片泵
single active failure　单一能动故障
single channel analysis　单通道分析
single channel impeller pump　单流道叶轮泵
single cylinder reciprocating pump　单缸往复泵
single diaphragm pump　单隔膜泵
single disc flat gate valve　单平板闸阀
single entry pump　单吸泵
single failure　单一故障
single failure criteria　单一故障准则
single flow turbine　单流程汽轮机
single gate disc　单闸板
single line diagram　单线图
single lobe pump　单凸轮泵
single opening exhaust valve　单口排

气阀
single pass weld 单道焊
single phase flow 单相流
single plunger pump 单柱塞泵
single port nozzle 单孔喷嘴
single random failure 单一随机故障
single stage regulator 单级调节器
single suction pump 单吸泵
single-failure-proof 防单一故障的
single-welded joint 单缝接头
sinking pump 潜水泵
sintered 烧结的
sintered flux 烧结焊剂
siphon break 虹吸阻断
siphon filter 虹吸滤池
sipping device （燃料破损检查）啜吸装置
siren tone generator 警报器，报警器
site acceleration 当地加速度，厂区加速度
site boundary 厂区边界
site installation 现场安装
site investigation 场地勘察
site personnel 厂区人员
site security system 厂区保安系统
site service 现场服务
site specific missile 场地特定飞射物
site supervision 现场监督
site-specific 特定厂址的，厂址特有的
siting 选址
skeleton 骨架，基干，纲要
sketch 草图
sketchy 大概的
skid 装置，成套设备，滑轨，滑槽
skill 技能
skill lab training 技能实验室培训
skimmer 撇沫器，去浮器
skimming 撇去浮渣
skip 跳，跳跃
skip sequence 跳焊
skirt 裙，裙边，裙座
skirt mounted pump 灯笼架支承泵
skirt support 裙式支座，支撑裙
sky-shine 天空照射
slab 厚平板，厚片，混凝土路面
slag bath 渣池
slag detachability 脱渣性
slag fluidity 熔渣流动性
slag forming constituents 造渣剂
slag inclusion 夹渣
slave cycler 从动计数器
slave cycler count and decoder 从动计数译码器
slave cycler input module 从动输入组件
sleeve 套管，套筒
sleeve end plug 套管端塞
slenderness ratio 长径比
slide valve 滑阀
slide way 滑道
sliding 滑行的，变化的
slightly 稍微地，轻微地
slightly enriched uranium 低富集铀
slightly enriched uranium dioxide 低富集二氧化铀
sling 吊索，钩悬带，投掷
sling block 吊绳滑轮
sling rod 吊杆

slip ring brush assembly　集电环电刷组件,滑环电刷组件
slit Y-type cracking test　Y 形坡口裂纹试验
slope-controlled arc welding power source　多特性弧焊电源,线性控制弧焊电源
sloping　坡度,倾斜的,有坡度的
slot　切槽
slot weld　槽焊
slow transient　缓慢瞬态
sludge　污泥,泥渣
sludge collector　(蒸汽发生器)泥渣收集器
sludge control system　泥渣控制系统
sludge dewatering　泥渣脱水
sludge dry　泥渣干燥
sludge incineration　泥渣焚烧
sludge pressure filtration　泥渣压滤
sludge thickening　污泥浓缩,泥渣稠化
sluice　水闸,泄水,冲洗
slumping　混凝土流动性试验,坍塌
small break　小破口
small load pump　低负荷泵
small-break LOCA (SLOCA)　小破口冷却剂流失事故
smooth over　缓解
smoothing　修平
smoothly　顺利地
snap switch　快速开关
snubber　阻尼器
sociality　社交
socket　插座
socket joint　套管接头
socket weld　承插焊接
sodium　钠
sodium vapor lamp　钠蒸气灯
soft control　软操作器,软操作
soft control device　软操作装置
soft control displays application software　软操作显示应用软件
soft control input devices　软操作输入装置
soft rubber lined pump　软胶衬里泵
software　软件
software development tool and libraries software　软件开发工具和数据库软件
software hazards analysis　软件风险分析
software license consistent with scope of Ovation hardware　与 Ovation 硬件范围相应软件许可证
software maintenance　软件维护
software package　软件包
soil　土壤
soil corrosion　土壤腐蚀
soil friction　土壤摩擦力
soil passive pressure　被动土压力
soil spring　土弹簧
soil-structure interaction　土与结构互相作用
solder　软钎料
solder cream　钎料膏
solder iron　烙铁
solder paste　钎料膏
solder preform　钎料预成形
solder-ability　钎焊性

soldered joint　钎焊接头
solderer　焊工
soldering　软钎焊
soldering flux　钎剂，焊剂
soldering seam　钎缝
solenoid　电磁线圈
solenoid motor brake　电磁式电机制动器
solenoid operator　电磁式操作机构
solenoid valve　电磁阀
solid (handling) pump　固体输送泵
solid bottom tray　实底桥架
solid conductor　单股导线，实心导体
solid laser　固体激光器
solid tetrahedral elements　实心四面体单元
solid waste treatment and storage system　固体废物处理和储存系统
solid wire　实心焊丝
solidification　固化
solidification crack　凝固裂纹
solidification mode in weld-bead　焊缝结晶形态
solidified slag　固熔渣
solids concentration　(内含)固态物的浓度
soluble boron concentration　可溶硼浓度
solution　解决方法
solution heat treatment　固熔热处理
somewhat　一点儿，稍微，有点，多少
sonic chock　声速节流圈，音速阻流器
sort out　整理，安排
sort out consignments　安排发运
sound limiting booth　隔音(电话)间
sound-powered system　声力电话系统
source　放射源，起源，源项
source file　源文件
source inspection　(放射)源检查，源项检查
source program　源程序
source range　源量程
source range channel　源量程(测量)通道
source range detector　源量程探测器
source range high neutron flux reactor trip　源量程中子注量率高停堆
source range neutron flux monitor　源量程中子注量率探测器
source range neutron flux multiplication signal　源量程周期信号
source range neutron flux preamplifier panel　源量程中子注量率前置放大器配电板
source rod　中子源棒
source term　源项
source verification　源项验证，原产地验证
source verification activities　原产地验证活动
source-term evaluation　源项评价
sourcing of raw material　原材料采购
spacer grid　定位格架
span　跨距，(仪表)量程
spanner　扳手，扳钳
spare capacity　备用容量
spare containment penetration　安全壳备用贯穿件

spare part 备件
spare penetration 备用贯穿件
spare penetration test connection 备用贯穿件试验接口
spare penetration test connection sparger 备用贯穿件试验接口鼓泡器/喷头
sparger 喷淋器,喷头
sparger arms 鼓泡器臂
sparger hole 鼓泡器孔
sparger hub 鼓泡器集管/集鼓
spark-proof valve 防静电(火花)阀门
spatial neutron kinetics 空间中子动力学
spatial power oscillation 功率空间振荡
spatially dependent parameter 空间相关参数
spatter 飞溅,溅污
special computer for radiation monitoring 辐射监测专用计算机
special construction technique 特殊施工技术
special inspection 专门视察
special lifting devices 专用提升装置
special monitoring 特殊检查,专门监测
special monitoring system (SMS) interface 特殊监测系统接口应用软件
special monitoring system cabinet 专用监测系统机柜
special spanners and hexagon bar wrench 专用扳手和六角棒扳手
special topics 专题
special waste water mixed drainage system 特种废水混合排放系统
speciality 专业
specialize 专门研究,专攻
specialize in 专营
specialized standard 专业标准
specific 明确的,特殊的
specific activity 比活度
specific burnup 比燃耗
specific conductivity 比电导率
specific gravity 比重
specific ion electrode 离子选择电极,特种离子电极
specific power 比功率
specific radioactivity 比活度
specific training 专项培训
specific training plan 专项培训计划
specification 规范,规格,说明书
specified test 规定的测试
specify 规定
specimen 样本
spectacle 展览物,景象
spectacle blind 视窗盲板,带双圈的盲板
spectacle flange 双孔法兰
spectral hardening 谱硬化
spectrophotometer 分光光度计
spectrophotometric method 分光光度法
spectroscopy 光谱,光谱学
speed reducer 减速器
spell out 明确指明,阐明
spent fuel 乏燃料
spent fuel assembly handling tool 乏燃

料组件装卸工具
spent fuel assembly shipping cask 乏燃料组件运输容器
spent fuel cask loading pit 乏燃料运输容器装料池
spent fuel cask washdown pit 乏燃料运输容器清洗池
spent fuel cooling pump 乏燃料池冷却泵
spent fuel cooling pump discharge to IRWST 乏燃料冷却泵至内置换料水箱的出口
spent fuel cooling pump suction from IRWST 内置换料水箱至乏燃料冷却泵入口
spent fuel examination facility 乏燃料检验装置
spent fuel handling tool 乏燃料操作工具
spent fuel pit cooling system 乏燃料池冷却系统
spent fuel pool 乏燃料池
spent fuel pool cooling and cleanup system 乏燃料池冷却及净化系统
spent fuel pool cooling heat exchanger 乏燃料池冷却热交换器
spent fuel pool cooling pump 乏燃料池冷却泵
spent fuel pool cooling water heat exchanger 乏燃料池冷却水热交换器
spent fuel pool demineralizer 乏燃料池/除盐装置
spent fuel pool emergency makeup isolation 乏燃料池应急补水隔离
spent fuel pool emergency makeup isolation valve 乏燃料池应急补水隔离阀
spent fuel pool emergency makeup lower isolation 乏燃料池应急补水低(水位)隔离
spent fuel pool filter 乏燃料池过滤器
spent fuel pool heat exchanger bolt tightener 乏燃料池热交换器螺栓紧定棘轮扳手
spent fuel pool level instrument 乏燃料池液位计
spent fuel pool level instrument isolation 乏燃料池液位计隔离
spent fuel pool level sensor 乏燃料池水位传感器
spent fuel pool purification pump 乏燃料池净化泵
spent fuel pool to cask washdown pit isolation 乏燃料池至乏燃料运输容器冲洗池的隔离
spent fuel rack insert for burnable poison rod assembly 乏燃料(贮存)格架为可燃毒物棒组件设置的插入物
spent fuel shipping cask 乏燃料运输容器
spent fuel shipping cask crane 乏燃料装运容器吊车
spent fuel storage area 乏燃料贮存区域
spent fuel storage pool 乏燃料贮存池
spent fuel storage rack 乏燃料贮存格架
spent fuel storage rack system 乏燃料

贮存格架系统
spent fuel system (SFS) 乏燃料系统
spent fuel system gates 乏燃料系统水闸门
spent resin 废树脂
spent resin collection tank 废树脂收集箱
spent resin tanks(with internal eductors and screens) 废树脂箱(带内置喷射器和滤网)
spherical shell 球形壳体
spider assembly 星形组件
spigot joint 扩口套管接头
spike 长钉,道钉
spiking 钉尖
spin test 旋转试验
spiral 螺旋形的,螺旋,盘旋
spiral welding 螺旋焊
spiral wound gasket 缠绕式垫片
splashing 飞溅,喷溅
splice box 分线盒,电缆与明线的连接箱
split contract 大块分包合同
split-wing type 分翼型
spool 短管,芯轴
spool piece (管道)调整段,补偿段
spot welding 点焊
spot welding robot 点焊机器人
spray 喷雾,喷淋
spray additive tank 喷淋添加剂箱
spray ejector 喷淋喷射器
spray header 喷淋集管
spray heat exchanger 喷淋热交换器
spray line 喷淋管道
spray nozzle 喷淋管嘴,喷雾管嘴,喷嘴
spray pump 喷淋泵
spray rate 喷涂率
spray shield 喷淋防护板
spray torch 喷炬(枪)
spray water valve 喷水阀
sprayed-on material 喷涂材料
sprayer (稳压器)喷雾器
sprayer valve 喷射阀,喷淋阀
spray-fused coating 喷涂层
spray-fusing 喷熔
spraying material 喷涂材料
spread factor/coefficient of spreading 铺展系数
spread footing 扩展底座
spread-ability 延展性
spread-ability test 延展性试验
spring diaphragm reducing valve 弹簧薄膜式减压阀
spring hanger 弹簧吊架
spring line (安全壳)起拱线
spring plate 弹簧座
spring preload 弹簧预载(力)
spring-loaded lift piston type 弹簧加载活塞式
spring-loaded safety valve 弹簧式安全阀
sprinkler 喷淋器,喷头
spurious 伪造的,虚假的,欺骗的
spurious closure 假关闭
spurious failure 虚假失效
spurious trip 假停堆,假跳闸/脱扣
square butt welding 无坡口对接焊

square cut　垂直切割
square root of the sum of squares (SRSS)　平方和的平方根
squeezing test　压扁试验
squib valve　爆破阀
squib valve controller cabinets　爆破阀控制器机柜
squirrel-cage　鼠笼式
squirrel-cage induction motor　鼠笼式感应电动机
squirrel-cage motor　鼠笼式电动机
stability　稳定性
stability index　稳定性指数
stabilize　稳定,使稳定,加固
stabilized electrical supply　稳压电源,稳定的电力供应
stack　烟囱,总管,立管
stack cutting　叠板切割
stack monitor　烟囱监测仪
staffing　人员配备
staffing and training　人员配备和培训
stage　级,阶段,时期,场所
stage casing　分段壳体
stage casing pump　分段式多级泵
staging area　集结区域
stainless steel　不锈钢
stainless steel electrode　不锈钢焊条
stainless steel filler rod　不锈钢填充棒,不锈钢焊料
stainless steel pump　不锈钢泵
stainless steel-wrought material　不锈钢锻件
stair　楼梯
stairway　楼梯
stairwell　楼梯间
stakeholder　受益者,保管者
stalemate,deadlock,impasse　僵局
stamping　冲压
stamping part　冲制件
stand　立场,立足点,看台,架子;经受
stand in the way　阻碍
standard design　标准设计
standard for comment　标准征求意见稿
standard penetration test (SPT)　标准贯穿件试验
standard pump　标准泵
standard safety analysis reports　标准安全分析报告
standard source　标准源
standard system table　标准体系表
standard thermal design procedure (STDP)　标准热工设计方法
standardization　标准化
standby　备用
stand by circulating pump　备用循环泵
standby diesel and auxiliary boiler fuel oil system　备用柴油机和辅助锅炉燃油系统
standby diesel generators　备用柴油发电机
standby gas treatment　备用气体处理
standby pump　备用泵
standing　商业状况,名望
standoff　支撑的,支座的,平衡的,抵消
standoff assembly　支撑组件,平衡组件
standpipe　竖管,管体式水塔

standpipe and hose system 消火栓系统
standpipe system 消防立管系统
standstill 停顿,停滞
star connection 星型连接
start up 启动
starting weld tab 引弧板
startup feedwater 启动给水
startup feedwater check valve 启动给水止回阀
startup feedwater control 启动给水控制
startup feedwater control valve (SFCV) 启动给水控制阀
startup feedwater isolation 启动给水隔离
startup feedwater isolation valve (SFIV) 启动给水隔离阀
startup feedwater line 启动给水管线
startup feedwater line penetration 启动给水管线贯穿件
startup feedwater piping penetration 启动给水管道贯穿件
startup feedwater system 启动给水系统
startup radios with base station (通信)无线基站
startup rate 启动速率
state oriented emergency operational procedure 状态导向应急操作规程
state-owned corporation 国有公司
statepoint 状态点
static analyse 静态分析
static balance 静平衡
static characteristics 静态特性
static characteristics derivation 静态特性演绎
static component 静态分量,静态组分
static cone penetration test 静力触探试验,静锥穿刺试验
static exciter 静止励磁调节器
static parameter 静态参数
static pressure 静态压力
static transfer switch 静态切换开关
statically misaligned RCCA RCCA 静态错对中
station blackout 全厂失电
stationary armature 固定衔铁,传递衔铁
stationary batteries 固定式蓄电池组
stationary gripper coil 固定钩爪线圈
stationary gripper latch 固定钩爪插销
stationary pole 固定磁极
stationary pump 固定式泵,固定泵
stationary seal ring 静密封环
statistical analysis 统计分析
stator 定子
stator flange 定子法兰
stator shell 定子外壳
status inputs from field equipment 就地设备状态输入
steady state 稳态
steady state operation 稳态运行
steady-state and transient power operation 稳态和瞬态功率运行
steady-state error 稳态误差
steady-state fluctuation 稳态波动
steady-state operation condition 稳态

运行工况
steady-state vibration 稳态振动
steam (bypass) dump system 蒸汽旁路排放系统
steam and powere conversions system 蒸汽与动力转换系统,二回路系统
steam cushion 汽垫
steam cycle system 蒸汽循环系统
steam dryer 蒸汽干燥器
steam dump 蒸汽排放
steam dump block 蒸汽排放隔离
steam dump control 蒸汽排放控制
steam dump control system 蒸汽排放控制系统
steam dump system 蒸汽排放系统
steam dump valve 蒸汽排放阀
steam explosion 蒸汽爆炸
steam flow 蒸汽流量
steam generator(SG) 蒸汽发生器
steam generator blowdown 蒸汽发生器排污
steam generator blowdown & demineralizer differential pressure trans mitter 蒸汽发生器排污和除盐床压差变送器
steam generator blowdown drain and recirculation pump 蒸汽发生器排污疏水和循环泵
steam generator blowdown EDI stack & brine conductivity transmitter 蒸汽发生器排污电渗析立管和炉水电导率变送器
steam generator blowdown electrodeionization package 蒸汽发生器排污电除盐组合件
steam generator blowdown heat exchanger 蒸汽发生器排污热交换器
steam generator blowdown line penetration 蒸汽发生器排污管贯穿件
steam generator blowdown system isolation 蒸汽发生器排污系统隔离
steam generator doghouse 蒸汽发生器鼓形罩
steam generator feedwater control system 蒸汽发生器给水控制系统
steam generator insulation package 蒸汽发生器保温包
steam generator layup system 蒸汽发生器保养系统
steam generator PORV 蒸汽发生器汽动卸压阀
steam generator PORV silencer 蒸汽发生器汽动卸压阀消声器
steam generator relief valve 蒸汽发生器卸压阀
steam generator safety valve 蒸汽发生器安全阀
steam generator tube leakage 蒸汽发生器传热管泄漏
steam generator wet layup 蒸汽发生器湿保养
steam hammer 汽锤
steam header pressure 蒸汽集管压力
steam header pressure controller 蒸汽集管压力控制器
steam jacket type 保温式,汽套式
steam line break 蒸汽管道破裂
steam line condensate drain control 蒸

汽管线冷凝水疏水控制
steam line condensate drain isolation 蒸汽管线冷凝水疏水隔离
steam line isolation 蒸汽管道隔离
steam line nitrogen supply isolation 蒸汽管线氮气供应隔离
steam loss 漏汽量,漏汽
steam manifold 蒸汽总管,蒸汽汽包
steam quality 蒸汽干度,蒸汽品质
steam restrictor 蒸汽限流器
steam trap 蒸汽疏水装置
steam turbine stop valve 主汽门
steam vent duct 蒸汽排放管
steam-compensated injection (堆芯补水箱)蒸汽补偿注射
steaming 蒸发
steam-line 蒸汽管线
steel containment 钢安全壳
steel grating 钢隔栅,钢栅栏
steel plate 钢板
steel plate/steel sheets 薄钢板
steel plate pretreatment 钢板预处理
steel sheet 薄钢片,薄钢板
steel shoe 钢挡板
steel sleeve 钢套管
steel structure 钢结构
steel support beam 钢支承梁
steep performance curve 阶跃形性能曲线
stellite clad journal 钨铬钴合金堆焊的轴颈
stem 阀杆
stem packing 阀杆填料
stem ring 阀杆环
stem seal 阀杆密封
step brazing/step soldering 分级钎焊
step change 阶跃变化
step insertion of reactivity 反应性阶跃引入
step length 步距,步长
step load change 阶跃负荷变化
step load decrease 阶跃负荷减少
step load increase 阶跃负荷升高,阶跃负荷增加
step load reduction 阶跃负荷降低
step open 阶跃开启
step-and-touch potentials 跨步电位和接触电位压
stepdown transformer 降压变压器
stepladder 直梯
stepless speed control 无级调速控制
step-nozzle 梯级管嘴(RNS与热段相连部分)
stepup transformer 升压变压器
step-wise 逐步地,一步一步地
stick build (模块制造)分块建造
stick to,adhere to 坚持
stiffener 加强勒
stiffening ring 加强环
stiffness 刚度,硬度
stiffness matrice 刚度矩阵
stipulate 规定
stirrup 箍筋
stochastic effect 随机效应
stock 原料,库存,托盘,树干
stone chipping 小石子,石头碎片
stop 止挡板,停止
stop check valve 截止止回阀

stop valve 截止阀，断流阀，关断阀
stop watch 秒表
stopping-off agent 阻流剂
storage 贮存，储存，存储，仓储
storage facility 储存设施
storage oscilloscope 存储式示波仪
storage rack 储存格架
storage reservoir 储水池
storage tank 储水箱
storage tank water level 储水箱水位
stored program contreol 存储程序控制
storm drain collector 雨水排水收集器
storm drain system 暴雨疏排系统
storm runoff 暴雨水量
storm surge banked-down 风暴潮减水波动
storm surge banked-up 风暴潮增水波动
storm water pump 暴雨排水泵
straight beam method 直射法超声波探伤
straight beam ultrasonic examination 直射波超声波检查
straight weld 直焊
straighten 矫正
straightening 校直，整直
straightforward 直接的，明确的
straightway check valve 直通单向阀，直通止回阀
strain 应变，张力，使紧张
strain fatigue 应变疲劳
strain fatigue cycle 应变疲劳循环
strain gage 应变片
strain rate 应变速率
strainer 管道过滤器，滤网
strand 钢绞线
stranded electrode 绞合焊条
strap 钢带，皮带，用带缚住绞线
stratification 层化，分层
stratigraphy 地层
straw pulp pump 草纸浆泵
stream 流，溪，流动，流注
strength 强度
strength of coating 涂层强度
strength property 强度特性
strength weld 强度焊
stress 加载，应力
stress concentration 应力集中
stress corrosion 应力腐蚀
stress corrosion cracking 应力腐蚀开裂
stress intensification factor 应力增强因子
stress intensity 应力强度
stress intensity factor 应力强度因子
stress or fatigue 应力或疲劳
stress range 应力变化范围
stress redistribution 应力重分布
stress relaxation 应力松弛
stress relief 消除应力
stress relief annealing crack (SR crack) 消除应力退火处理裂纹
stress relieving by annealing 退火消除应力
stretch 张拉，伸展，伸长
stretching gallery 张拉廊道
stria depth 割纹深度

strike　冲击,撞击,侵袭
strike and damage probability　撞击与损坏概率
striking end (of an electrode)　焊条引弧端
stringent　严厉的,迫切的,严酷的
stringent requirements　严格要求
strip　条,带,钢索
strip footing　条形基础,(安全壳)钢索张拉起点
stripper　脱气塔
stripping　萃取,脱模
strive to　力求,努力
strobe　闸门,滤波,滤光
strobe light assemble　闪烁灯组件
stroke　撞击,打击,敲打
stroke testing　行程试验,撞击试验
stroke time　全行程时间,冲击时间
stroke-tested　冲击试验的
stroking the valve　炸开(爆破阀)
strong-back　定位板
strontium group　锶系元素
structural analysis　结构分析
structural component　结构部件
structural criteria　结构准则
structural deterioration　结构劣化
structural element　结构单元
structural grid　结构格架
structural margin　结构安全裕量
structural member　构件
structural module　结构模块
structural shield　结构屏蔽
structural weld　结构焊
structure　构筑物,结构
structure, system and component (SSC)　构筑物、系统和部件
strut　支撑杆,支柱,压杆
stub　短管,残余,断株,短柱
stub end　短端头,残端
stuck rod criterion　卡棒准则
stuck-open　卡在开位
stud　螺栓
stud tensioner　螺栓拉伸机
studhole plug handling tool　螺孔塞装卸工具
stuffing box　填料箱
sub compartment pressurization　子隔间加压
sub component　子部件
subatmospheric containment　负压安全壳
sub-channel　子通道
subcontractor　分包商
subcoold temperature　欠热度,过冷度
subcooled blowdown　过冷喷放,欠热喷放
subcooled temperature of close valve　关阀过冷度,关阀欠热度
subcooled temperature of open valve　开阀过冷度,开阀欠热度
subcritical　亚临界,次临界
subcritical condition　亚临界状态,次临界状态
subcritical core　次临界堆芯
subcriticality　次临界
subdivide into　再(细)分成
subgrade　地基,路基
sub-item　子项

subject matter expert(SME) 专业工程师
subject to 以…为准
sublet 转包,分包
submerasion 浸没
submerged arc welding (SAW) 埋弧焊
submerged arc welding machine 埋弧焊机
submerged motor pump 潜水电泵(排污泵)
submersion pump 潜水泵
submission of document 文件递交
submit 提交
submit tender 提交标书
submittal 提交
suborder 分订单
subsection (ASME 规范)分卷,子节
subsequent 随后的
subsequently 随后
subsidence 陷落,下沉,沉淀
subsidiary 子公司
subsonic velocity 次音速
substantial 坚固的,实质的,真实的,充实的
substation 变电所
sub-stitutionary oxidation 置换氧化
substrate 基底,基础
substrate material 基底材料
substructure 子结构
subsystem 子系统
sub-zero valve 低温阀门
suction 吸入口
suction barrel 吸入口吊篮
suction isolating valve 入口隔离阀门
suction nozzle 入口管嘴
suction piping 入口管线/管道
suction strainer 入口过滤器
suction valve 入口阀
suction valve guard 入口阀保护装置
suction well 吸水井
suitable medium 适用介质
suitable temperature 适用温度
sulfated battery plate 硫酸蓄电池板
sulfur(sulphur) 硫
sulfur hexafluoride breaker 六氟化硫断路器
sum of the aux/annex exhaust fan flow 辅助/附属排风机总流量
sump discharge containment isolation IRC 地坑排水安全壳内隔离
sump 集水坑,地坑
sump pump 地坑泵,污水泵
super cavitation pump 超汽蚀泵
super high pressure valve 超高压阀门
superelement 超级单元
superimpose 添加,附加
superimposed back pressure 超附加背压力
super-pressure pump 超高压泵
supersonic pump 超声速泵
supersonic velocity 超音速
supervise 监督,管理
supervised area 监督区
supervisor workstation 值长工作站
supplementary 补充的,增补的
supplementary loaded safety valve 带补充载荷的安全阀

supplier 供方,供货商
supplier evaluation 供方评价
supplier evaluation, qualification and selection 供货商评估,资质鉴定和选择
supplier surveillance 供货商监督
supply and drainage system 给排水系统
supply chilled water temperature 冷冻水供水温度
supply fan 送风机
support 支承,支撑,支座
support block 支承块
support bracket 超级支撑架,大牛腿
support column 支承柱
support for reactor head vent piping 反应堆顶盖放气管道支承
support lug 支承凸条
support pad 支承垫
support point 支撑点
support shirt 支撑套,(反应堆压力容器)支撑裙
supporting plate 支承板
supports & hangers 支吊架
supports and restraints 支承和约束
suppressant 抑制性的,制止的,遏抑的
suppression of reactor power oscillation 反应堆功率振荡抑制
suppressor 抑制器
surcharge effect 超载效应
surface activity 表面活度
surface carburization 表面渗碳
surface contamination survey meter 表面污染测量仪
surface crack 表面裂纹
surface dose rate 表面剂量率
surface examination 表面检查
surface radioactivity contamination 表面放射性污染
surface roughening 表面粗糙化处理
surface roughness 表面粗糙度
surface wash pump 表面清洗泵
surfacing by fire cracker welding 躺板极堆焊
surfacing electrode 堆焊焊条
surfacing with band-electrode 带极堆焊
surge 波动,奔放,涌动;电涌
surge capacity 波动容量
surge line insulation 波动管保温
surge line piping 波动管
surge load 波动载荷
surge nozzle 波动管接管,波动管管嘴
surge protection 冲击(电压)保护,电涌保护,波动保护
surge suppressor 电涌抑制器
surge tank 波动箱
surveillance 监视,监督
surveillance capsule (反应堆压力容器)辐照监督管
surveillance personnel 监督人员
surveillance plan 监督计划
surveillance program 监督大纲
surveillance requirement (SR) 监测要求,监督要求
survey 检查,调查
survey table 调查表

suspend 暂停,临时取消
suspended solids 悬浮颗粒
suspension of work 停工
sway 摇摆,摇动,支配
sway strut 摇摆支撑杆
swelling 肿胀
swelling rate 肿胀速率
swing 摆动,回旋,晃动,摇摆
swing disk type check valve 摆动板式止回阀
swing gate piston pump 摆动门活塞泵
swing type 摆动式,旋启式
swinging vane pump 旋叶泵,摆动滑片泵
swirl 旋转,打旋,旋状物
swirl vane separator 旋叶式分离器
switch over 切换
switchboard (distribution) 配电盘
switchbox 配电箱
switchgear 开关设备,开关柜
switchgear & main electrical equipment fire protection system 电气开关设备及主要电气设备消防系统
switchgear breaker(reactor trip) (反应堆停堆)开关设备断路器
switchgear room 电气开关室,开关设备间
switching and tagging area 隔离标记区
switchyard 开关站
symbol 标志,象征,符号,记号
symmetrical section 对称截面
symmetry 对称性,匀称
symptom 征状,征兆
symptom-based functional guideline 基于征状的功能导则
synchronization 同步
synchronized operation 同步操作
synchronous 同步的
synthetic heat-affected zone test 综合热影响区模拟试验
system 系统
system description 系统说明,系统描述
system diagnostic 系统诊断的,系统诊断
system generation 系统生成
system inquiry 系统询价
system load 系统负荷
system operation pressure 系统运行压力
system packages 系统包
system resistance 系统阻抗,系统阻力
system response 系统响应
system software 系统软件
system specification documents 系统说明书,系统规格书文件
systematic approach to training (SAT) 系统性培训方法
system-level actuation 系统级驱动,系统级触发
system-related 系统有关的
systems required for safe shutdown 安全停堆要求的系统

T

T headed bar T 形镦头钢筋
tab 焊舌,突舌

table 表格,工作台
table in main control room 主控制室工作台
tack 大头钉,平头钉,航向,增补
tack weld 点焊
tack welding 平头焊接,点焊,定位焊
tack-expanded 定位胀,平头胀
tag 标签,标牌,签条
tag no. 标牌号
tagging 编标签码,加标牌
tagging room (主控制室)签票室
take precedence over 优于,先于
take the matter up 汇报一下情况,商量一下
tandem sequence 纵列多丝埋弧焊
tank 水箱
tank drainage pump 罐柜抽泄泵,水箱排水泵
tank lorry ball valve 槽罐车球阀
tank system 水箱系统
tap 水龙头,分接头,抽头
tap changer 抽头切换开关,抽头变换器
tape reader 磁带读取器,读带机
tape recorder 磁带录音机
taper 开坡口
tapered valve 锥形阀
target 靶,目标,指标
target isotope 靶同位素
target material 靶材料
targeted axial offset 目标轴向偏移
targeted time 规定的时间
tariff 关税
task 任务
task action plan 任务行动计划
task analysis 工作任务分析
task element 工作任务单元
teaching 教学
team collaborator 团队合作者
team member 团队成员
tearing strength 撕裂强度
technical discipline 技术训练
technical interface 技术接口
technical specification 技术规格书
technicality of the electrode 焊条工艺性
technique (专门)技术,(艺术上的)技巧,技能
technological coating 工艺镀层
technology-intensive 技术密集型的
tectonic 构造的,建筑的
tedious 繁琐的
tee type cracking test T 形裂纹试验
tee T 接头
telegraphic transfer 电汇
telephone desk 电话桌
telephone FAX interface card 电话传真接口卡
telephone MODEM interface card 电话 MODEM 接口卡
telephone outdoor enclosure 电话机户外附件
telephone set 电话机
telephone switchboard 电话交换台
teleprinter 电传打印机
telescopic tube 伸缩套筒
telescoping sipping system 拉杆式啜漏系统

teletype　电传打字机
tellurium group　碲系元素
temper　回火
temperature at discharging condensate　凝结水出口温度
temperature coefficient　温度系数
temperature resistance　耐热性，热阻
temperature sensor　温度传感器
temperature transient　温度瞬变
temporary　临时的，暂时的
temporary backing　临时垫板
temporary containment washdown isolation valve　临时的安全壳冲洗隔离阀
temporary instrument isolation valve　临时仪表隔离阀
temporary strainer　临时过滤器
temporary wiring　临时接线
ten path selector　10路选择器
tender　投标
tender/bid for the project　投标争取这个项目
tendering party　招标方
tendon　钢筋束
tensile and impact test　拉伸和冲击试验
tensile restraint cracking test　拉伸约束裂纹试验(TRC试验)
tensile strength　抗拉强度
tensile stress　拉伸应力
tensile stress limit　拉伸应力限值
tensile test　拉伸试验
tensile test data　拉伸试验数据
tension　紧张，张力，拉力
tension link　活节拉杆
tension ring　拉紧环
tension structure on top of the CRDMS　控制棒驱动机构顶部拉紧结构
tentatively　暂时地，初步地
term　期限，条件，条款，价格，费用
term cable support　终期电缆支撑
term of agreement　合同期限
term of validity　有效期
terminal　终端，接线端子
terminal box　端子盒
terminal objective　最终目标
termination　终端，端子
terms & conditions development　编写条款与条件
test　考试，试验
test and distribution cabinet　试验和分配柜
test conditions transient　试验工况瞬态
test coupler　试验连接器
test piece　试件
test prerequisite　试验前提条件
test pressure　试验压力
test program　试验大纲
test records　试验记录
test report　试验报告
test specifications and test procedures　试验技术规格书和试验规程
test specimen　试验样品
test weld　试验焊，试焊
testing program　测试程序，试验大纲
text editor　文本编辑
that's fair enough　言之有理

the exchange of needed goods　互通有无
the first party　甲方
the insured　投保人
the insurer　保险人
the proper authorities　主管部门
the same line of business　同一行业
the second party　乙方
theoretical density (T. D.)　理论密度
theoretical flowing capacity　理论排量，理论流量
theoretical surface　理论表面
theoretically　理论上
there from　从那里，从那一点
thereafter　自那以后，以后，之后
therein　在其中
thereof　由此，因此
thereto　此外，又
thermal　热的
thermal aging　热老化
thermal aging effect　热老化效应
thermal annealing　退火
thermal barrier　热屏障
thermal capture　热中子俘获
thermal conductivity　热导率
thermal cutting　热切割
thermal cycling　热循环
thermal design flow　热工设计流量
thermal diffusion coefficient　热扩散系数
thermal efficiency　热效率
thermal expansion　热膨胀
thermal feedback　(反应性)热反馈
thermal gradient　热梯度
thermal growth　热增长，热生长
thermal inertia　热惯性
thermal insulation　保温层
thermal lag　热滞后
thermal load　热载荷，热负荷
thermal margin　热工裕量
thermal mass flowmeter　量热式质量流量计
thermal neutron　热中子
thermal neutron flux　热中子注量率
thermal plume　热烟羽
thermal power　热功率
thermal relay　热继电器
thermal shield　热屏蔽
thermal shock　热冲击
thermal sleeve　热套
thermal spraying　热喷涂
thermal stability of flux　钎剂热稳定性
thermal stratification　热分层
thermal stratification, cycling, and striping (TASCS)　热分层、热循环和热振荡
thermal stress　热应力
thermal striping　热振荡
thermal transient　热瞬态
thermal weldability　热焊接性
thermal-hydraulic design　热工水力设计
thermal-hydraulic transient　热工水力瞬态
thermalization　热能化，热化
thermally treated　热处理的
thermally-saturated　热饱和的
thermit crucible　热剂坩埚，热熔坩埚

thermit mold　热剂铸模,热熔模铸
thermit powder　热剂,热熔粉剂
thermit rail welding　钢轨热剂焊,热熔轨道焊
thermit reaction　热剂反应
thermit repair welding　热剂补焊
thermit slag　热剂熔渣
thermit steel　热剂钢水
thermo unit　热量单位
thermocouple　热电偶
thermocouple alloy　热电偶合金
thermocouple calibrator　热电偶校准器
thermocouple column　热电偶导管柱
thermocouple conduit seal　热电偶套管密封
thermocouple level switch　热电偶液位开关
thermocouple reference junction box　热电偶参考接线盒
thermodynamic　热力学的
thermo-luminescence detector　热释光探测器
thermo-luminescence dosimeter (TLD)　热释光剂量计
thermophoresis　热迁移,热泳
thermostat control　恒稳调节控制器
thick sludge pump　稠泥浆泵,沉积物泵
thimble　顶针,心环,嵌环,套管,支撑环
thimble cell　冷壁栅元,套管栅元,支撑环栅元
thimble guide tube　套管导向管
thimble plug　阻力塞
thimble plug assembly　阻力塞组件
thimble screw　套管螺栓
thin film　薄膜
thin gate valve　薄型闸阀
thinskirt　薄壁裙(筒)
thread　螺纹,线,细丝
threaded fastener　螺纹紧固件
threaded fitting　螺纹接头
three dimensional　三维的
three loop XL Westinghouse 17×17 robust fuel　西屋公司三环路加长型17×17
three phase arc welding　三相电弧焊
three second gust speed　三秒强风速度
three valve manifold　三阀歧管,三阀集管
three way type　三通式
three way valve　三通阀
three-dimensional flux map of the reactor core　堆芯中子注量率三维分布图
three-pin socket　三芯插座
threshold　门槛,阈值
threshold dose　阈值剂量
throat diameter　喉径
throat size　喉管尺寸
throttle　节流
throttle device　节流装置
throttle valve　节流阀
throttling orifice　节流孔板
throughput　吞吐量,产量,生产能力
through-wall crack　贯穿裂纹(穿透壁厚的裂纹)
throughway valve　直通阀

thrust bearing 推力轴承
thrust collar 推力环
thrust collar nut tool 推力环螺母紧定工具
thrust collar tool (for shaft movement adjustment) 推力环工具(用于轴运动的调整)
thrust collar withdrawal tool 推力环拆卸工具
tidiness 清洁
tie bus between ECS load center 发动机控制系统负荷中心间的连接母线
tier (阶梯)层次,看台
tier 2 material 第二层次文件,第二层次材料
ties 拉筋
tight lattice 稠密栅格
tightener 紧固物,收紧器,张紧装置
tile 瓷砖,瓦片
tilt 倾斜
time average application software 时间平均应用软件
time constant 时间常数
time delay circuit 延时电路
time division multiplex 分时多路
time history analyses 时程分析
time out 超时
time slice 时间片
timed delay 定时延迟,时间延迟
time-history analyzer 时程分析仪
time-history integration method 时程积分法
timing 时间选择
timing signal 时序信号
tiny drag slow shut check valve 微阻缓闭止回阀
tip 提示
titanium 钛
title 标题
title block 标题栏
TMI action plan 三哩岛行动计划
to a great extent 在很大程度上
to my regret 很遗憾
toe crack 焊趾裂纹
toe of weld 焊缝边缘裂纹
toilet 盥洗室
tolerance 允许误差,公差
tone generator system 音频发生器系统
tool monitor (used for checking tool and small packages for radiation) 工具监测仪(用于工具和小型物件的辐射监测)
tool storage rack 工具存放架
top grid 顶部格架
top inlet pump 顶端吸入泵
top nozzle 上管座
top plate (上管座)顶板,上框板
top shell 上部筒体
top suction pump 顶端吸入泵
topographic factor 地形因素,地形因子
torch 焊炬
torch brazing/ torch soldering 火焰钎焊
tornado 龙卷风
tornado damper 防龙卷风阀
tornado loading 龙卷风荷载

tornado missile 龙卷风飞射物
torque 扭矩,转矩,力矩
torque motor 力矩电动机,陀螺修正电动机,罗盘矫正电动机
torsional strength 抗扭强度
total enthalpy rise factor 总焓升因子
total head 总扬程
total investment 总投资额
total iron 总铁
total organic carbon (TOC) analyzer 总有机碳分析仪
total peaking factor 总峰值因子
total power peaking factor 总功率峰值因子
total pressure 全压,总压力
total quality management (TQM) 全面质量管理
total tensile strain 总拉伸应变
toughness 韧性,韧度,抗断强度
toxic 有毒的
T-pattern three way T形三通
trace 跟踪,追踪,查找
trace element 微量元素
tracer 跟踪装置
tracer ball 示踪球
track/store module 跟踪/储存组件,跟踪/储存模块
track ball 跟踪球
traction-type elevator 牵引式升降机
trade study 商贸(对策)研究
trade-off 折中(办法,方案)
traffic control 通行控制
train 序列,系列
trainee 学员
trainer 教员
trainer of synthetic weld 焊工综合培训器
training engineer 培训工程师
training implementation approach 培训实施方法
training information management system (TIMS) 培训信息管理系统
training inspection engineer 培训监督工程师
training management 培训管理
training material 培训材料
training module 培训单元
training needs analysis 培训需求分析
training objective 培训目的
training plan 培训计划
training policy 培训政策
training principle 培训原则
training program 培训大纲
training program development 培训大纲开发
training record 培训记录
training regulation 培训制度
training responsibility 培训责任
training scope 培训范围
training setting 培训方式
training system 培训体系
train-the-trainer 卓越内部培训师培训
transaction 交易,业务
transcrystalline crack 穿晶裂纹
transfer 运输,输送
transfer canal 运输通道
transfer expenses 搬迁费,安家费
transfer line 输送管

transfer of records 记录移交
transfer pump 输送泵
transfer switch 切换开关,转换开关
transfer tank 输送箱
transfer tube 运输通道
transferred arc 转移弧
transformer 变压器
trans-granular fracture 穿晶断口
trans-granular stress corrosion cracking (TGSCC) 穿晶应力腐蚀开裂
transient 瞬时现象,瞬态,瞬时的
transient analysis 瞬态分析
transient and postulated accident conditions 瞬态和假想事故工况
transient curve 瞬态曲线
transient event 瞬态事件
transient heat transfer 瞬态热传导
transient operation 瞬态运行
transient response 瞬态响应
transient ring 过渡环
transient temperature 瞬态温度
transient welding stress 焊接瞬时应力
transistor arc welding power source 晶体管弧焊电源
transit time 过渡时间,渡越时间
transition cone 锥形过渡端,过渡锥
transition zone 过渡区
translational speed 平移速度,移动速度
transmission 传动
transmission antenna 发射天线
transmitter 变送器
transport theory 迁移理论
transportation 运输
transportation route 运输路线
transverse 横向的
transverse crack 横向裂纹
transverse seam 横焊缝
transverse stiffness 横向刚度
transverse submerged arc welding 横列双丝并联埋弧焊
transverse weld/cross-weld 横向焊
trap 疏水器,捕集器,存水弯管
trash rack 拦污栅
travel speed 穿行速度
traveler 行走机构
trays 托架
treatment 处理
trefoil tube hole 三叶状管孔
trend group software configuration 趋势组软件配置
trial 试验性的
triangular pitch 三角排列
triaxial acceleration sensor 三轴向加速度传感器
trimming 去毛刺
triode gun 三极电子枪
trip (反应堆)事故保护,停堆,跳闸,脱扣,断路
trip delay time 脱扣延时
trip of turbine generator 汽轮发电机跳闸,脱扣
trip reactivity insertion (紧急)停堆反应性引入
trip setpoint 停堆整定值
tripping 脱扣,断路,跳闸
trisodium phosphate (TSP) 磷酸三钠

tritium sampler　氚取样器
trochoid rotor pump　次摆线转子泵
trochoid screw pump　次摆线螺杆泵
trolley　小车
trolley frame　小车架
trough　槽
trough duct　(电缆)沟槽
truck bays of the radwaste building　放射性废物厂房卡车停放区
trunk line　干管,总管,中继线
trunnion　吊耳
trustworthy　值得信赖的
tube　管,传热管
tube bundle　管束
tube expansion　胀管
tube lane　(U 形管)管廊
tube pitch　管间距
tube plate　管板
tube rolling　管子滚压,辊碾
tube sheet　管板
tube spacing　管间距
tube stream　管程
tube support plate　传热管支承板
tubes plugged　堵管
tube-side　管侧
tubewell pump　管式深井泵
tubing　装管,制管,管道,管子
tubular　管
tubular type axial flow pump　管流泵,直管轴流泵
tundish　漏斗
tungsten carbide surfacing electrode　碳化钨堆焊焊条
tungsten electrode　钨电极
tungsten inclusion　夹钨
tungsten inert-gas welding machine　钨极惰性气体保护弧焊机
tunnel　通道
turbidimeter　浊度计
turbidity　混浊度
turbine　汽轮机,透平
turbine building　汽轮机厂房
turbine building and water treatment building fire protection system　汽轮机厂房和水处理车间及其内部设备消防系统
turbine building HVAC system　汽轮发电机厂房暖气通风空调系统
turbine bypass　汽轮机旁通
turbine bypass control valve　汽轮机旁路控制阀
turbine bypass system　汽轮机旁路系统
turbine control system　汽轮机控制系统
turbine control valve　汽轮机调节阀
turbine controls and protective function　汽轮机控制和保护功能
turbine driven centrifugal pump　汽动离心泵
turbine extraction steam system　汽轮机抽汽系统
turbine gland sealing system　汽轮机轴封系统
turbine gland sealing system and exhaust blower　汽轮机轴封系统和抽气机
turbine inlet valve　汽轮机进汽阀

turbine island　汽轮机岛，常规岛
turbine missile　汽轮机飞射物
turbine overspeed transient　汽轮机超速瞬态
turbine power and frequency regulating system　汽轮机功率和频率调节系统
turbine protection system　汽轮机保护系统
turbine pump　汽动泵
turbine regulation and control　汽轮机调节与控制
turbine roll test　汽轮机冲转实验
turbine runback module　汽轮机降功率组件
turbine stop valve　汽轮机截止阀
turbine throttle valve　汽轮机节流调节阀
turbine thrust bearing　汽轮机推力轴承
turbine trip　汽轮机脱扣，汽轮机紧急停机
turbine vent and drain system　汽轮机排放和疏水系统
turbo-generator　汽轮发电机组
turbulence of sediment　沉积物紊流
turbulent flow　紊流，湍流
turn out　结果是，原来是
turn out，produce，fabricate　生产
turning gear　盘车装置
turnkey contract　交钥匙承包合同
turnkey contractor　交钥匙承包商
turret type piston pump　阀箱式活塞泵，炮塔式活塞泵
twin electrode　双芯焊条
two group theory　双群理论
two out of two voting hardware　2 取 2 硬件
two phase flow　两相流
two stage regulator　两级调压器
two-out-of-four logic　四取二逻辑
two-out-of-three logic　三取二逻辑
two-sided access　双侧通道
two-way valve　二通阀
type　类型
type A test　A 类试验
type B test　B 类试验
type C test　C 类试验
type model　型号
type of connection　连接形式
type of construction　结构形式，建筑类型
type test　型式试验，例行试验
typical cell　典型栅元

U

ultimate　最终的，最后的，根本的
ultimate bending strength　抗弯强度极限
ultimate compressive strength　抗压强度极限
ultimate heat sink　最终热阱
ultimate heat sink cooling tower and water storage tank system　最终热阱冷却塔与储水箱系统
ultimate safety earthquake　极限安全地震
ultimate strain　极限应变
ultimate strength　极限强度
ultimate stress　极限应力

ultimate tensile strength 抗拉强度极限
ultrahigh vacuum pump 超高真空泵
ultrasonic decontaminating tank 超声波去污槽
ultrasonic decontamination 超声去污
ultrasonic examination 超声检查
ultrasonic inspection 超声波探伤
ultrasonic liquid level meter 超声波液位计
ultrasonic soldering 超声波钎焊
ultrasonic test 超声试验
ultrasonic testing examination 超声检测
unanimously 一致同意的，无异议的，全体一致的
unannounced inspection 突击检查
unaugmented critical flow 未增加的临界流，未提高的临界流
unborated water 无硼水，未含硼水
unborated water source 非硼化水源
uncertainties in the calibration 标定的不确定性
uncertainty 不确定性
uncertainty analysis 不确定性分析
unchokeable pump 不堵塞泵
uncompensated ionization chamber 非补偿电离室，未补偿电离室
uncontrolled release 失控排放
uncontrolled rod withdrawal accident 控制棒失控抽提事故
undeniable 不可否认的
under coating 结合层
under water pump 潜水泵
underbead crack 焊道下裂纹
underclad cracking 堆焊层下裂纹
undercut 咬边
underestimate 低估，把…的价值估计过低，估计不足
underground wiring 地下配线，地下布线
underload 低载，轻载
under-mentioned 下述的
undertake 从事
undertaking 任务，保证，许诺
undervoltage 欠电压
undervoltage trip attachment （汽轮发电机）欠压脱扣附件
underwater cutting 水下切割
underwater cutting blowpipe 水下割炬
underwater gas shielded arc welding 水下气体保护电弧焊
underwater lighting device 水下照明装置
underwater plasma arc welding 水下等离子弧焊
underwater transfer tube 水下运输管道
underwater welding 水下焊
underwiring 地下配线
undiluted weld deposits 未稀释焊缝熔敷金属
unfortunately 抱歉
unidentified leakage 未识别的泄漏，不可识别泄漏
uniform corrosion 均匀腐蚀
unilateral 单方面的，单边的

uninterruptible　不间断的
unit　单元,单位,机组
unit heater　单元式加热器
unit heaters providing AP1000 equipment class　供 AP1000 D 级功能设备的单元
unit loading and unloading　机组增减负荷
unit shutdown　机组停运
unit weight　重度,单元重量
unless otherwise mentioned(noted,stated)　除非另有说明(说明,陈述)
unless otherwise noted　除非另外注明
unless otherwise specified　除非另有规定
unless otherwise stated　除非另外声明
unload　卸货
unmixed zone　未混合区
unplanned dilution　非计划稀释
unresolved safety issues　未解决的安全专题
unrodded　没有装控制棒
unseal type　非封闭式,非密封式
unsealed source　开放源,非密封源
unstabilized austenitic stainless steels　未钝化奥氏体不锈钢
up to date　最新
upender　倾翻机
upfront　坦率地,公开的,预先
upgrade　提高,提升
uphill welding　仰焊
uplift　提升
upper annulus drain　上部环廊排水沟
upper compartment　上部隔间
upper core internal　上部堆内构件
upper core plate　堆芯上板
upper core plate inserts　堆芯上板插入件
upper core support assembly　上部堆芯支撑组件
upper core support structure　堆芯上部支撑结构
upper head　上封头,顶盖
upper internals storage stand　压紧筒存放架,上部堆内构件存放架
upper limit　上限
upper plenum　(燃料棒)上贮气腔
upper shelf energy　上平台能量
upper shelf fracture energy　上平台断裂能量
upper shell　(压力容器)上筒体
upper shell course　(压力容器)上筒体接管段
upper support column　上支撑柱
upper support plate　上部支撑板
upset load combination　异常载荷组合
upsetting　镦粗
upstream　上游
uranium dioxide　二氧化铀,UO_2
urgently　迫切地
urinalysis　尿分析
usability of the electrode　焊条工艺性
use,utilize,apply to,employ　使用
use-as-is　原样照用
usual practice　惯例,通行做法
U-tension test　U 形拉伸试验
utility bill　公共费用
utility program　业主大纲

utility steering committee　用户指导委员会
U-tubes　U 形管
U-tube bundle　(蒸汽发生器)U 形管束
U-tube heat exchanger　U 形管式热交换器,U 形管式蒸汽发生器
U-tube manometer　U 形管式气压计
UV-Vis spectrophotometer　紫外分光光度计

V

vacuum brazing　真空钎焊
vacuum breaker　真空破坏器
vacuum breaker valve　真空破坏阀
vacuum cleaner　(真空)吸尘器
vacuum degasifier　真空脱气器
vacuum pump　(水力喷射)真空泵
vacuum　真空度
valid　有效的
validation　确认,培训效能检查
value　价值,值,评价
valve　阀门
valve alignment　阀门排列,阀门调节
valve and breaker status monitoring application software　阀门和断路器状态检测应用软件
valve and fitting　阀门和配件
valve arm　气阀摇臂
valve ball　阀球
valve body　阀体
valve bonnet　阀盖,阀帽
valve box　阀箱
valve box cover　阀箱盖
valve box type piston pump　阀箱式活塞泵
valve cage　阀箱
valve carrier　阀座
valve chest　阀箱
valve deck plate　阀盖板
valve deck plate type piston pump　阀盖板式活塞泵
valve disc　阀盘,阀瓣
valve driving arm　阀驱动臂
valve gallery　阀廊
valve grid　阀挡
valve guide　阀门导向器
valve handle　阀门手柄
valve lapping machine　阀门研磨机
valve lifting device　阀门开启装置
valve motion arm　阀门驱动臂
valve operating rod　阀门操纵杆
valve packing　阀门填料
valve plate type piston pump　阀板式活塞泵
valve position　阀位
valve position meter　阀位计
valve positioner　阀门定位器
valve pot type piston pump　阀罐式活塞泵(美)
valve rod　阀杆
valve seat　阀座,阀盘
valve seat recess　阀座槽
valve spring　阀簧
valve spring plate　阀簧压板
valve stem　阀杆
valve stem leakoff　阀杆引漏
valve stem position　阀杆位置
valve stop　阀挡

valve tail rod　阀导杆
valve test panel　阀门试验控制屏
valve trim　阀门芯件(与介质接触的内件)
valve type bucket　阀式桶体,阀式铲斗
valve type piston pump　阀形活塞泵
valve yoke　平阀框
valve-by-valve basis　逐个阀门地
valve-controlled hydraulic ram　阀控液压油缸
valveless diaphragm pump　无阀隔膜泵
valveless pump　无阀泵
valveless vibration pump　无阀振动泵
vanadium　钒
vane　(涡轮)叶片
vane-axial type fan　轴流式风机
vapor　蒸汽
vapor condenser　蒸汽冷凝器
vapor extractor　抽汽器
vapor phase　汽相
vapor phase soldering　蒸汽钎焊
variable　变量,可变的
variable area flow converter　可调流量转换器
variable area flowmeter　可调流量计
variable area ratio jet pump　可变比喷射泵
variable capacity pump　变流量泵,可调流量泵
variable capacity screw pump　流量可调螺杆泵
variable category　变量种类
variable classification　变量分级
variable frequency drive(VFD)　变频器
variable output pump　变量泵
variable speed pump　变速泵
variable stroke reciprocating pump　变行程往复泵
variable voltage & variable frequency speed regulation 变压变频调速
various combinations of reactor coolant pump　反应堆冷却剂泵的不同组合
vary　变化
vault　房间,拱顶,地下室
velocity head　速度压头,风压
velocity head flow probe　速度压头流量探测器,风压流量探头
velocity pressure　速度压头,风压
velocity pressure exposure coefficient 风压高度变化系数,风压方位系数
vendor　供货单位,供货商
vendor code plate rating　供货商名录等级,供货商名录分类
vent　通风,排气
vent cabinet　通风柜
vent condenser　排气冷凝器
vent pipe　通风管
vent stack　通风烟囱,通风总管
vent type thermometers with specific gravity　带比重修正的排气式温度计
vent valve　放气阀
ventilate　通风
ventilated protective suit　通风防护服
ventilated suit　通风衣,气衣
ventilation cabinet　通风柜

ventilation control room system　室内通风控制系统
ventilation　通风，唤起
ventilation shroud　通风罩
ventilation system　通风系统
venue　（指定的）集合地点
verification　验证
vertical deep well pump　立式深井泵
vertical disc type butterfly valve　垂直板式蝶阀
vertical down welding　垂直下焊
vertical lift check valve　立式升降止回阀
vertical pipe　垂直管，立管
vertical position welding　垂直位置焊
vertical type sump pump　立式地坑泵
vertical velocity　垂直速度
very soft silty clay　极软的淤泥土/粉质黏土
VES MCR air delivery line flowrate instrument　VES 主控制室空气输送管路流量仪表
VES MCR DP sensing instrument　VES 主控制室 DP 感应仪表
vessel　容器
vessel head　压力容器封头
vessel head multi-stud tensioner　压力容器顶盖多头螺栓拉伸机
vessel manway　容器人孔
vessel support bracket insulation　容器支架保温层
viable　可行的
viable option　灵活的选择
vibration　振动
vibration analyzer　振动分析仪
vibration frequencies　振动频率
vibration measurement program　振动测量大纲
vibration of the shaft　轴承振动
vibration testing　振动试验
vibratory arc surfacing　振动电弧堆焊
video display units　视频显示单元
video teleconference (VTC)　视频会议
viewing window　观察窗
vinyl paint　乙烯基漆
violate a contract　违反合同
virtually　实际上，事实上
viscose spinning pump　纺丝泵
viscous slag　黏性熔渣
visual alarm　可视警报，光报警
visual alerts　视觉报警
visual display　荧屏显示，视觉显示
visual examination　外观检查
visual inspection　目视检查
visual inspection report　目视检查报告
visual signal　视觉信号
vital　重大的，紧要的，不可缺少的
vital area　重要（设备）区
vital instrumentation power supply　重要仪表电源
vitreous slag　玻璃状焊渣
void　无效的
void content　空泡份额，空泡含量
void ratio　空隙比
void swelling　空泡膨胀
void volume　空腔体积
volatile　挥发，挥发性的，易变的
voltage　电压，电位，电势

voltage generator-dcs testing （用于 dcs 试验的）电压发生器
voltage of slag bath　渣池电压
voltage ratio　变压比
voltage regulating transformer　调压变压器
voltage tolerance　容许电压偏差
volume average effective stress　体积平均有效应力
volume change　容积变化
volume control tank　容控箱
volume coverage　体积覆盖
volume defect　体积缺陷
volumetric examination　体积检查
volumetric flask　容量瓶
volumetric power density　体积功率密度
volumetric ultrasonic shear wave test　体积超声剪切波试验
volumetrically inspect　体积法检测
vortex air pressure　旋涡中心气压
vortex pump with shrouded impeller　封闭叶轮旋涡泵
vortex suppression plate　涡流抑制板，漩涡抑制板
voting logic module　表决逻辑组件
V-type piston pump　V形活塞泵
VUS aftercooler/moisture separator　VUS后置冷却器/汽水分离器
VUS air compressor　VUS空气压缩机
VUS air dryer　VUS空气干燥器
VUS compressed air filter　VUS压缩空气过滤器

W

wafer check valve　对夹式止回阀
wafer plate valve　对夹平板阀
waiver　放弃，弃权证书，让步，特许
walking beam pump　步进梁式泵
walkway　走道
wall mounted pump　壁式泵
wall panel information station　大屏幕信息显示板
wall panel information system　大屏幕信息系统
wall panel information system panel in main control　主控制室的大屏幕信息系统显示
wall panel　墙板，墙屏
wall speaker　壁挂扬声器
want，need，demand　需求
warning sign　警告标志
warranty　保证书
washer　洗涤器，垫圈
wastage　耗蚀，消耗量，废物
waste accumulation room　废物集中（存放）间
waste collecting tank　废物收集箱
waste decay tank　废物衰变箱
waste gas after-cooler　废气后置冷却器
waste gas compressor　废气压缩机
waste gas pre-cooler　废气前置冷却器
waste gas sampling box　废气取样箱
waste holdup and monitor tank　废液滞留和监测箱
waste monitor tank　废物监测箱
waste paper pulp pump　废纸浆泵
waste sampling box　废物取样箱
waste sampling cabinet　废物取样隔间

waste sump 废液地坑
waste valve 排污阀
waste water flow 废水流量
water bath microprocessor with digital read out 带数字化读取的水浴微处理机
water bill 水费
water chemistry 水化学
water collection trough 收集水槽
water consumption 用水量
water content 含水量
water conveying pump 运水泵
water curtain type dry underwater welding 水帘局部干式水下焊
water distribution bucket （安全壳冷却）水分配桶
water distribution system （安全壳冷却）水分配系统
water distribution weir （安全壳冷却）水分配围堰
water distribution weir system （安全壳冷却）水分配围堰系统
water film 水膜
water glass 水玻璃
water hammer 水锤
water heater 水加热器，热水器
water inlet assembly 进水口组件
water intake and discharge 取排水
water inventory 水装量
water jet method underwater arc cutting 喷水式水下电弧切割
water level 水位
water level transmitter 水位变送器
water line corrosion 水管腐蚀
water proofing material 防水材料
water quality 水质
water recirculation 水的再循环
water reduced coating (acrylics) 减水涂层(丙烯酸酯类)
water removal 除水
water seal 水封
water seal gate valve 水封闸阀
water solid 水实体
water spray system 水喷雾系统，喷水系统
water sprinkler 水喷淋器，水喷头
water stop 止水，断水
water supply pump 供水泵
water supply system 供水系统
water treatment 水处理
water treatment building HVAC system 水处理车间暖通空调系统
water vapor arc welding 水蒸气保护电弧焊
water-closing type arrestor 水封式回火防止器
waterproofing 防水的
water-quenched 水淬
water-steam equilibrium 水-汽平衡
watertight floors and wall 防水楼板和墙壁
watertight 不漏水的，不透水的
watt-hour meter 电表
wattles power meter 无功功率表
wattmeter 功率表
wave soldering 波峰钎焊
wave soldering machine 波峰钎焊机
wave-like deformation 波浪变形

way of business 经营方式
wear 磨损
wear allowance 允许磨损量
weaving speed of wire 焊丝横摆频率
web-based learning（WBL） 网络化学习
wedge 楔，楔形物
wedge diaphragm valve 楔式隔膜阀
wedge disc 楔盘，楔片，楔板
wedge gate valve 楔式闸阀
weight fraction 重量份额
weight percent 重量百分比
weighting factor 权重因子
weir 堰，围堰
weir diaphragm valve 堰式隔膜阀
weir-type 堰型
weld 焊接
weld assembly 焊接部件
weld cladding 堆焊层
weld crack 焊接裂纹
weld decay 敏化区腐蚀
weld defect 焊接缺陷
weld gauge 焊缝检测规
weld inspector 焊接检验员
weld junction（英）/bond line（美） 熔合线
weld metal 焊接金属
weld metal crack 焊缝裂纹
weld seam 焊缝
weld seam depth-to-width ratio 焊缝深宽比
weld temperature field 焊接温度场
weld thermal cycle 焊接热循环
weld zone 焊接区
weldability 可焊性
welded construction 焊接结构
welded joint 焊缝
welded part 焊件
welded structure 焊接结构
welder 焊工，焊机
welder qualification 焊工资格评定
welder qualification test 焊工考试
welder qualified certification 焊工合格证
welders training 焊工培训
welding backing 焊接衬垫
welding booth 焊接工作间
welding by both sides 双面焊
welding by one side 单面焊
welding circuit 焊接回路
welding condition 焊接条件
welding consumables 焊接消耗材料
welding deformation 焊接变形
welding equipment 焊接设备
welding expert system 焊接专家系统
welding flame 焊接火焰
welding flux 焊剂
welding glass 滤光镜片
welding goggles 气焊眼镜
welding gun 焊枪
welding head 焊接机头
welding heat source 焊接热源
welding image processing 焊接图像处理
welding imperfection 焊接欠缺
welding machine 焊机
welding material 焊接材料
welding mechanics 焊接力学

welding nozzle　焊嘴
welding operator　焊接操作工
welding operator qualification　焊接操作员资格评定
welding parameter　焊接工艺参数
welding parameter recorder　焊接参数记录仪
welding post　焊接工位
welding power source　焊接电源
welding procedure qualification　焊接工艺评定
welding procedure qualification test　焊接工艺评定试验
welding process　焊接过程
welding residual deformation　焊接残余变形
welding residual stress　焊接残余应力
welding robot play back　焊接机器人示数
welding robot　焊接机器人
welding robot station　焊接机器人工作站
welding sequence　焊接工序
welding set　焊接设备
welding shop　焊接车间
welding station　焊接工位
welding stress　焊接应力
welding symbol　焊缝代号,焊接标记
welding technique　焊接技术
welding technology　焊接工艺
welding tip　焊嘴
welding tractor　焊车
welding variable　焊接变量
welding wire/welding rod　焊丝
welding with flux backing　焊剂垫焊
weldment　焊接,焊件
well intentioned　善意的,好心(做)的
well type manometer　管式气压计
well-advised　深思熟虑的
well-defined and specific information　明确具体的信息
well-established　已被认可的,已建立的
Well's wide plate test　韦尔斯宽板拉伸试验
Westinghouse alarm presentation system software　西屋报警显示系统软件
wet annular burnable absorber　水冷环型可燃毒物体
wet bulb air temperature　湿球温度
wet deposition　湿沉降
wet layup　(蒸汽发生器)湿保养
wet method underwater welding　湿法水下焊
wet pipe　湿管
wettability　润湿性
wetted coverage　潮湿覆盖面
wetting　受潮
wetting angle　润湿角
wetting balance test　润湿称量试验
wheeled truck　车轮架
whereby　借以
whirl-pool pump　旋涡泵
whole body contamination monitor　全身污染检测仪
whole body counting system　全身计数系统

whole lot 整批货
whole-body counter 全身计数器
wide range 宽量程
wide range pressurizer level steam space instrument root 稳压器汽空间宽量程液位计根部
wide-range cold leg temperature 宽量程冷段温度
wide-range hot leg temperature 宽量程热段温度
willingness 意愿,愿望
winch 绞盘,卷扬机
winchester disk 文切斯特硬盘
wind loading 风荷载
wind profile 风断面
wind tunnel testing 风洞试验
wind up the transaction 成交
wind up,terminate,pack up 结束
wind-borne missile 风卷带的飞射物
winding 绕组,线圈,缠绕
windmill(driven) pump 风车驱动泵
wing valve 圆盘导翼阀,翼形阀
wing with valve 有阀翼板
wire 导线
wire/cables 电缆/电线
wire feeder 送丝机构
wire fence 铁丝网
wire reel 焊丝盘
wireless base station 无线基站
wireless paging remote terminal unit 无线寻呼远程终端机组
wireless telephone master control unit 无线电话主控制器
wireless telephone switch 无线电话交换机
wireless telephone system 无线电话系统
wireway 线槽
wire-wireless switching controller 有线-无线转换控制器
wiring 布线,配线
with reference to 关于
with respect to 就…而言,关于,根据
with the exception of 除…外
withdraw 撤销
withdraw an offer 撤销报盘
withdrawable 可拔出的,可取出的
withdrawal limit (控制棒)提棒限值,提升限值
withdrawal speed (rate) 提棒速度
withdrawn 抽出,提升,提棒
without engagement 不受约束,无约束
withstand 承受,抵挡
withstand voltage 耐压,承压
witness point 见证点
WLS connection penetration 低放废物系统(WLS)连接贯穿件
work breakdown structure 任务分解结构,工作分解结构
work coating 工作包覆层
work distance 工作距离
work leader 工作负责人,工头
work on 从事
work order 工单,工作指令,工作票
work out a contract 制订合同
work schedule 工作进度表
working authorization 工作许可证

working instruction 操作指南，工作细则
working medium 工作介质
working pressure 工作压力
working temperature 工作温度
workload 工作负荷
workplace 工作场所
workstations and processors 工作站和处理器
worm gear actuator 涡轮传动装置，涡轮驱动装置
worm gear reducer 涡轮减速器
worm hole 蛀孔，蚀孔
wrapper 封套
wrench 扳钳，扳手，猛扭
written examination 书面考试
wrought 熟铁
wrought austenitic stainless steel stock 锻造奥氏体不锈钢的坯料
wrought material 锻造材料

X

xenon 氙
xenon buildup 氙积累
xenon concentration 氙浓度
xenon distribution 氙分布
xenon equilibrium 平衡氙
xenon oscillation 氙震荡
xenon poisoning 氙中毒
xenon reconstruction 氙重积累
xenon transient 氙瞬态
xenon-induced spatial oscillation 氙致空间(功率)振荡
X-ray industrial television inspection X 射线工业电视探伤
X-ray radiographic inspection X 射线探伤
X-ray stress analysis X 射线应力分析
x-y pattern x-y 模式

Y

yard fire protection water distribution system 厂区消防水分配系统
yard main 厂区主管，厂区总管
y-globe type 直流式
yield strain 屈服应变
yield strength 屈服强度
yoke 支架，轭，套，结合
yoke assemble 轭状组件
yoke nut 阀杆螺母

Z

z-direction window type restraint cracking test z 向窗口约束裂纹试验
zero power 零功率
zinc 锌
zinc addition tank 锌添加箱
zinc injection package 注锌组合件
zircalloy cladding tube 锆合金包壳管
zircalloy-4 锆-4 合金
zirconium 锆
zirconium based alloy 锆基合金
zirconium-steam reaction 锆-水蒸气反应
zirconium-water reaction 锆-水反应
ZIRLO (用于 AP1000 型燃料棒包壳的)锆铌合金材料
zone-selected controller 区域选择控制器
zoning 分区

Ⅱ. 设备采购常用词汇与术语

a matter of 有关…的事
a wide range of validity and high accuracy 宽幅有效和高精度
above-mentioned 上述的
absentee 不在者；缺席者，缺勤者
accept an offer 接受报盘
accept an order 接受订单
accessory 附件
accommodate 顺应，调节；迁就，迎合，接受
accommodation 住宿设施
according to 照，依照，据，据…所说
accordingly 随着；相应地
account for 考虑
achieve consensus 达成一致，圆满解决
acquiesce 默许，默认，勉强同意
addendum 附录，补遗
address 致力于
advance payment 预付款
adversary 对手，敌手
affiliate 分支机构，分会；隶属企业
aforementioned 前面提到的
aforesaid 前面所说的；上述的
agenda 议事日程，会议事项
agree about 达成一致
agree to terms 达成协议
agreement 协议
alienate 使疏远
all in all 总的说来
all risks 一切险
allowance 折扣
along with… 随着…
alternative 一种替代选择，代替品
ambiguity 含糊；意义不明确
amendment, modification, alteration 变更，更改，修正
amicably 友好地
an offer without engagement 虚盘
anticipate 预期，指望
apart from 除了…以外，此外
apology 道歉
applicability 适用性
applicable 适用的
apply the same metaphor 用同样的比喻
approach 方法，手段
appropriate 恰当的，合适的
approval 批准；认可
approximate 大概的
arbitration 仲裁
arrange a deal 达成交易
as a gesture of 作为…表示
as already noted 如上所述
as far as…concerned 就…而言
as noted above 如上所述，正如(上面)所指出的那样
as regards 关于，至于
as requested 按照要求
as to 关于

assertive　断言的，肯定的
assessment　评定，估定
associate　伙伴，同人
associate with　把…同…联系起来
assortment　种类
asymmetrical　不对称的
at one's discretion　凭某人自行处理
at your earliest convenience　在你方便时，尽早
attach great attention on　非常重视…
attain, reach　达到
attendant　出席人，参加人
attorney　代理人，代言人
auditing　审计
authorization, permission, permit, release　认可，批准，许可
authorized approval　权威认证
auxiliary　辅助的
availability　供应能力
available　可得到的，有用的
avoid　避免
balance　剩余部分
be authorized to do　被授权，有权
be aware of　了解，知道，觉得
be bound by　受…的约束
be compatible with　与…协调的，相容的，可两立的
be concerned with　涉及，参与，干预
be consistent to(with)　一致的，协调的，相容的，不矛盾的
be corresponding to　相当的，对应的，符合…的
be due to do sth　定于…做某事
be engaged in　从事
be engaged in a discussion　参与讨论
be entitled to　有…资格，被授予权利
be fully efficient　全面投产
be grateful to　感谢…
be greatly indebted to　非常感激…
be in a position to do sth　能，有能力做某事
be in accord with　与…一致的，符合…的
be in production　投入生产
be in the position to do　有资格(权利)做决定，能够做
be in this line　从事这个行业
be liable for　对…应负(法律)责任，有义务的
be more likely to　很有可能
be not inclined to　不打算
be not prepared to　不准备
be not ready yet to　还不准备
be of great significance　具有重要意义
be operational　投产，投入使用
be referred to as　叫做，称为，被认为是
be responsible for　负责
be subject to　须经，受支配，从属于
be suitable to　合适的;适宜的，适当的;相当的
be under sb responsibility　由…负责
be unlikely to…　不太可能(做)…
be well under way　正在顺利进行
be well-versed in　熟悉，精通
beforehand, in advance, upfront　事先，预先
behavioral　关于行为的

beyond one's expectation　超出某人的预期
bid award　确定中标单位，发标
bid evaluation　评标
bidder　投标方
bidding　电厂招投标
board　董事会
bottom line　底线
breach　违反，破坏
breach of contract　违约
breach party　违约方
breakthrough　突破
bring about　引起
bring the price down　降价
broaden, enlarge　加宽，放阔，使扩大
broadly, widely, extensively　广泛地
buyer　需方，买方
by and large　从全体看来
by virtue of　凭借，因为，由于
C&F(cost and freight)　到岸价(成本＋运费)
cancel an order　取消订单
cancellation　取消
carry on　从事
cash　现金
casual　偶然的，碰巧的;临时的
certify, attest to, demonstrate, prove　证明
characterize　表现…的特征，刻画…的性格
charter a ship　租船
chronological　按时间顺序的
cif(cost, insurance and freight)　到岸价(成本＋保险＋运费)
circulate　流转，流行，传播
circulating capital　流动资金
claim　索赔，要求，请求
clarify misunderstanding　消除误解
clarity　清晰
clause　条款
clear up, clarify, make sure, identify　澄清，弄清
close this deal　做成这笔生意
code　规则，法规
coincide with　与…一致，相合，符合，相符，相巧合
collection　托收
come into effect, come into force　被实施，生效
come into operation　投入运行
come quite a long way　进展不小
come to a compromise　达成妥协
come to an agreement　达成协议
come to terms with　与…达成协议
come up　的确出现，发生
come up with, put forward, bring up　提出
comment　评语，评论
commission　佣金
commit to　承诺
commitment　许诺，诺言
common ground　共同立场，一致点
commonplace　平常事情
commuting allowance　交通补助
competitive　有竞争性的
complement, replenish　补充
complication　麻烦情况
comply with　符合，与…一致，依从，

照办
comply with, meet 满足
composure 冷静,沉着
comprehensive, overall 全面的
compromise 妥协,调和,和解
conceal 隐藏
concede 让步,允许,承认
concession 让步,迁就,让与
conciliation 调解
conciseness 简明
conclude (条约等的)缔结,议定
conclude a contract 缔结合同
conclude the business 达成交易
concreteness 具体
confidential 保密的
confirm 证实,证明(陈述、主张等)有根据
confirm an order 确认订单
confirmation, acknowledgement 确认
conform to 与…一致
consent 同意
consequently 所以,因此
considerable 明显的,值得考虑的
consideration 体贴,考虑,报酬
consignee 收货人
consigner, consignor 发货人
consolidate 巩固
constantly 不断地,经常地
construction funds 建设资金
consult with 商量
contemplate 考虑,思考
contentious 有争议的
contingency 意外事件
contract 合同
contractual party 缔约方
contribute 出力,做出贡献
controversial 争议的
cooperative 合作的
coordinator 协调员
corrective action processing and tracking 纠正行动处理和跟踪
correspond to 对应,与…一致,符合
correspondence 商函
count on 指望,依靠
counter offer 还盘(买方),买方还价
counter proposal 反建议
counter-counter offer 再换盘(卖方)
counterpart 对方,对手
counter-sign 会签
courtesy, etiquette, protocol 礼仪,得体
cover, insure 保险
cover quite a lot of ground 取得了不小的进展
coverage 险别
covering letter 伴函
credit terms 信贷条件
criterion 标准,准则
crucial 关键的,极紧要的
crux 要点,难点,要害
cultivated 有修养的,文雅的
customary 习惯的,通常的
cut a deal 达成协议
D/A(document against acceptance) 承兑交单
D/D(remittance by bank's demand draft) 票汇
D/P(document against payment) 付款

交单
date of issue　开证日期
deadline　截止时间，工期
dealing　生意，业务
decline, reject, turn down　拒绝
dedicate　专门用于
deferment　延期
define　规定
definitely　的确，一定，绝对
delineate　描写，描绘
delivery　运送，投递，传送
demarcate　画界，区别
demonstration　技术示范
departure　背离，违背
deposit　（买方支付的）定金，保证金
deserve　值得
designate　指定，规定，限定
designate, appoint, assign　指定
desperate　急切需要的
despite　尽管
destine　预定，注定
detail　细节，详告
devastating　破坏性的
development, format and review　（指询价单的）编制，格式和审核
deviation disposition requestion (DDR)　偏差处理要求
devote one's attention to　集中精力…
differ from　与…不同，与…有区别
discrepancy　不一致，不符
discuss, talk over　讨论，磋商
dismantle　拆除，拆卸，解体
disposal　支配权，（自由）处置权
dispute　争议
distinction, difference　差别
distinctly　独特地
distinct　明显的，明确的，显著的
distinguish　区别，辨别，识别
diversified　多种形式的
divert into　转化到，转移到
do something reciprocal in return　礼尚往来，有来有往
document submittals, approvals and tracking　文件提交，批准和跟踪
documentary draft　跟单汇票
down payment　定金，入门费
draft, draw up　起草
draft at sight　即期汇票
draft contract　合同草案
drag on, prolong, make longer　拖延，持续
draw up a contract　起草合同
due　预定的，到期的
dynamic　有活力的，不断进取的
easy-going　轻松的，舒适的
effect a contract　实施合同，合同生效
effectiveness　有效性
effectuation　发生，生效
electricity rate　电费
element　要素，纲要，成分
elucidate　阐明，说明
embark on　着手，开始从事
emphasize　强调
enclose　随信付上，付入
enclosed please find　随函附上
enclosure　附件
encounter　遇到
endeavor to　全力以赴

endowment insurance 养老保险
engage in 经营，从事
enhance，strengthen 加强
enter into/establish business relations 建交
enter into an agreement 签置协议
enterprise 企业集团
entity 实体
entrust 委托
environmentally friendly 环保的
equality 平等
equivalent dose 当量剂量
especially 尤其
evaluate 对…估价，对…作评价
evaluation report 评估报告，评价报告
event oriented emergency operational procedure 事件导向应急操作规程
eventually 最后，终于
exaggerate 夸大，浮夸
examination & approval 审查/批复
exceptionally 特别地
exclusive agent 独家代理
exclusively 专门地，排他地
exclusivity clause 排他条款
execute a contract 履行合同
execute an order 执行订单
executive 总经理，董事
exemption 豁免
expand 扩大，扩展
expediting techniques and tool 催货的技巧和工具
expertise 评定，鉴定(报告)，专业鉴定
export-oriented 出口型的
extend 拓宽，拓展
extend/renew an agreement （协议)延续/续约
extend a warm welcome to 向…致以热烈欢迎
extend an offer 延长报盘
extend business 扩大业务，扩大经营
extremely 相当
facilitate 使容易，使顺当，助长，促进
fail to honor the contract 不履行合同
fall into 分成，落入
fall short of 没有达到
fall within the scope of 属于…范围
familiarize 熟悉
feasibility 可行性
feature 特色，特点
financial position，financial standing 财务状况
financing 财务
firm 公司
flexibility 灵活性
free on board 离岸价(船上交货价)
foreign direct investment(FDI) 外国直接投资
forenamed 前面举出的，前面所举的
formalize 使成正式，正式化
forthcoming 即将来到的
forthwith 立即，立刻
foster 促进(发育)，鼓励，扶植
foster cooperation and trust 促进发展合作与互信
free from particular average 全损险(即平安险)

freight costs 运费
fruitful 富有成效的
fulfill an order 履行订单
gas bill 气费
get down to brass tacks / get down to business 进入正题
get nowhere 不会有任何结果
get stuck 纠缠
get the message 领会用意
get to the bottom of 寻根问底
get to the heart of 找到问题的关键所在
give a briefing on 就…作一简单介绍
give and take 讨价还价
give away 泄露
give ground 退让
give in 让步
go into details 进行细节的探究
gpg(general program) 总大纲(指三大控制大纲)
gracious 有礼貌的，得体的
grant 给予，授予
grant permission 准许，批准
gross weight 毛重
ground (工作)有所进展
guarantee 保证
hands-on experience 实践经验
hash out 深入讨论
have full authority 完全有权
have good connections with/in 有好的渠道，有好的关系
have in common 共同之处
have no alternative but… 只能
have no intention of 没有…意向
have not much choice 没有选择余地
head office 总部，总公司
headquarters 总部
hereby 特此，兹
herein 此中，于此，据此看来
hereinafter 在下文
hereunder 据此
herewith 与此一起
hierarchical 等级的
hierarchy 等级制度
hold /conduct(meeting) 举行(会议)
hold consultations with 与…进行协商/磋商
hold…covered 给…保险
hypothesis, assumption 假设
identical 共同的，一致的
identify 弄清，识别，验明
identify common areas of interest and establish rapport 寻找共同之处，建立融洽关系
ill-advised 失策的，愚蠢的
illustrate, elaborate, elucidate, spell out 说明，阐明
impact, affect, influence 影响
implement, put into effect 实施
implementation 执行，履行，贯彻，落实
imply 暗示
improvise 临时准备，临时凑成，即席创作
in accordance with 依照，与…一致
in charge of 负责
in compliance with 遵照
in conformity to 遵照，与…相适应，

与…一致
in consideration of　考虑到，由于
in current use　在使用中
in due course　在适当的时候，到一定的时候
in due time　按时，在通常的时间内
in duplicate　一式两份
in good agreement with　符合…，照…
in line with　与…相符
in mass production　投入批量生产中
in motion　正在运动
in one lot　一次性，全部
in proportion to　按…的比例，…成比例，与…相称
in regard to　关于
in terms of　依照，按照
in that case　既然是那样
in the context of　在…背景下
in the form of　以…形式
in the interest of　符合…利益
in the light of　根据
in triplicate　一式三份
in view of　鉴于
incentive　刺激，鼓励
incident　事件
inconspicuously　不太显眼地
incur　引起，遭到
indemnification　赔偿
indemnify　赔偿
indicate　指示，表示，指出
indisputable　无可争辩的
inevitable　不可避免的，不可逃避的；必然的
inferior to　低于，小于
inquiry　询盘(买方)，询价
insofar as　在…的情况下，在…范围内
installment　分期付款
insufficient　不够的，不足的，不充分的，不适当的
insurance policy　保险单
intangible　不现实的
internal department procedure (IDP)　部(处)内管理程序
introduce　引进，引入
investment　投资
invoice　发票
invoice processing　发票处理手续
involuntary　不知不觉的，无意识的
iron out　解决，处理
irrational arguments　非理性的争论
irrespective of　不管，不顾，不论，不考虑
issue　发给，发行，放出，(争论)问题，议题
it should be noted here that…　应该注意
it will be appreciated if…　如果…，本人将不胜感激
item　条款，项目
itemize　开列
justification for conciliation　妥协的理由
kind of　些许，有些
know-how　专门技能，知识，技巧
knowledge-intensive　知识密集型的
launch (a new product)　推出(一个新产品)

lay on　安排
legal force　法律效力，法律约束力
letter of credit(L/C)　信用证
liability　责任，义务
liable to　易于…的，有…倾向的
liaison　联络
license　准许，发许可证给…
licensee　接受授权者(接受方)
licensor　授权者(转让方)
likelihood　可能性
likewise　同样
line of business　业务范围
linkage close　附带条件的成交
linking strategy　附加条件的战略
litigation　打官司，涉讼，起诉
load　装，加载
localization　本地化
lodge a claim　提出索赔
long-term loan　长期贷款
mail transfer(M/T)　信汇
maintain composure　保持冷静
make(take)a note of　记录(下)
make a concession　作出让步
make a corresponding move　开始相应的行动
make a deal　成交
make a point of　强调，重视
make a presentation on /give a brief account of　介绍，阐述
make a prompt move　立即行动
make a sacrifice　作出牺牲
make an offer　报价
make concession　作出让步
make enquire about　咨询，询问，查询
make it　做到，办成功
make some headway　取得了一些进展
management　管理部门，经理部
managerial expertise　管理经验
managerial personnel　管理人员
managing director　总经理，总裁
manifold　多方面的，多样的
manufacturer　制造商
manufacturing supervision　监造
mark the occasion　庆祝这个盛会
master contract　主合同
materialize　实现，使具体化
mechanic　机械的
mediate　协调，调停
medium-term loan　中期贷款
memorandum　备忘录
memorandum of understanding　谅解备忘录
methodically　有条不紊地，有组织地，有次序地
minimize　使减到最少
minute　纪要，记录
misunderstanding　误会，误解，曲解
mitigate　减轻，减缓
mode of transport　运输方式
modify, revise, amend　修改，修正
monitor　监督，监视，监控
motion　(议会中的)提议
motivation　动机
move on to　继续
move on, proceed with　进行
multi-purpose　多用途的
mutual benefit　互利
needless to say　不用说

negotiable　可协商的，可兑现的
negotiation　协商，谈判，交涉，协定
negotiation consciousness　谈判意识
net weight　净重
nominate　指定，任命
nonbinding　非约束力的
nonconforming item　不符合项
non-conforming processing, tracking and resolution　不符合项处理，跟踪和解决
non-negotiable　不可兑现的
norm, standard　标准，规范
note that…　注意
notify　通知
notwithstanding　虽然，尽管…仍…
objective　目的，目标，任务
observe　遵守
obtain, secure　获得
occasional　偶然的，不经常的
occurrence　（故障、事件的）发生，出现
offense　冒犯，得罪
offer　报价
offer firm　报实盘，报实价
offer subject to our final confirmation　报价以最后我方确认为准
offeree　受盘人（买方）
offeror　发盘人（卖方）
on balance　总的来说，总而言之
on the subject of　关于
on the whole　总的说来，总而言之
once for all　一次性地
open a saving account　开设活期存款户头
opening bank　开户行
opening offer　首次报价（开盘）
operate on CIF terms　按到岸价格成交
opponent　对手，竞争者
oriental　东方的
out of line with　与…偏离，与…不相符
outage, shutdown　停运，停机，停堆
outcome, result　结果
outlet　批发商，销售公司
outline　叙述（要点），概括地论述
outstanding　未解决的，未完成的
overdue　过期的，过期未付的
overestimate　估计过高，过分评价
overview　全貌
packages bids　投标承揽项目的各个部分，打包投标
paramount　首要的，最高的，至上的
participant　参加者，与会代表
particular average　单独海损（即水渍险）
partner　伙伴，配手，搭档
pay on dispatch of the goods　发货付款
penalty　惩罚
pending　悬而未决的，未定的
perfect, improve　改善，改良
peril　危险
persuade, talk into　劝说，说服
pertinent　相关的，有关的
pertinent information　相关信息
pictorial　有插图的，用图表示的
place an order　提交订单
please rest assured　请放心
point out　指出
potential　潜在的

PRC　公司级管理程序
preamble　序言
precaution　预防措施
precedent　惯例，先例，范例
preferential　优惠的
preferential terms　优惠条件
preferential treatment　优惠待遇
preliminary　初步的
premise　前提，所在地
premium　保险金
prerequisite　先决条件，前提
president　院长，总统
presume　假设
prevention　预防
previously　预先，在前
primordial　理所当然的
principal　委托人，委托方
prioritize　分清重点，明确轻重缓急
priority　优先(权)
procedure　程序，手续，方法
proceed to, take up　动手，开始，着手
procrastinate　拖延，耽搁
product receipt and inspection　产品接收和检验
profile　简况，简介
profit margin　利润，利润率
progress, come along　进展
progressively　渐进地，逐渐地
project expenses　工程费用
project on a turnkey basis　交钥匙工程
project proposal　项目建议书
promote　促进
proportion　比例
proposal　提议，建议，提案
propose claims　提出索赔
propose, move　提议，动议
prospect　前途，希望，远见，潜在的客户
prospective　预期的，将来的
protocol　草案
protracted　拖延的
provision　(法律)条款，规定，措施，储备
punctually　准时地
purchase order format and review　采购订单格式和审核
purchasing department　采购部
put forward　提出
put into operation　投入运行
quality observation report(QOR)　质量观察报告
quotation　报价，报盘
quote　报价(虚盘)
raise, collect, make up　筹集，自筹
rather than　而不
reach an agreement　达成协议
recognize　承认
reduction　减少
refer to　参考，参照，叫某人参看，叫某人查询
reference　证明，提及，参考
regardless of　不管，不顾，不拘，不注意
registered capital　注册资金
registration　注册，登记
regularly　经常地
reimburse　报销，补偿
reiterate　重申

related to 与…有关
reliable 可靠的
remain composure 保持冷静
remain in force 保持有效
remain in good terms with 与…保持良好关系
remit 汇寄，付(款)，开发(支票)
remittance 汇款，汇款额，支付(金额)
render 提出，提供
render advisory service 提供咨询服务
renew 续签合同
renew an offer 更新报盘
re-occurrence (事件的)再次发生
repair and replacement parts 修复和更换部件
repair 修理
repayment 还款
replacement 替代，替换
report to 向…汇报
representative 代表，代理(人)
requisite 必要的，必不可少的
resolve 解决，处理
resolve conciliation 调停
resort to arbitration 诉诸仲裁
respecting 关于
rest with 在于，归于
restate 重申，再声明，重新陈述
restriction 限制
result from 由…引起(产生，造成)
result in (引起)导致
retrieval 取回，恢复
return on investment(ROI) 投资回报
reveal 揭露
rewarding 有益的，值得做的
rework 返工
ridiculously 不合理地
room to maneuver 回旋余地
royalty (专利权)使用费
rudimentary 基本的
run through 浏览，从头到尾过一遍
sample survey 抽样调查
scarce 稀少的
schedule development, integration and management 进度编制，综合和管理
scope 范围，余地
scrap 废弃，报废
screening criteria 审批标准
see to that 注意做到
seismic 地震的
self-reliance 自主化
seller 供方
sentimentality 感情用事
services 服务
session 会议，会议的一次(一届)
set out to do something (带着某种目的的)开始做某事
set out to prepare 着手准备
settle a claim 理赔
settle a dispute 解决纠纷
settle an account 结账
settlement 解决方法
shipper 托运人
shortage 短缺
sign, take on 签订
sign a contract 签订合同
signify 表示

simply put　简而言之
simultaneously　同时地
site service　现场服务
sketchy　大概的
slightly　稍微地，轻微地
smooth over　缓解
smoothly　顺利地
sociality　社交
solution　解决方法
somewhat　一点儿，稍微，有点，多少
sort out　整理，安排
sort out consignments　安排发运
sourcing of raw material　原材料采购
specialize　专门研究，专攻
specialize in　专营
specification　规范，规格，说明书
specify　规定
specimen　样本
spell out　明确指明，阐明
stalemate, deadlock, impasse　僵局
stand in the way　阻碍
standardization　标准化
standing　商业状况，名望
standstill　停顿，停滞
state-owned corporation　国有(营)公司
stick to, adhere to　坚持
stipulate　规定
straightforward　直接的，明确的
stringent requirements　严格要求
strive to　力求，努力
subcontractor　分包商
subdivide into　再(细)分成
subject to　以…为准
sublet　转包，分包
submit　提交
submit tenders　提交标书
subsequent　随后的
subsequently　随后
subsidiary　子公司
substantial　显著的，相当大的
superimpose　添加，附加
supervise　监督，管理
supplementary　补充的
supplier evaluation, qualification and selection　供货商评估，资质鉴定和选择
supplier surveillance　供货商监督
survey　检查，调查
suspend　暂停，临时取消
system inquiry　系统询价
telegraphic transfer　电汇
take precedence over　优于，先于
take the matter up　汇报一下情况，商量一下
targeted time　规定的时间
tariff　关税
technical interface　技术接口
technique　(专门)技术，(艺术上的)技巧，技能
technology-intensive　技术密集型的
tedious　繁琐的
tender /bid for the project　投标争取这个项目
tendering party　招标方
tentatively　暂时地，初步地
term　期限，条件，条款，价格，费用
term of agreement　合同期限

term of validity　有效期
terms & conditions development　编写条款和条件
that's fair enough　言之有理
the exchange of needed goods　互通有无
the first party　甲方
the insured　投保人
the insurer　保险人
the proper authorities　主管部门
the same line of business　同一行业
the second party　乙方
theoretically　理论上
there from　从那里，从那一点
thereafter　自那以后，以后，之后
therein　在其中
thereof　由此，因此
thereto　此外，又
timing　时间选择
tip　提示
to a great extent　在很大程度上
to my regret　很遗憾
trace　跟踪，追踪，查找
trade-off　折中(办法，方案)
transaction　交易，业务
transfer expenses　搬迁费，安家费
transportation　运输
trial　试验性的
trustworthy　值得信赖的
turn out　结果是，原来是
turn out, produce, fabricate　生产
unanimously　一致同意的，无异议的，全体一致的
undeniable　不可否认的
underestimate　低估，把…的价值估计过低，估计不足
under-mentioned　下述的
undertake　从事
undertaking　任务，保证，许诺
unfortunately　抱歉
unilateral　单方面的，单边的
unless otherwise mentioned (noted, stated)　除非另有说明(说明，陈述)
unless otherwise specified　除非另有规定
unload　卸货
up to date　最新
upgrade　提高，提升
urgently　迫切地
use, utilize, apply to, employ　使用
use-as-is　原样照用
usual practice　惯例，通行做法
utility bill　公共费用
valid　有效的
value　尊重
vary　变化
venue　(指定的)集合地点
viable　可行的
viable option　灵活的选择
violate a contract　违反合同
virtually　实际上，事实上
vital　重大的，紧要的，不可缺少的
void　无效的
want, need, demand　需求
warranty　保证书
water bill　水费
way of business　经营方式
well intentioned　善意的，好心(做)的

well-advised　深思熟虑的

well-defined and specific information　明确具体的信息

well-established　资深的

whereby　借以

willingness　意愿，愿望

wind up the transaction　成交

wind up, terminate, pack up　结束

with reference to　关于

with respect to　就…而言，关于，根据

with the exception of　除…外

withdraw　撤销

withdraw an offer　撤销报盘

without engagement　不受约束，无约束

work on　从事

work out a contract　制定合同

Ⅲ. 设备监造常用词汇与术语

AC arc welding power source 交流弧焊电源
AC plasma arc welding 交流等离子弧焊
AC/DC arc welding power source 交直流两用弧焊电源
acceleration voltage 加速电压
accept without modification 原样验收
acceptance 接受，验收
accessory 附件
acetylene cylinder 乙炔瓶
acetylene generator 乙炔发生器
acicular ferrite 针状铁素体
acid covering 酸性药皮
acid electrode 酸性焊条
acid slag 酸性渣
acidity 酸度
acoustic emission testing 声发射检测
activation temperature range of flux 钎剂活性温度范围
active filler metal 活性钎料
active metal brazing alloy 活性钎料
adaptor 贯穿件，适配器
adaptive control welding 适应控制焊接
adhering slag 挂渣
adhesive strength 结合强度
age hardening 时效硬化
ageing 老化
agglomerated flux 团块状焊剂，陶质焊剂，烧结焊剂
air arc cutting 空气电弧切割
air plasma arc cutting 空气等离子弧切割
air plasma water injection arc cutting 水注入空气等离子弧切割
air tight test 气密性检验
air-acetylene welding 空气乙炔焊
allotment 分批，分配
allowable stresses 许用应力
alloying 渗合金
alloying constituent 合金剂
alternate wire-feed system 变速送丝方式
alternating current arc welding 交流电弧焊
alumetized steel electrode 渗铝钢焊条
aluminium 铝
aluminum alloy arc welding electrode 铝合金焊条
amorphous filler metal 非晶态钎料
anchoring 抛锚效应
angle beam method 斜射法超声波探伤
angle joint 角接接头
angular distortion 角变形
anneal 退火
applicability 应用范围
appurtenance 配件，附件
arc brazing 电弧钎焊
arc cutting 电弧切割
arc scratch 电弧擦伤

arc stabilizer　稳弧剂
arc surfacing　电弧堆焊
arc voltage feedback controlling arc welding　弧压反馈控制电弧焊
arc welding　电弧焊
arc welding generator　弧焊发电机
arc welding machine using a consumable electrode　熔化极弧焊机
arc welding machine using a non-consumable electrode 不熔化极弧焊机
arc welding power source　弧焊电源
arc welding rectifier　弧焊整流器
arc welding robot　弧焊机器人
arc welding transformer　弧焊变压器
arc welding with consumable　熔化电弧焊
arc welding with nonconsumable　不熔化极电弧焊
argon　氩
argon arc welding machine　氩弧焊机
arrangement drawing　分布图，布置图
as appropriate　根据情况
as built　竣工
as-built drawing　竣工图
assemble　装配，组装
assembly　装配
assembly drawing　组装图
as-welded (AW)　焊态
atomic hydrogen welding　原子氢焊
atomic hydrogen welding apparatus　原子氢焊机
audit plan　监察计划
audit report　监查报告
auditee　受监查方，受审核方
austenitic electrode　奥氏体焊条
automatic oxygen cutting　自动气割
automatic surfacing　自动堆焊
automatic welding　自动焊
availability　可用率
axial bend test　纵弯试验
backfire　逆火
backing　焊接衬垫
backward deformation　反变形
balanced pressure torch　等压式焊炬
ballast rheostat　镇定变阻器
bar type cracking test　圆棒裂纹试验
bare terminal (of an electrode)　焊条夹持端
base metal　母材
base-plate　垫板
basic covering　碱性药皮
basic electrode　碱性焊条
basic slag　碱性渣
basicity　碱度
batch/lot　批，一次投料量；一次生产量
battelle type underbead cracking test　巴特尔焊道下裂纹试验
bead　焊道
bead crack　焊道裂纹
beam current　电子束电流
beam power　电子束功率
beam power density　电子束功率密度
bedplate　垫板
benchmark　基准
benchmarking　对标，对准
bending strength　抗弯强度
bevel cutting　坡口切割

beveling 倒角，倒棱，坡口切
bias electrode 偏压电极
billet 钢坯
binder 粘接剂
blank 毛坯，坯料
blocking 垫块
blow hole 气孔
blow pipe 吹管，喷焊器，焊炬
bolting 螺栓
bond area 熔合区
bond coating 结合层
bonded flux 粘结焊剂
bottom head 下封头
bottom shell 下部筒体
brace metal 钎缝金属
braced joint 钎焊接头
braided electrode 编织焊条
branch T saddle joint T形管接头
brazability 钎焊性
braze cream 钎料膏
braze paste 钎料膏
brazed rate 钎着率
brazer 钎焊
braze-welding 钎接焊
brazing 硬钎焊
brazing (soldering) blowpipe 钎炬
brazing.(soldering) operation 钎焊操作
brazing (soldering) process 钎焊过程
brazing filler metal 硬钎料
brazing filler metal erosion 钎料流失，钎料熔蚀
brazing flux 钎剂，焊剂
brazing in controlled atmosphere 保护气氛钎焊
brazing lamp 喷灯
brazing process variables 钎焊参数
brazing retort 钎焊盒
brazing seam 钎缝
brazing temperature 钎焊温度
brazing time 钎焊时间
breaking strength 断裂强度
brittle fracture 脆性断裂
brittle temperature range 脆性温度区间
buildup 局部堆焊
buildup welding 堆焊
built-up member 组装件
burden 焊剂层
burring 去毛刺
burn through 烧穿
burst pressure 冲破压力
butt and lap joint 搭接对接接头
butt welding 对焊
buttering parts 打底焊件
BWRA cracking test for austenitie steel BWRA 奥氏体钢裂纹试验
calibration 标定
canned-motor RCP 屏蔽电机式反应堆冷却泵
carbon 碳
carbon arc air gouging 碳弧气割
carbon arc cutting 碳弧切割
carbon arc welding 碳弧焊
carbon equivalent 碳当量
carbon-dioxide arc welding 二氧化碳气体保护电弧焊
carburization 渗碳处理，碳化

carburizing 渗碳处理，碳化
carburizing flame 碳化焰
cast iron 铸铁
cast iron electrode with steel core 钢芯铸铁焊条
cast nodular 球墨铸铁
casting part 铸件
ceramic spraying 陶瓷喷涂
certificate of compliance 合格证
chemical analysis 化学分析
chevron cracking 锯齿形裂纹
chipping hammer 敲渣锤
chroming 镀铬
chromium 铬
chromium equivalent 铬当量
circular-groove cracking test 环形槽裂纹试验
circular-path cracking test 圆形镶块裂纹试验
circumferential weld 环焊
cladding part 堆焊件
cleanliness 清洁
clearance between focal point and (plate) surface 离焦量
clearance fillability test 填缝性试验
cleavage fracture 解理断口
CO_2 arc welding 二氧化碳气体保护电弧焊
CO_2 arc welding machine 二氧化碳弧焊机
CO_2 arc welding with thick wire 粗丝二氧化碳亚弧焊
CO_2 arc welding with thin wire 细丝二氧化碳亚弧焊
coarse grained region 粗晶区
coating flux 涂料
coating material 涂料
cobalt 钴
cobalt base alloy surfacing electrode 钴基合金堆焊焊条
coefficient of heat flow concentration 热流集中系数
cold crack 冷裂纹
cold cracking susceptibity coefficient 冷裂敏感系数
cold soldered joint 虚钎，虚焊
cold wall type vacuum brazing 冷壁真空钎焊炉
collar 焊筋
colophony flux 松香钎剂
combined cutting and welding torch 焊割两用炬
combined plasma arc 联合型等离子弧
combustion intensity 燃烧强度
combustion rate 燃烧速度，燃烧率
commissioning 调试，试运行
commissioning program(CP) 调试大纲
compatibility 兼容性
compliance 一致性
compressive strength 抗压强度
computer aided design for welding structure 计算机辅助焊接结构设计
computer-aided welding process programming 计算机辅助焊接工艺设计
concave 凹面
concession 让步，妥协，特许

conformatory test　验证试验
conformity　合格，符合
conical shell　锥形筒体
connector　接头
consignment　交运货物
consignment note　发货通知书
constant-current arc welding power source　恒流弧焊电源
constant-voltage arc welding power source　恒压弧焊电源
constant wire-feed system　等速送丝方式
constituted component　组装件
constraint stress　约束应力
constricting nozzle　压缩喷嘴
consumable electrode　可熔电极
consumable nozzle　熔嘴
consumable wire　熔丝
contact welding　接触焊
contact-reaction brazing　接触反应钎焊
continuous cooling transformation　连续冷却转变图
continuous laser welding　连续激光焊
continuous weld　连续焊
contraction deformation　收缩变形
contraction stress　收缩应力
controlled atmosphere arc welding　充气室中电弧焊
conventional load voltage　常负载电压
cooling time　冷却时间
copper shoe　铜滑板
copper-alloy arc welding electrode　铜合金弧焊条
copper-plating welding wire　渡铜焊丝
core specific enthalpy rise　堆芯比焓升
corner double seam　锁缝角接头，双缝角
corrective action　纠正措施
corrosion resistance　抗腐蚀强度
cover glass　防护镜片
crack　裂纹
crack propagation rate　裂纹扩展速率
cracking sensibility　裂纹敏感性
cracking test　裂纹试验
crater　弧坑
crater crack　弧坑裂纹
criterion of cold cracking　冷裂判据
critical crack size　临界裂纹尺寸
critical current of double arcing　双弧临界电流
critical flow　临界流
critical stress intensity factors　临界应力强度因子
cruciform cracking test　十字接头裂纹试验
cruciform weld　十字焊缝
crush resistance　抗压强度
cu-cooling plate　铜冷却板
cut face　切割面
cut line　切割线
cutting blowpipe　割炬
cutting gun　割枪
cutting nozzle　割嘴
cutting oxygen　切割氧
cutting robot　切割机器人
cutting shoulder　切口上缘
cutting speed　切割速度

cutting stream　切割氧
cutting tip　割嘴
cutting torch　割炬
cylinder cap　阀罩
cylinder manifold　汇流排
cylinder valve　气瓶阀
DC arc welding power source　直流弧焊电源
de-brazing　脱钎，脱焊
deburring　去毛刺
defect　缺陷
deficiency　不足，亏空
deflection deformation　挠曲变形
degree of demonstration　证实程度，验证度
degree of free for robot　机器人运动自由度
dehydrogenation　去氢
delayed crack　延迟裂纹
Delong's diagram　德龙组织图
deoxidizer　脱氧剂
dependability　可信性，可信赖度
dephosphorization　脱磷
deposition efficiency　沉积效率
depth of cutting veins　割纹深度
depth of slag bath　渣池深度
descaling　去氧化皮
desiccant　干燥剂
desiccation　干燥作用
desiccator　干燥器
design review　设计审查
de-soldering　脱钎，脱焊
desoxydation　脱氧
destructive test　破坏性检验
desulphurization　脱硫
developer　显影剂
deviation permit　偏离许可
die formed　冲制成型
differential pressure　差压
diffusible desoxydation　扩散脱氧
diffusible hydrogen　扩散氢
diffusible hydrogen remained at 100℃　100℃残余扩散氢
diffusible oxydation　扩散氧化
diffusion brazing　扩散钎焊
dimensional diagram　尺寸图
dimple fracture　韧窝断口
diode gun　二极电子枪
dip brazing　浸渍钎焊
dip soldering　浸渍钎焊
direct current arc welding　直流电弧焊
direct current suppressor　直流分量抑制器
discrepancy　偏差，违背
dislocation　错边
dishing　冲压
disposition　处理，安排
disposition of nonconforming items　不符合项处理
distance between welding wires　焊丝间距
divergent nozzle　快速割嘴
document submittals　文件提交
double arcing　双弧现象
double coated electrode　双层药皮焊条
double tension test　双重拉伸试验
downhill welding　俯焊
drawled part　拉制件

dropping characteristic arc welding power source　下降特性弧焊电源
dross　结瘤，浮渣
dry flashback arrestor　干式回火防止器
dry method underwater welding　干式水下焊
dual gas plasma arc cutting　双层气体等离子弧切割
ductile fracture　延性断裂
ductility-dip crack　失延裂纹
duty cycle　负载持续率
dye penetrant inspection　着色探伤
dye penetrant test　着色渗透检验
dynamic (static) pressure　动(静)压
dynamic characteristic　电源动特性
earth clamp　地线夹头
eccentricity (of an electrode)　焊条偏心度
eddy current inspection　涡流检查
effectiveness　有效性
elastoplastic fracture mechanics　弹塑性断裂力学
electric arc spraying　电弧喷涂
electric welder　电焊机
electric welding machine　电焊机
electrical test　电气试验
electricity grid　输电网
electrode　电极，焊条
electrode for under-water cutting　水下切割用焊条
electrode for welding spheroidal graphite cast iron　球墨铸铁焊接用焊条
electrode holder　焊钳，电极夹
electrode keep time of slider　焊丝停摆时间
electrode setback　电极内缩长度
electrode with cast core wire　铸芯焊条
electro-gas (arc) welding machine　气弧立式电焊机
electro-gas (arc) welding　气电弧立式焊
electromagnetic inspection　电磁探伤
electromagnetic lens　电磁透镜
electron beam brazing　电子束钎焊
electron beam cutting　电子束切割
electron beam deflection coils　电子束偏转线圈
electron beam welding (EBW)　电子束焊
electron beam welding machine　电子束焊机
electron gun　电子枪
electroslag process stability　电渣工艺稳定性
electroslag surfacing　电渣堆焊
electroslag welding (ESW)　电渣焊
electroslag welding machine　电渣焊机
electroslag welding with consumable nozzle　熔嘴电渣焊
electroslag welding with plate electrode　板极电渣焊
electroslag welding with tube electrode　管极电渣焊
electroslag welding with wire electrode　丝极电渣焊
elongated cavity　加长气孔
end of manufacturing report　完工报告

epitaxial solidification　联生结晶
erosion　溶蚀
evenness of cutting surface　切割面平面度
excessive penetration　塌陷
exercisability　可操作性
exhaustiveness　(论述等)详尽，彻底
expanded part　胀件
expediting plan　催交计划
external characteristic　(电源)外特性
fabrication weldability　工艺焊接性
face bend testing　面弯试验
face of cut　切割面
failure analysis of weldments　焊件失效分析
fatigue fracture　疲劳断口
fatigue strength　疲劳强度
faying face　钎焊面
feedhead　(铸件的)冒口
ferreous　铁的
ferrite　铁素体
ferrite side-plate　铁素体侧板条
ferritic　铁素体的
ferritic electrode　铁素体焊条
ferrule　套环
fettling　去毛刺
fibrous zone　纤维区
field of weld temperature　焊接温度场
field welding　现场焊
filler material　焊接填充材料
filler metal　填充金属，焊条
fillet　钎角
filter glass　滤光镜片
final inspection　最后检查
findings　调查结论
fine grained region　细晶区
finger (cracking) test　指形裂纹试验
finishing　精加工，完工
fire cracker welding　躺焊
FISCO (type) cracking test　FISCO 裂纹试验
fish eye　白点
fishbone cracking test　鱼骨形裂纹试验
fissure　狭长裂纹
fittings　装配
flame cutting　氧气切割
flake　白点
flame cone　焰芯
flame envelope　外焰
flame gouging　火焰气刨
flame heating efficiency　火焰热效率
flame planer　门式切割机
flame spraying　火焰喷涂
flame straightening　火焰矫正
flange　法兰
flanged connection　法兰连接
flanged coupling　法兰连接
flare tube fitting　外喇叭口套管接头
flaring　扩口
flashback　回火
flashback arrestor　回火防止器
flat lock seam joint　平面锁缝接头
flat position weld　平焊
flattening　压扁，压平
flaw　缺陷
flaw detection sensitivity　探伤灵敏度
flexural strength　抗弯强度

flow diagram　工艺流程图
flow distribution baffle　流量分配板
flow soldering　波峰钎焊
fluidity of the slag　熔渣流动性
fluorescent penetrant inspection　荧光渗透探伤
flush lap joint　平面搭接头
flushing　冲洗
flux　焊药，焊剂
flux activity　钎剂活性
flux backing　焊剂垫
flux coated brazing rod　药皮钎料，药皮焊条
flux for narrow-gap submerged arc welding　窄间隙埋弧焊焊剂
flux inclusion　钎剂夹杂
flux layer　焊剂层
flux paste　钎剂膏
flux-cored colder wire　药芯钎料丝
flux-cored electrode　管状焊条，药芯焊条
fluxer　钎剂涂敷器
flux-hopper　焊剂漏斗
focal length　焦距
focal spot　焦点
focussed diameter of the beam　焦斑直径
folded joint　锁缝接头
foreign elements　杂质元素
forging part　锻件
forming part　成型件
fractography　断口金相
fracture　断口
fracture analysis diagram　断裂分析图
fracture apperance　断口形貌
fracture mechanics　断裂力学
fracture test　断口试验
fracture toughness　断裂韧度
free deformation　自由变形
fresh fuel　新燃料
fuel assembly　燃料组件
fuel element　燃料元件
full penetration weld　全透焊
full vacuum electron beam welder　高真空电子束焊机
furnace brazing　炉中钎焊
furnace soldering　炉中钎焊
fused flux　熔炼焊剂
fusible backing　保留垫板
fusion type plasma arc welding　熔透型等离子弧焊
fusion zone　母材熔化区
gamma-ray inspection　γ射线探伤
gas cavity　气孔
gas cutting　气割
gas cutting machine with magnetic wheels　磁轮式气割机
gas cylinder　气瓶
gas flux　气体钎剂
gas forming constituents　造气剂
gas jet laser cutting　喷气激光切割
gas laser　气体激光器
gas pore　气孔
gas shielded arc welding　气体保护电弧焊
gas shielded arc welding machine　气体保护弧焊机
gas welder　气焊工

gas welding　气焊
gasification speed　气体发生速度
gauge pressure　表压
G-BOP weld metal crack test　G-BOP 焊缝金属裂纹试验
girth weld　环焊
glide plane fracture　滑移面断口
gouging blowpipe　表面割炬
grain boundary ferrite　晶界铁素体
granular bainite　粒状贝氏体
graphite type electrode　石墨型电极
gravity coefficient of coating　药皮重量系数
gravity welding　重力焊
grinding　磨光
hammering　锻造，锻打
handling　搬运，装卸
hard facing　表面硬化
hard surfacing　表面硬化
hardened steel　淬火钢
hardened zone　硬化区
hardfacing　耐磨堆焊
hardfacing electrode　耐磨堆焊焊条
hardness　硬度
healing effect　愈合效应
heat resisting steel　耐热钢
heat treatment for dehydrogenation　去氢热处理
heat-affected zone　热影响区
heat-affected zone crack　热影响区裂纹
helium　氦
helium-arc welding　氦弧焊
helmet　面罩
high cellulose (type) electrode　高纤维素型焊条
high frequency spraying　高频感应喷涂
high iron oxide type electrode　氧化铁型焊条
high manganese steel surfacing electrode　高锰钢堆焊焊条
high speed welding flux　高速焊剂
high titania (type) electrode　高钛型焊条
high toughness super low hydrogen electrode　高韧性超低氢焊条
high-current plasma arc welding　大电流等离子弧焊
high-speed cutting　高速切割
high-speed nozzle　高速割嘴
hoisting　起重，提升
holding time of brazing　钎焊保温时间
horizontal bend test　横弯试验
horizontal position weld　水平焊
hot crack　热裂纹
hot extrusion　热挤压
hot gas soldering　热风钎焊
hot straining embrittlement　热应变脆化
hot wall type vacuum brazing furnace　热壁真空钎焊炉
hot wire MIG welding　热丝 MIG 焊
hot wire plasma arc surfacing　热丝等离子弧堆焊
hot wire submerged-arc welding　热丝埋弧焊
hot wire TIG welding　热丝 TIG 焊

hot-ductility test　热塑性试验
H-type cracking test　H 形裂纹试验
hydraulic test　水压试验
hydrogen　氢
hydrogen embrittlement　氢脆
hydrogen test　测氢试验
hydrogen-induced crack　氢致裂纹
hydrostatic test　水压试验
icicle　毛刺
ID grit blasting　内部喷丸处理
identification　标记，编码，标志
identification　认出，识别，鉴定
identification tagging　编码，标记
IIW cracking test　IIW(国际焊接学会)裂纹试验
ilmenite type electrode　钛铁矿形焊条
immersion method　液浸法超声波探伤
impact test of haz　热影响区冲击试验
impeller　叶轮
implant test　插入试验，植入试验
impulsed laser welding　脉冲激光焊
inclusion　夹杂
inclusion content　杂质含量
incomplete fusion　未熔合
incomplete penetration　未焊透
incompletely filled groove　未焊透
incompletely filled weld　未焊满
induction brazing　感应钎焊
industry /economic sector　工业行业/经济部门
inert-gas backing　惰性气体衬垫
inflammable limit of the gaseous　混合气体可燃浓度限值
infrared brazing (soldering)　红外线钎焊
ingot　铸块，锭
inherent strain method　固有应变法，本征应变法
inherent strain zone　固有应变区，本征应变区
inherent stress　固有应力，本征应力
initial diffusible hydrogen　初始扩散氢
injector　喷射器
injector torch　射吸流焊炬
inner cone　焰芯
insert　嵌条
inset T joint　嵌入 T 形接头
inspection　检查，检验
inspection plan　检验计划
inspection report　检查报告
installation　安装，装配
installation instruction　安装说明
installation(erection)drawing　安装图
instantaneous concentration heat source　瞬时集中热源
interchangeability　可换性
intercrystalline crack　晶间裂纹
interfacial region　钎缝界面区
intergranular fracture　晶间断裂
intermediate weld　间歇焊
intermediate-current plasma arc welding　中电流等离子弧焊
intermittent weld　间歇焊
internal flame　内焰
inverter welding power source　逆变式焊接电源
iron oxide type electrode　氧化铁型焊条

iron soldering　烙铁钎焊
ISO 9000 family　ISO 9000 系列标准
item　物项
job batching　分批加工
joggled lap joint　平面搭接头
joint design　接头设计
kerf　切口
kerf angle　切口角
kerf width　切口宽度
keyhole effect　小孔效应
keyhole-mode welding　小孔型等离子弧焊
killed steel　镇静钢
Kinzel test　肯泽尔(Kinzel)弯曲试验
knife line attack　刀状腐蚀
lack of fusion　未熔合
lack of penetration　未焊透
ladle　钢水包
ladle analysis　熔炼分析，桶样分析，炉前分析
lamellar segregation　层状偏析
lamellar tearing　层状撕裂
laminated part　轧件
lap joint　搭接接头
lapped T-joint/flanged T-joint　搭接 T 形接头
laser beam cutting　激光切割
laser beam welding　激光焊
laser brazing (soldering)　激光钎焊
laser cutting (LC)　激光切割
laser welding　激光焊
laser welding equipment　激光焊机
lath ferrite　条状铁素体
lath martensite　板条马氏体
layout drawing　布置图
lead time　产品从设计到交货的时间
leak test　密封性检验，泄漏试验
least developed countries　欠发达国家
legibility　字迹清楚
lehigh restraint cracking test　里海抑制裂纹试验
lick side seam jointt　平面锁缝接头
light soldering (brazing)　光束钎焊
lime titania type electrode　钛钙型焊条
lime type covered electrode　碱性包覆焊条
linear elastic fracture mechanics　线性弹性断裂力学
linear heat source　线热源
linear porosity　线性气孔率
linearity　线性度
liquation crack　液化裂纹
liquation of filler metal　钎料熔析
liquid crystal test　液晶试验
local deformation　局部变形
local dry underwater welding　局部干式水下焊
local stress　局部应力
local stress relieving　局部消除应力，局部应力释放
lock corner joint　锁缝角接头
lone of cut　切割线
long slag　长渣
longitudinal crack　纵向裂纹
longitudinal-bend test　焊道纵向弯曲试验
lot　批次，批量
lot number　批号

low hydrogen type electrode　低氢型焊条
low hydrogen type flux　低氢型焊剂
low poison flux　低毒焊剂
low pressure acetylene generator　低压乙炔发生器
low temperature steel electrode　低温钢焊条
low temperature stress relieving　低温(拉伸)消除应力
low-current plasma arc welding　小电流等离子弧焊
low-fume and low-toxic electrode　低烟尘低毒焊条
machined part　机加工件
macrofractography　断口宏观分析
macrograph　目视检查
magnetic flux　磁性焊剂
magnetic particle inspection　磁粉探伤
main arc　主弧
maintenance　维修
malleable (cast)iron　(可)锻(铸)铁
management review　管理评审
manganese　锰
manipulating　搬运，装卸
manual arc welder　手工电焊工
manual arc welding machine　手工弧焊机
manual electroslag welding　手工电渣焊
manual oxygen cutting　手工气割
manual welding　手工焊
mark　标记
marking　标记，编码，标志
martensite-austenite constituent　m-a 组元
martensitic　马氏体的
maximum hardness test　最高硬度试验
mean solidification rate　平均结晶率
mechanical stress relieving　机械消除应力
mechanical test　机械试验，力学试验
median shell　中间筒体
melting characteristics of an electrode　焊条熔化性能
metal active-gas arc welding　活性气体保护金属极电弧焊
metal arc welding　金属极电弧焊
metal inert-gas welding machine　金属极惰性气体保护弧焊机
metal spraying　金属喷涂
metallurgical structure　金相结构
metallurgical weldability　冶金焊接性
microalloying　微量合金化
micro-crack/micro-fissure　微裂纹
micrograph　显微检查
micro-plasma arc welding　微束等离子弧焊
micro-plasma welding equipment　微束等离子弧焊机
misalignment　错边
mismatching deformation　错边变形，失配变形
mixed gas arc welding　混合气体保护电弧焊
mixing chamber　混合室
mixing ratio　混合比
model for quality assurance　质量保证

模式
modules of water glass　水玻璃模数
mold for thermit weld　热剂铸模，热熔模铸
molten metal bath dip brazing　金属浴钎焊
molybdenum　钼
momentary cooling rate　瞬时冷却速度
Monel electrode　蒙乃尔焊条
multi-pass weld　多道焊
multiple wire submerged arc welding　多丝埋弧焊
Murex hot cracking test　缪雷克丝热裂纹试验
narrow gap welding　窄缝焊
narrow-gap electroslag welding　窄间隙电渣焊
narrow-gap submerged arc welding　窄间隙埋弧焊
NC cutting machine　数控切割机
neatness　清洁
neutral flame　中性焰
nickel　镍
nickel base alloy covered electrode　镍基合金焊条
nickel equivalent　镍当量
nitriding　氮化，渗氮（处理）
nitrogen　氮
nitrogen-arc welding　氮弧焊
nominal (rated) pressure　额定压力
non-conforming processing　不符合项处理
non-consumable electrode　不熔化电极
non-destructive examination　无损检验
nondestructive test　无损试验
non-transferred arc　非转移弧
non-vacuum electron beam welder　非真空电子束焊机
notch　缺口
notch bend test　缺口弯曲试验
notification　通知
nozzle　喷嘴
numerical-control cutting（NC）　数控切割
nuts　螺母
onsite weld　现场焊
open arc welding　明弧焊
open circuit voltage (no load voltage)　空载电压
operability　可使用性，可操作性
operating (service，working) pressure　工作压力
operating characteristics of an electrode　焊条使用性
orbital welding　轨道式焊
order sheet　订单
organization structure　组织结构
orifice diameter　节流孔板直径
orifice throat length　节流孔板孔长
orifice throat ratio　节流孔板节流比
oscillator　频率振荡器
outline drawing　轮廓图
overhaul　大修
overheated structure　过热组结构
overheated zone　过热区
overlap　焊瘤
oxide melting slag　氧化物型熔渣
oxidizing flame　氧化焰

oxy-acetylene flame 氧乙炔焰
oxy-acetylene torch 氧乙炔焊炬
oxy-acetylene welding 氧乙炔焊
oxy-arc cutting 氧气电弧切割
oxy-coal gas flame 氧煤气焰
oxygen 氧
oxygen cutting 氧气切割
oxygen evaporator 液氧气化器
oxygen lancing/oxygen lance cutting 氧气切割
oxygen plasma arc cutting 氧等离子切割
oxygen-free flux 无氧焊剂
oxy-hydrogen flame 氢氧焰
oxy-hydrogen welding 氢氧焊
packaging 包，打包
parent metal 母材
Paris formula 帕瑞斯(Paris)公式
partial melting region 部分熔化区
partial penetration weld 离子渗透焊
partial vacuum electron beam welder 低真空电子束焊机
pass sequence 焊道工序
passivation 钝化
path repeatability 轨迹可重复性
pattern recognition for welding 焊接图像识别
peak temperature 峰值温度
pearlitic heat resistant steel electrode 珠光体耐热钢焊条
penetration inspection 渗透探伤
percentage of moisture for covering 焊条药皮含水量
performance test 性能试验
permanent backing 保留垫板
permeability 渗透性
perveance 导流系数
phosphorus 磷
photo-electric tracing cutting 光电跟踪切割机
physical separation 实体隔离
pickling 酸洗，腐蚀
pillow test 枕形气密检验
pilot arc 维弧
pilot arc current 维弧电流
pinhole 针尖状气孔
pipe flame cutting machine 火焰切管机
pit 凹坑
plain glass 防护镜片
plane heat source 面热源
planeness of cutting surface 切割面平面度
plasma (welding) torch 等离子弧焊枪
plasma arc 等离子弧
plasma arc cutting 等离子弧切割
plasma arc powder surfacing 粉末等离子弧堆焊
plasma arc surfacing 等离子弧堆焊
plasma arc welding 等离子弧焊
plasma arc welding machine 等离子弧焊机
plasma gas/orifice gas 等离子气
plasma jet 等离子流
plasma MIG welding 等离子-混合气体保护焊
plasma spraying 等离子喷涂
plastic spraying 塑料喷涂

plating　电镀
plug weld　管塞焊
pneumatic test　气动试验
point heat source　点热源
polygonization boundary　多边化边界
polygonization crack　多边化裂纹
porosity　密集气孔，多孔性
porous slag　多孔焊渣
powder cutting　氧溶剂切割，粉末切割
powder dispenser　粉剂罐
powder lancing　溶剂氧切割
powder washing　氧熔剂表面修整
powdered filler metal　粉状钎料
power capacity　电厂发电能力
precedent desoxydation　先期脱氧
precipitation　沉淀
precipitation desoxydation　沉淀脱氧
prefabrication (piping)　预制（管道）
preformed filler metal　成形钎料
preheat flame　预热火焰
preheat oxygen　预热氧
premise　厂房
pressure regulator/gas regulator　减压器
pressure test　耐压检验，压力试验
pre-stressing　预应力
preventive action　预防措施
primary combustion　一次燃烧
primary solidification structure　一次结晶组织
probe　探头
procedure　程序
process　过程，工艺
process qualification　工艺评定
processed material　流程性材料
procurement　采购
product liability　产品责任
production permit　生产许可
program　程序
programme　大纲
project coordination　工程协调
provisional acceptance report　临时验收报告
ptp repeatability　点位可重复性
pulsed arc exciter　脉冲激弧器
pulsed arc stabilizer　脉冲稳弧器
pulsed arc starter　脉冲引弧器
pulsed arc welding machine　脉冲弧焊机
pulsed arc welding power source　脉冲弧焊电源
pulsed arc welding rectifier　脉冲弧焊整流器
pulsed electron beam welding　脉冲电子束焊
pulsed plasma arc welding　脉冲等离子弧焊
purchaser　买方
pure iron electrode　纯铁焊条
pure nickel cast iron electrode　纯镍铸铁焊条
purifier　净化器
qualification certificate　资格评定证书
qualification process　鉴定过程
qualified　鉴定合格
quality assurance program　质量保证大纲，质保大纲

quality audit　质量监查
quality audit observation　质量监查与观察结果
quality evaluation　质量评价
quality loop　质量环
quality of cut face　切割面质量
quality planning　质量策划
quality policy　质量方针，质量政策
quality system　质量体系
quality-related costs　质量相关成本
quartz powder cutting　氧-石英砂切割
quasi-cleavage fracture　准解理断口
quasi-stationary temperature field　准稳定温度场
quench　淬火
radical zone　放射区
radiographic inspection/ radiography　射线探伤
radiographic test　射线试验
range of welding current regulation　焊接电流调节范围
rated duty cycle/standard service　额定负载持续率
rated welding current　额定焊接电流
reaction flux　反应钎剂
reconditioning　修理，修补
recurrence　再发生
reference　编号
reheat crack　再热裂纹
reinforcement　加强
rejection　报废，拒收
relative humidity　相对湿度
release　交付，产品的放行
repair　返修，修补
requirement society　要求
residual hydrogen　残余氢
residual stress analysis　残余应力测定
resistance brazing　电阻钎焊
restraint coefficient　约束系数
restraint intensity　约束强度
rework　返工
rigging　搬运，装卸
rigid restraint cracking test　刚性约束裂纹试验（RRC 试验）
ripple　结晶层状线
rising characteristic arc welding power source　具有上升特性的弧焊电源
Robertson crack arrest test　罗伯逊止裂试验
robot working space　机器人工作空间
rolling　辊轧
root　根部（焊接）
root bend test　根部弯试验
root concavity　根部凹陷
root crack　根部裂纹
rosin flux　松香钎剂
rotating arc welding　旋转电弧焊
rough machining　粗加工
roughness　粗糙度
round bar cracking test　圆棒裂纹试验
running characteristics of an electrode　焊条使用性
run-off tab　引出板
run-on tab　引弧板
safety　安全性
safety instruction　安全细则
salt bath dip brazing (soldering)　盐浴钎焊

salt melting slag 盐型熔渣
salt-oxide melting slag 盐-氧化物型熔渣
sand blast 喷砂
sandwich filler metal 夹层状钎料
scarfing 火焰表面清理
Schaeffler's diagram 舍夫勒组织图
SCR arc welding rectifier 晶闸管弧焊整流器
seal weld 密封焊
sealant 封孔剂
sealing 封孔处理
seamless part 无缝件
seawater corrosion resistant steel electrode 耐海水腐蚀钢焊条
secondary combustion 二次燃烧
secondary solidification structure 二次结晶组织
segmented circular groove cracking test 分块圆周形槽热裂纹试验
segregation 偏析，熔析（冶金）
self-adjusting arc welding 自调节电弧焊
self-fluxing alloy powder 自熔剂合金粉末
self-fluxing brazing alloy/self-fluxing filler metal 自钎剂钎料
self-inspection 自检
self-releasing slag 自动脱落焊渣
semi-automatic welding 半自动焊
semi-volatile covering 气渣联合保护型药皮
series submerged arc welding 横列多丝埋弧焊
service delivery 服务提供
service weldability 使用可焊性
set pressure 整定压力
shape cutting 仿形切割
shear lip zone 剪切唇区
shearing strength 剪切强度
shielded gas plasma arc cutting 保护气体等离子弧切割
shielded metal arc welding (SMAW) 保护电弧焊
shipment/shipping/dispatch 装货，装运
short circuiting arc welding 短路过渡电弧焊
short slag 短渣
shot-blasting 喷丸处理
shoulder meltability/ melting degree of shoulder 上缘熔化度
shrinkage cavity （铸件的）缩孔
side bend test 侧弯试验
silicon 硅
silicon arc welding rectifier 硅弧焊整流器
single pass weld 单道焊
single port nozzle 单孔喷嘴
single stage regulator 单级调节器
single-welded joint 单缝接头
sintered flux 烧结焊剂
sketch 草图
slag bath 渣池
slag detachability 脱渣性
slag fluidity 熔渣流动性
slag forming constituents 造渣剂
slag inclusion 夹渣

slit type cracking test 狭长形坡口裂纹试验
slope-controlled arc welding power source 多特性弧焊电源，线性控制弧焊电源
slot weld 槽焊
smoothing 修平
socket joint 套管接头
socket weld 承插焊接
solder 软钎料
solderer 焊工
solder cream 钎料膏
solder paste 钎料膏
solder preform 钎料预成形
solder-ability 钎焊性
solder iron 烙铁
soldered joint 钎焊接头
soldering 软钎焊
soldering flux 钎剂，焊剂
soldering seam 钎缝
solid laser 固体激光器
solid wire 实心焊丝
solidification crack 凝固裂纹
solidification mode in weld-bead 焊缝结晶形态
solidified slag 固熔渣
solution heat treatment 固熔热处理
source verification activities 原产地验证活动
spare part 备件
spent fuel 乏燃料
spigot joint 扩口套管接头
spiking 钉尖
spiral welding 螺旋焊
spot welding 点焊
spot welding robot 点焊机器人
spray rate 喷涂率
spray torch 喷炬（枪）
spray-fused coating 喷涂层
spray-fusing 喷熔
spraying material 喷涂材料
spread factor/coefficient of spreading 铺展系数
spread-ability 延展性
spread-ability test 延展性试验
square cut 垂直切割
squeezing test 压扁试验
stack cutting 叠板切割
stainless steel electrode 不锈钢焊条
stakeholder 受益者，保管者
stamping 冲压
stamping part 冲制件
starting weld tab 引弧板
steel plate/steel sheets 薄钢板
steel plate pretreatment 钢板预处理
steel shoe 钢挡板
step brazing/ step soldering 分级钎焊
stiffener 加强肋
stopping-off agent 阻流剂
straight beam method 直射法超声波探伤
straight weld 直焊
straighten 矫正
strain rate 应变速率
stranded electrode 绞合焊条
strength of coating 涂层强度
strength weld 强度焊
stress corrosion cracking 应力腐蚀开

裂
stress intensity factor 应力强度因子
stress redistribution 应力重分布
stress relaxation 应力松弛
stress relief 消除应力
stress relief annealing crack (SR crack) 消除应力退火处理裂纹
stress relieving by annealing 退火消除应力
stria depth 割纹深度
striking end (of an electrode) 焊条引弧端
strong-back 定位板
structural weld 结构焊
subcontractor 分包商
submerged arc welding (SAW) 埋弧焊
submerged arc welding machine 埋弧焊机
submittal 提交
suborder 分订单
sub-stitutionary oxidation 置换氧化
sulphur (sulfur) 硫
support shirt 支撑套
surface crack 表面裂纹
surface examination 表面检查
surface roughening 表面粗糙化处理
surface roughness 表面粗糙度
surfacing by fire cracker welding 躺板极堆焊
surfacing electrode 堆焊焊条
surfacing with band-electrode 带极堆焊
surveillance plan 监督计划
suspension of work 停工
synthetic heat-affected zone test 综合热影响区模拟试验
TEE type cracking test T形裂纹试验
tack weld 点焊
tandem sequence 纵列多丝埋弧焊
taper 开坡口
tearing strength 撕裂强度
technicality of the electrode 焊条工艺性
technological coating 工艺镀层
temper 回火
temporary backing 临时垫板
tender 投标
tensile restraint cracking test 拉伸约束裂纹试验(TRC试验)
tensile strength 抗拉强度
test piece 试件
test program 试验大纲
test report 试验报告
test specimen 试验样品
test weld 试验焊
thermal cutting 热切割
thermal efficiency 热效率
thermal spraying 热喷涂
thermal stability of flux 钎剂热稳定性
thermal stress 热应力
thermal weldability 热焊接性
thermit crucible 热剂坩埚，热熔坩埚
thermit mold 热剂铸模，热熔模铸
thermit powder 热剂，热熔粉剂
thermit rail welding 钢轨热剂焊，热熔轨道焊
thermit reaction 热剂反应

thermit repair welding　热剂补焊
thermit slag　热剂溶渣
thermit steel　热剂钢水
threaded fitting　螺纹接头
three phase arc welding　三相电弧焊
tidiness　清洁
titanium　钛
toe crack　焊趾裂纹
top shell　上部筒体
torch　焊炬
torch brazing/ torch soldering　火焰钎焊
torsional strength　抗扭强度
total investment　总投资额
total quality management　全面质量管理
toughness　韧性，韧度，抗断强度
tracer　跟踪装置
trainer of synthetic weld　焊工综合培训器
transcrystalline crack　穿晶裂纹
transferred arc　转移弧
transgranular fracture　穿晶断口
transient welding stress　焊接瞬时应力
transistor arc welding power source　晶体管弧焊电源
transition zone　过渡区
transverse crack　横向裂纹
transverse submerged arc welding　横列双丝并联埋弧焊
transverse weld/cross-weld　横向焊
traveler　行走机构
trimming　去毛刺
triode gun　三极电子枪
tube expansion　胀管
tube plate　管板
tubular　管
tungsten carbide surfacing electrode　碳化钨堆焊焊条
tungsten electrode　钨电极
tungsten inclusion　夹钨
tungsten inert-gas welding machine　钨极惰性气体保护弧焊机
twin electrode　双芯焊条
two stage regulator　两级调压器
ultimate bending strength　抗弯强度极限
ultimate compressive strength　抗压强度极限
ultimate tensile strength　抗拉强度极限
ultrasonic inspection　超声波探伤
ultrasonic soldering　超声波钎焊
ultrasonic test　超声试验
under coating　结合层
underbead crack　焊道下裂纹
undercut　咬边
underwater cutting　水下切割
under-water cutting blowpipe　水下割炬
underwater gas shielded arc welding　水下气体保护电弧焊
underwater plasma arc welding　水下等离子弧焊
underwater welding　水下焊
unmixed zone　未混合区
uphill welding　仰焊

upper support plate 上部支撑板
usability of the electrode 焊条工艺性
U-tension test U形拉伸试验
vacuum 真空度
vacuum brazing 真空钎焊
validation 确认
vanadium 钒
vapor phase soldering 蒸汽钎焊
verification 验证
vertical down welding 垂直下焊
vertical position welding 垂直位置焊
vibratory arc surfacing 振动电弧堆焊
viscous slag 黏性熔渣
visual examination 外观检查
vitreous slag 玻璃状焊渣
voltage of slag bath 渣池电压
waiver 让步，特许
washers 垫圈
water curtain type dry underwater welding 水帘局部干式水下焊
water glass 水玻璃
water jet method underwater arc cutting 喷水式水下电弧切割
water vapor arc welding 水蒸气保护电弧焊
water-closing type arrestor 水封式回火防止器
wave soldering 波峰钎焊
wave soldering machine 波峰钎焊机
wave-like deformation 波浪变形
weaving speed of wire 焊丝横摆频率
weld assembly 焊接部件
weld crack 焊接裂纹
weld decay 敏化区腐蚀
weld defect 焊接缺陷
weld gauge 焊缝检测规
weld inspector 焊接检验员
weld junction(英)/bond line（美） 熔合线
weld metal crack 焊缝裂纹
weld seam depth-to-width ratio 焊缝深宽比
weld temperature field 焊接温度场
weld thermal cycle 焊接热循环
weld zone 焊接区
weldability 焊接性
welded construction 焊接结构
welded joint 焊缝
welded part 焊件
welded structure 焊接结构
welder 焊工，焊机
welder qualification 焊工资格评定
welder qualification test 焊工考试
welder qualified certification 焊工合格证
welders training 焊工培训
welding backing 焊接衬垫
welding booth 焊接工作间
welding circuit 焊接回路
welding consumables 焊接消耗材料
welding deformation 焊接变形
welding equipment 焊接设备
welding expert system 焊接专家系统
welding flame 焊接火焰
welding flux 焊剂
welding glass 滤光镜片
welding goggles 气焊眼镜
welding gun 焊枪

welding head 焊接机头
welding heat source 焊接热源
welding image processing 焊接图像处理
welding imperfection 焊接欠缺
welding machine 焊机
welding mechanics 焊接力学
welding nozzle 焊嘴
welding operator qualification 焊接操作员资格评定
welding operator 焊接操作工
welding parameter recorder 焊接参数记录仪
welding position 焊接工位
welding power source 焊接电源
welding procedure qualification 焊接工艺评定
welding residual deformation 焊接残余变形
welding residual stress 焊接残余应力
welding robot 焊接机器人
welding robot play back 焊接机器人示数
welding robot station 焊接机器人工作站
welding sequence 焊接工序
welding set 焊接设备
welding shop 焊接车间
welding station 焊接工位
welding stress 焊接应力
welding tip 焊嘴
welding tractor 焊车
welding wire/welding rod 焊丝
weldment 焊接，焊件
Well's wide plate test 韦尔斯宽板拉伸试验
wet method underwater welding 湿法水下焊
wettability 润湿性
wetting angle 润湿角
wetting balance test 润湿称量试验
whole lot 整批货
wire feeder 送丝机构
wire reel 焊丝盘
work coating 工作包覆层
work distance 工作距离
wormhole 条虫状气孔
wrapper 封套
wrought 熟铁
xenon 氙
X-ray industrial television inspection X射线工业电视探伤
X-ray radiographic inspection X射线探伤
X-ray stress analysis X射线应力分析
z-direction window type restraint cracking test z向窗口约束裂纹试验
zinc 锌
zirconium 锆

Ⅳ. 核电培训常用词汇与术语

American Society for Training and Development(ASTD) 美国训练发展协会
annual training plan 年度培训计划
assessment 考核
assessor 考核人员
attendance 考勤，出席，到场
attitude 态度，行为
authorization 授权
auxiliary operator (AO) 辅助操纵员
basic safety authorization training (BSA) 基本安全授权培训课程
certification 毕业
classroom 教室
classroom training 课堂培训
competence 工作能力
computer-based learning(CBL) 计算机化学习
corporate learning strategy 企业的学习策略
course 课程
courseware 教材
creativity 创新性
curriculum 课程
distance learning 远距/程学习
educational administration 教务处
educational improvements 教学改进
e-learning 电子化学习
elite course 精品课程
enabling objective 分解目标
entry level requirement 入学水平要求
evaluation 评价
evaluation tool 评价工具
excitement 趣味性
external 外部
faculty 师资
feedback 反馈
finance 财务
full scope simulator (FSS) 全范围模拟机
go out training 外出培训
individual training plan(ITP) 个人培训计划
industry safety(IS) 工业安全
information management system 信息管理系统
instructor 教员
instructor lesson guide 教案
interactivity 互动性
internal 内部
internal training 内部培训
job 岗位工作
job analysis 岗位工作分析
job and task analysis(JTA) 岗位工作和任务分析
job competencies analysis(JCA) 岗位全面工作能力分析
job shadowing training 岗位影子培训
know-how training(KH) 专业技能培训课程
knowledge, skill and attitude(KSA) 知识、技能和态度

lesson plan　课程计划
licensing　颁发执照
management　管理部门
management training(MTT)　管理培训课程
mock-up　模拟体
module plan　单元计划
monthly training plan　月度培训计划
motivating mechanism　激励机制
nuclear safety(NS)　核安全
online learning　在线学习
on-the-job training(OJT)　在岗培训
operating organization　营运部门，运营组织
operating personnel　运行人员
operation　运行
operational experience　运行经验
operational safety assessment review team(OSART)　运行安全评审团
other specific training(OST)　其他专项培训课程
part-time instructor　兼职教员
performance　业绩
performance assessment　业绩评价
plant department manager　核电厂部门经理
plant manager　核电厂经理
plant training manager　培训经理
procedure　程序
qualification　资格，资格评审
qualification program　人员资格评审培训大纲
reactor kinetics　反应堆动态学
reactor operator (RO)　反应堆操纵员
reactor power burst　反应堆功率突增
reactor power excursion　反应堆功率偏差
real practice　实践性
rewetting temperature　再湿温度
safety culture　安全文化
SAT analysis phase　SAT 的分析阶段
SAT design phase　SAT 的设计阶段
SAT development phase　SAT 的编制阶段
SAT evaluation phase　SAT 的评价阶段
SAT implementation phase　SAT 的实施阶段
schedule　进度
senior reactor operator(SRO)　高级操纵员
simulation　模拟
simulator training　模拟机培训
skill　技能
skill lab training　技能实验室培训
special topics　专题
specific training　专项培训
specific training plan　专项培训计划
subject matter expert(SME)　专业工程师
survey table　调查表
systematic approach to training(SAT)　系统化培训方法
task　工作任务
task analysis　工作任务分析
task element　工作任务单元
teaching　教学
technical discipline　技术训练

terminal objective　最终目标
test　考试
trainee　学员
trainer　教员
training engineer　培训工程师
training implementation approach　培训实施方法
training information management system(TIMS)　培训信息管理系统
training inspection engineer　培训监督工程师
training management　培训管理
training material　培训材料
training module　培训单元
training needs analysis　培训需求分析
training objective　培训目的
training plan　培训计划
training policy　培训政策
training principle　培训原则
training program　培训大纲
training record　培训记录
training regulation　培训制度
training responsibility　培训责任
training scope　培训范围
training setting　培训方式
training system　培训体系
train-the-trainer　卓越内部培训师培训
validation　培训效能检查
video teleconference(VTC)　视频会议
web-based learning(WBL)　网络化学习
workplace　工作场所
written examination　书面考试

Ⅴ. AP1000 核电厂设备名称

英文名称	中文名称
120/12 VAC hydrogen igniters transformers	120/12 伏交流氢点火器变压器
accumulator tank A	安全注射箱 A
accumulator tank B	安全注射箱 B
ancillary diesel generator packages	附属柴油机包
annex BLDG MCCs	附属厂房电机控制中心
annex building PWR XFMRs	附属厂房电力变压器
ASME safety class 1,2 & 3 valve	ASME1、2、3 级阀门
AUX BLDG MCCs	辅助厂房电机控制中心
AUX building power transformers	辅助厂房电力变压器
auxiliary building lighting transformers	辅助厂房照明变压器
batteries	蓄电池
battery chargers	蓄电池充电器
battery monitors	蓄电池监测器
building lighting fixtures	厂房照明装置
burnable poison rod assembly handling tool	可燃毒物棒组件操作工具
cable tray	电缆桥架,电缆托架
cask wash down pit level instrument	运输容器冲洗池液位仪表
cathodic protection	阴极保护
cold leg insulation package	冷段保温包
communication/CCTV	通信/闭路电视
compressor	压缩机
cont. recirc. isolation valve	连续循环隔离阀

英文名称	中文名称
containment air baffle	安全壳空气导流板
containment H_2 concentration instrument	安全壳氢气浓度仪表
containment lighting XFMRs	安全壳照明变压器
containment MCCs	安全壳电机控制中心
containment polar crane	安全壳环吊
containment power transformers	安全壳电力变压器
containment pressure instruments	安全壳压力仪表
containment spray ring headers	安全壳喷琳环集管
containment vessel	安全壳容器
control rod drive shaft handling tool	控制棒驱动轴装卸工具
control rod drive unlatch tool	控制棒驱动拔闩工具
core makeup tank A	堆芯补水箱 A
core makeup tank B	堆芯补水箱 B
crane/elevator/lift	吊车/电梯/起吊设备
CRDM external cables	控制棒驱动机构外部电缆
CRDM internal cables	控制棒驱动机构内部电缆
CRDMs	控制棒驱动机构
damper actuators	挡板控制器
damper	风阀,阻尼器,节气闸
DAS processor cabinet	DAS 处理器机柜
DAS squib valve controller cabinet	数据采集系统爆破阀控制器机柜
DDS data highway	设计开发系统数据总线
DDS processor cabinets	设计开发系统处理器机柜
dedicated safety panel ＃1	1 号专用安全控制盘
dedicated safety panel ＃2	2 号专用安全控制盘

英文名称	中文名称
diesel GEN BLDG lighting XFMRs	柴油发电机厂房照明变压器
diesel GEN BLDG MCCs	柴油发电机厂房电机控制中心
diesel generator building power transformers	柴油发电机厂房电力变压器
distributed processing cabinet	分布式处理机柜
distribution panel	配电盘
diverse actuation system panel	多样化驱动系统控制盘
doors and hatche	门及闸门
ECS switchgear buses	发电机控制系统开关母线
electrical penetration	电气贯穿件
electronic cabinets	电子机柜
equipment module	设备模块
excore detector	堆外探测器
facilities specialties	设施专用产品
fans	风扇
FHS fixed underwater light	燃料操作系统固定水下照明灯
FHS portable underwater light	燃料操作系统移动式水下照明灯
field mounted instrument	现场安装仪表
fire control cabinet	消防控制机柜
floor module	地板模块
floor seals	地板密封
forklift	叉式升降机,铲车
fourth stage ADS valve	第 4 级自动泄压系统阀
fuel handling machine	燃料装卸机
fuel transfer conveyor	燃料转移运输机
fuel transfer tube	燃料运输管

英文名称	中文名称
ganged support hangers	成组的支吊架
heat exchangers	热交换器
heater	加热器
hot leg insulation package	热段保温包
IIS guide tubes	堆内测量系统导管
IIS processing cabinets	堆内测量系统处理机柜
in core instrument	堆内测量仪表
in core thimble assemblies	堆芯内套管组件
intermediate range neutron flux preamplifier panel	中间量程中子注量率前置放大器配电板
inverter	逆变器
irradiated sample handling tool	辐照样本装卸工具
IRWST injection isolation valve	安全壳内置换料水箱注射隔离阀
lighting panel	照明配电盘
load cell (IC) with linkage	相互联系的负荷单元
load test fixture inside containment	安全壳内负荷试验装置
load test fixture outside containment	安全壳外负荷试验装置
local panel	就地控制盘
low voltage distribution panel	低压配电盘
low voltage load centers	低压负荷中心
mach shop DIST XFMR	机械车间配电变压器
machine tool	机械工具
main feedwater check valve	主给水逆止阀
main feedwater control valve	主给水控制阀
main feedwater isolation valve	主给水隔离阀

英文名称	中文名称
main steam line isolation valve	主蒸汽管道隔离阀
MCR RO WS A desk	主控制室反应堆操纵员工作站 A 桌
MCR RO WS B desk	主控制室反应堆操纵员工作站 B 桌
MCR SRO/SHIFT foreman desk	主控制室反应堆高级操纵员/值长办公桌
MCR wall panel displays	主控制室墙面大屏幕
MCR/RSW transfer switch panel	主控制室/远程停堆工作站转换开关盘
MIM reactor coolant pump instrument	金属碰撞检测设置在反应堆冷却剂泵的仪表
MIM reactor vessel upper head instrument	金属碰撞检测设置在反应堆压力容器上封头的仪表
MIM steam generator instrument	金属碰撞检测设置在蒸汽发生器的仪表
misc AHUs	其他空气操作单元
misc. I&C orifice plate	其他仪控节流孔板
motor control center	电机控制中心
mounting plate	安装盘
new fuel assembly handling tool	新燃料组件装卸工具
new fuel ELEV & hoist	新燃料电梯/吊车
new fuel JIB crane	新燃料悬臂吊车
new RCC handling tool	新控制棒束装卸工具
on-site diesel generator A packages	厂内柴油发电机 A 包

英文名称	中文名称
packages-radiation monitoring	包—放射性监测
passive RHR heat exchanger	非能动余热排出热交换器
PCCWST water level instrument	非能动设备冷却水贮存箱水位仪表
PCS flow instrument	主冷却剂系统流量仪表
pipe hanger	管道支吊架
piping assembly	管道组件
plumbing material and equipment	铅管材料及设备
PMS cabinets	保护与安全监督系统机柜
PMS QDPS MCR qualified display units	保护与安全监督系统数据处理系统主控制室经鉴定的显示器装置
power operated relief valve	电动泄压阀,动力释放阀
power range high voltage distribution boxes	功率量程高压配电箱
pressurizer	稳压器
pressurizer heater controller	稳压器加热器控制器
pressurizer insulation	稳压器保温
pressurizer safety valve	稳压器安全阀
PRHR HX flow	非能动余热排出热交换器流量
printer station	打印机站
pump	泵
PZR HTRS control GRP MCCs	稳压器加热器控制 GRP MCCs
QDPS thermocouple reference junction panel	合格 1E 级数据处理系统热电偶参比端盘
radwaste BLDG MCC	放射性废物厂房电机控制中心
radwaste building power transformer	放射性废物厂房电力变压器

英文名称	中文名称
RCP bearing water temperature instrument	反应堆冷却剂泵轴承水温计
RCP pump speed instrument	反应堆冷却剂泵泵速仪表
RCP stator temperature instrument	反应堆冷却剂泵定子温度计
RCP switchgear	反应堆冷却剂泵开关装置
RCP variable speed drive units	反应堆冷却剂泵变速驱动装置
RCP vibration instrument	反应堆冷却剂泵振动测量仪表
reactor coolant loop piping	反应堆冷却剂回路管道
reactor coolant pump insulation package	反应堆冷却剂泵保温包
reactor integrated head package	反应堆一体化顶盖包
reactor lower internals	反应堆下部堆内构件
reactor trip switchgear	反应堆停堆开关装置
reactor upper internals	反应堆上部堆内构件
reactor vessel	反应堆压力容器
reactor vessel reflective insulation (all other)	反应堆压力容器反射绝热层(其他部分)
reactor vessel reflective insulation in reactor cavity (IVR)	反应堆腔内(用于堆芯熔融物压力容器内滞留)的反应堆压力容器反射绝热层
reactor vessel stud tensioner system	反应堆压力容器螺栓拉伸器系统
refueling machine	换料机
regulating transformers	调节变压器
residual heat removal pump A	余热排出泵 A
residual heat removal pump B	余热排出泵 B
rod control cabinet	控制棒控制柜,棒控机柜
rod control cluster handling tool	棒控束装卸工具

英文名称	中文名称
rod drive M/G set	控制棒驱动电动发电机组
rod drive M-G control cabinet	控制棒驱动电动发电机组控制机柜
rod position indication cabinet	棒位指示机柜
rod position indicators	棒位指示器
room module	房间模块
RSR WS desk	远程停堆工作站办公桌
RX coolant pump	RX 冷却剂泵
RXS non-threaded fasteners	RXS 非螺纹紧固件
RXS threaded structural fasteners	RXS 螺纹结构紧固件
service modules	服务模块
SG main steam line flow instrument	蒸汽发生器主蒸汽管线流量仪表
SG feedwater flow instrument	蒸汽发生器给水流量仪表
SG startup feedwater flow instrument	蒸汽发生器启动给水流量仪表
source range neutron flux preamplifier panel	源量程中子注量率前置放大器配电板
sparger	喷淋器,喷头
special monitoring system cabinet	专用监测系统机柜
spent fuel assembly handling tool	乏燃料组件装卸工具
spent fuel pool level instrument	乏燃料池液位仪表
spent fuel shipping cask crane	乏燃料装运容器吊车
squib valve controller cabinets	爆破阀控制器机柜
stairs	楼梯
startup feedwater control valve	启动给水控制阀
steam generator insulation package	蒸汽发生器保温包

英文名称	中文名称
steam generator	蒸汽发生器
storage rack	储存格架
structural module	结构模块
supports & hangers	支吊架
surge line insulation	波动管保温
switchboard	配电盘
switchbox	配电箱
system packages	系统包
tank	水箱
tie bus between ECS load centers	发动机控制系统负荷中心间的连接母线
trays	托架
turbine bypass control valve	汽轮机旁路控制阀
valve test panel	阀门试验盘
valve	阀门
VES MCR air delivery line flowrate instrument	VES 主控制室空气输送管路流量仪表
VES MCR DP sensing instrument	VES 主控制室 DP 感应仪表
vessel	容器
wall panel	墙板,墙屏
wire/cables	电缆/电线
workstations and processors	工作站和处理器

第二部分　术语缩写

Ⅰ. AP1000 核电厂常用缩写

英文简称	英文全称	中文名称
ABT	Automatic Bus Transfer	自动总线转换(自动母线切换)
AC	Alternating Current	交流电
AC	Acceptance Criteria	验收准则
ACC	Accumulator	安(全)注(射)箱,蓄压箱
ACL	Access Control List	通行控制清单
ACR	Automatic Cable Routing	自动路由器(自动电缆路由)
ADL	Applicable Document List	使用文件清单
ADSP	Automatic Depressurization System Sparger	自动卸压系统喷头
A-E	Architect-Engineer	建筑工程师
AESOP	Atomic Energy Simulation of Optimization	最优化原子能模拟机
ALARA	As-Low-As-Reasonably Achievable	合理可行尽量低
ALWR	Advanced Light Water Reactor	先进轻水反应堆
ANNEX	Annex Building	附属厂房
AOD	Argon-Oxygen Decarburization	氩氧脱碳装置
AOO	Anticipated Operational Occurrences	预计运行事件

英文简称	英文全称	中文名称
APY	Applicability Document	适用性文件
APYA	APY Administrator	APY 管理员
ASL	Authorized Supplier List	授权供货商清单
ATM	Atmosphere	大气压
ATP	Authorization to Proceed	批准开工
ATWS	Anticipated Transient Without Scram	未能紧急停堆的预期瞬态
ATWT	Anticipated Transient Without Trip	未能紧急停堆的预期瞬态
AUX	Auxiliary Building	辅助厂房
BA	Burnable Absorber	可燃吸收剂,可燃毒物
BEA	Bid, Evaluate, Award	投标,评标,中标
BEACON	Best Estimate Analyzer for Core Operations - Nuclear	堆芯核操作最佳估计分析器
BFS	Build For Shipment	装运厂房
BIS	Bidder Invitation Specification	招标说明书,邀标书
BLC	Boiler Level Control	锅炉水位控制
BLW	Boiling Light Water	沸腾轻水
BNI	Balance of Nuclear Island	核岛辅助设施
BOL	Beginning of Life	寿期初
BOM	Bill of Material	材料清单
BOP	Balance of Plant	核电厂其余部分(除核岛和常规岛外)
BPC	Boron Precipitation Control	硼的沉淀控制
BSEMT	Basement	地基,基础

英文简称	英文全称	中文名称
BTP	Branch Technical Position	部门技术岗位
BTU	British Thermal Unit	英制热单位
BWR	Boiling Water Reactor	沸水反应堆
C	Centigrade	摄氏度
CA	Containment	安全壳
CAD	Computer Aided Design	计算机辅助设计
CAE	Computer Aided Engineering	计算机辅助工程
CAGS	Compressed Air and Gas System	压缩空气和气体系统
CALC	Calculation	计算,统计
CAR	Corrective Action Requests	纠正(措施)行动要求
CAV	Cumulative Absolute Velocity	累积绝对速度
CB	Concentration of Boron	硼浓度
CBS	Cost Breakdown Structure	成本分类
CCB	Configuration Control Board	控制盘布置
CCP	Critical Channel Power	临界通道功率
CCWS	Component Cooling Water System	设备冷却水系统
CD	Core Design	堆芯设计
CDE	Common Desktop Environment	公共桌面环境
CDM	Certified Design Material	规范认可的设计材料
CDP	Common Design Principles	通用设计准则
CDWS	Chilled Water System	冷冻水系统
CEDE	Committed Effective Dose Equivalent	约定有效剂量当量
CEIDS	Cable, Equipment, Instrument Database System	电缆、设备、仪器数据库系统
CEO	Cognizant Engineering Organization	认知工程组织(被认可的工程机构)

英文简称	英文全称	中文名称
CERT	Constant Extension Rate Test	持久延伸率试验
CG	Critical Group	关键组
CHF	Critical Heat Flux	临界热流密度,临界热通量
CI	Conventional Island	常规岛
CIM	Component Interface Module	部件接口模块
CIPIMS	Construction Inspection Program Information Management System	施工监督大纲信息管理系统
CIV	Combined Intercept Valve (turbine)	混合式截止阀(汽轮机)
CIV	Containment Isolation Valve	安全壳隔离阀
CL	Column	柱子
CL	Cold Leg	冷段
CM	Construction Manager	施工经理
CMF	Common Mode Failure	共模故障
CMS	Configuration Management System	结构管理系统
CMT	Core Makeup Tank	堆芯补水箱
CMTR	Certified Material Test Report	材料鉴定试验报告
COB	Close of Business	交易结束
COD	Commercial Operation Date	商业运行日期
CofC	Certificate of Compliance	一致性证书
COL	Combined Operating License	联合运行许可证,一步法审批许可证
COLA	Combined Construction Permit-Operating License Application	建造、运行联合许可执照申请
COLR	Core Operating Limits Report	堆芯运行限制报告
CORS	Catalytic Oxidation Reduction System	催化降低氧化系统

英文简称	英文全称	中文名称
CPM	Counts per Minute	每分钟的计数
CPM	Critical Path Method	临界通道法
CPM	Customer Project Manager	客户项目经理
CPR	Critical Power Ratio	临界功率比
CPS	Computerized Procedure System	计算机程序处理系统
CPS	Counts per Second	每秒钟的计数
CRD	Control Rod Drive	控制棒驱动
CRDM	Control Rod Drive Mechanism	控制棒驱动机构
CRT	Cathode Ray Tube	阴极射线管
CSA	Construction Support and Advisors	工程支持顾问
CSA	Control Support Area	控制支持区域
CSDV	Condenser Steam Discharge Valve	凝汽器蒸汽排放阀
CSF	Critical Safety Function	临界安全功能
CST	Condensate Storage Tank	凝结水储存箱
CT	Construction Testing	工程测试
CTS	Correspondence Tracking System	信息跟踪系统
CV	Containment Vessel	安全壳
DAC	Design Acceptance Criteria	设计验收准则
DACR	Days Away Case Rate	日患病率
DBA	Design Basis Accident	设计基准事故
DBE	Design Basis Event	设计基准事件
DBT	Design Base Threat	设计基准威胁
DC	Design Certification	设计取证
DC	Design Commitment	设计承诺
DC	Direct Current (electrical power and signals)	直流(电力与信号)

英文简称	英文全称	中文名称
DCB	Design Certification Basis	设计许可证基准
DCC	Design Control Computers	设计控制计算机
DCD	Design Certification Departure	设计许可证偏差
DCD	Design Control Document	设计控制文件
DCP	Design Control Proposal	设计变更建议
DCPA	DCP Administrator	设计变更建议管理人
DCPI	DCP Initiator	设计变更建议提出方
DCR	Design Certification Rule	设计许可证规则
DCR	Design Change Review	设计变更审查
DCR	Deviation from Contract Requirements	偏离合同要求
DCS	Distributed Control System	分布式控制系统
DDE	Deep Dose Equivalent/External Dose Whole Body Exposure	深部剂量当量/全身外照射剂量
DEGB	Double Ended Guillotine Break	双端剪切断裂
D-EHC	Digital Electrohydraulic Control	数字化电动液压控制
DF-ET	Drift Flux Equal Temperature	等温漂移流量
DF-UT	Drift Flux Unequal Temperature	漂移流非等温
DG	Diesel Generator	柴油发电机
DI	Design Institutes	设计院
DID	Defense in Depth	纵深防御
DID	Detail Interface Document	详细接口文件
DMD	Document Management Directory	文件管理手册
DOR	Division of Responsibilities	责任划分
DP	Design Participation	设计参与

英文简称	英文全称	中文名称
DP	Differential Pressure	压差
DPU	Distributed Processing Unit	分布式处理单元
D-RAP	Design Reliability Assurance Program	设计可靠性保证大纲
DRPI	Digital Rod Position Indication	数字棒位指示
DRWM	Dynamic Rod Worth Measurement	动态棒价值测量
DSF	Document Submittal Form	文件提交格式
DSN	Design	设计
DTS	DCP Tracking System	设计变更提议跟踪系统
DVI	Direct Vessel Injection	压力容器直接注入
DWGS	Drawings	图纸
DXS	Decontamination System	去污系统
E&DCR	Engineering & Design Charge Request	工程设计收费请求
EAAMP	Equipment Aging Assessment and Management Program	设备老化评估与管理大纲
EBP	Evaluated Bid Price	评标价
ECCS	Emergency Core Cooling System	紧急堆芯冷却系统
ECD	NI Contract Effective Date	核岛合同生效日
ECT	Eddy Current Testing	涡流探伤检验
EDB	Engineering Data Base	工程数据库
EDMS	Electronic Data Management System	电子数据管理系统
EDS	Enterprise Data Servers	厂用数据服务器
EDS	Engineering Documents Standardization	工程文件标准

英文简称	英文全称	中文名称
EFPD	Effective Full Power Days	等效满功率天数
EFW	Emergency Feed Water	应急给水
EHC	Electric Heating Coil	电加热线圈
EHC	Special Process Heat Tracing System	工艺热跟踪专用系统
EIB	Enforcement and Inspection Bulletin	强制执行与检查通告
EIS	Environmental Impact Statement	环境影响报告
EL	Elevation	标高
ELEC	Electrical	电的
EMI	Electromagnetic Interference	电磁干扰
ENSA	Equipos Nucleares, S. A.	核设备有限公司
ENUSA	Empresa Nacional del Uranio, S. A.	国家铀资源有限公司
EOC	Effectiveness of Contract	合同生效
EOF	Emergency Offsite Facility	场外应急设施
EOMR	End of Manufacturing Report	制造完工报告
EPABX	Emergency Procedure Automatic Branch Exchange	应急规程自动交换处
EPC	Engineering, Procurement, Construction	设计,采购,建造
EPR	European Power Reactor	欧洲动力反应堆
EPS	Enterprise Project Structure	企业项目框架
EQ	Equipment Qualification	设备鉴定
EQTR	Equipment Qualification Test Report	设备鉴定试验报告

英文简称	英文全称	中文名称
ER	Environmental Report	环境报告
ERF	Emergency Response Facility	应急响应装置
ERG	Emergency Response Guidelines	应急响应指南
ESBWR	Economic and Simplified Boiling Water Reactor	经济简化型沸水反应堆
ESD	Electrostatic Discharge	静电释放
ESF	Emergency Safety Features	应急安全设施
ESF	Engineered Safety Features	专设安全设施
ESFAS	Engineered Safety Features Actuation System	专设安全设施触发系统
ESP	Early Site Permit	早期现场许可证
ESR	Electroslag Remelting	电渣再熔
ESSOW	Engineering Services Scope of Work	设计管理服务工作范围
EUM	English Units of Measurement	英制测量单位
EUP	Enriched Uranium Product	浓缩铀产品
EUR	European Utility Requirements	欧洲电力公司要求文件
EVMS	Earned Value Management System	收益值管理系统
EWN	Export Warning Notice	出口训诫
EWS	Emergency Water Supply	应急水供应
EXH	Exhaust	乏汽
EXT	Exterior	外部
F	Fahrenheit	华氏
F. O.	Fuel Oil	燃料油
FA	Fuel Assembly	燃料组件

英文简称	英文全称	中文名称
FAB	Fabricate	制造
FAC	Final Acceptance Certificate	最终验收证书
FAI	Fails As-Is	故障原状
FAT	Factory Acceptance Test	工厂验收试验
FBR	Feed Bleed Relief	给水泄放阀
FBTA	Function Based Task Analysis	基于功能的任务分析
FC	Fail Closed	故障关闭
FC	First Concrete	第一灌混凝土
FCD	Fuel and Core Design	燃料堆芯设计
FDA	Final Design Approval	最终设计批准
FDC	Final Design Certification	最终设计证书
FDDI	Fiber Distributed Data Interface	光纤数据分配接口
FDN	Foundation	奠基,基准,地基
FE	Flow Element (maybe an orifice)	流量部件(可能是一个孔板)
FEIS	Final Environmental Impact Study	最终环境影响研究
FHA	Fuel Handling Area	燃料装卸区
FIATECH	Fully Integrated and Automated Technology	整体自动技术
FID	Fixed Incore Detector	固定式堆内探测器
FIP	Fail in Position	没有就位
FIV	Flow Induced Vibration	流致振动
FLD	Function Logic Diagram	功能逻辑图
FLP	Fuel Load Permit	装料许可
FME	Foreign Material Exclusion	排除外国材料
FMEA	Failure Modes and Effects Analysis	失效模式和效应分析

英文简称	英文全称	中文名称
FNL	Final	最终的
FO	Fail Open	故障开
FOAKE	First of a Kind Engineering	首堆工程
FPCS	Fuel Pool Cooling and Cleanup System	燃料池冷却净化系统
FRD	Fuel Rod Design	燃料棒设计
FRLC	Fission Product Leakage Control	裂变产物泄漏控制
FSAR	Final Safety Analysis Report	最终安全分析报告
FSS	Full Scope Simulator	全范围模拟机
ft	feet	英尺
FT	Flow Transmitter	流量变送器
FTE	Full Time Equivalent	全时间当量
FTS	Fuel Transfer System	燃料输送系统
FTT	Fuel Transfer Tube	燃料输运通道
FUAT	Final Upgrade Acceptance Test	最终改进验收试验
FWC	Framework Contract	框架合同
FY	Fiscal Year	财政年度
GAD	General Arrangement Drawings	总体布置图
GB	Graphics Builder	制图人员
GDC	General Design Criteria	通用设计准则
GERS	Generic Equipment Ruggedness Spectra	通用设备粗糙度分布谱
GFI	Ground Fault Interrupter	接地故障断路器
GIF	Generation Ⅳ International Forum	第四代反应堆国际论坛
GIS	Geographical Information System	地理信息系统

英文简称	英文全称	中文名称
GPM	Gallons per Minute	加仑/分
GRCA	Gray Rod Control Assembly	灰棒控制组件
GSE	GSE System, Inc. (real time simulators)	GSE 系统工程公司(实时仿真机)
GSI	Generic Safety Issues	一般安全议题
GUI	Graphical User Interface	图形用户接口
HART	Highway Addressable Remote Transducer	高速可编址的远程传感器
HEPA	High Efficiency Particulate Air (filter)	高效微粒空气(过滤器)
HFE	Human Factors Engineering	人因工程
HFP	Hot Full Power	热态满功率
HFT	Hot Functional Test (prior to fuel load)	热态试验(燃料装载前)
HL	Hot Leg	热段
HP	Health Physics	保健物理
HP	High-Pressure	高压
HPC	High Performance Concrete	高性能混凝土
HRA	High Radiation Area (>100 mrem, <1 000 mrem)	高放射性区(> 100 mrem, <1 000 mrem)
HSE	Health Safety Environment	健康安全环境
HSI	Human-System Interface	人机接口
HTML	Hyper Text Markup Language	超文本链接标示语言
HTS	Heat Transport System	热传输系统
HTTP	Hyper Text Transfer Protocol	超文本传输协议

英文简称	英文全称	中文名称
HVAC	Heating, Ventilation, and Air Conditioning	供暖、通风和空调(暖通和空调)
HWP	Heavy Water Plant	重水堆电站
HX	Heat Exchanger	热交换器
I&C	Instrumentation and Control	仪表和控制
I/O	Input/Output	输入/输出
IAW	In Accordance with	按照
ICTN	Integral Clamp Top Nozzle	整体夹紧顶部接头
ID	Inside Diameter	内径
IDCOR	Industry Degraded Core Rule-making	堆芯降级工业标准制定
IDR	Intermediate Design Review	中期设计审查
IED	Index of Engineering Documents	设计文件索引
IFB	Invitation for Bid	投标
IFBA	Integral Fuel Burnable Absorbers	燃料一体化可燃毒物
IFC	Issued for Construction	许可建造
IFMG	Intermediate Flow Mixing Grid	中间流量搅混格架
IGSCC	Intergranular Stress Corrosion Cracking	晶间应力腐蚀裂纹
IHP	Integrated Head Package	一体化堆顶结构
ILRT	Integrated Leak Rate Test	整体泄漏率试验
IMB	Inside Missile Barrier	内部投射物屏障
IMQP	Integrated Manufacturing and Quality Plan	集成化加工与质量计划
IMS	Information Management System	信息管理系统

英文简称	英文全称	中文名称
INSP	Inspection	检查
IPS	Integrated Project Schedule	项目总体进度
IPS	Integrated Protection System	总体保护系统
IR	Independent Reviewer	独立审查方
IRC	Inside Reactor Containment	反应堆安全壳内
IRWST	In-Containment Refueling Water Storage Tank	安全壳内置换料水贮存箱
ISI	In-Service Inspection	在役检查
IST	In-Service Testing	在役试验
ITA	Inspections ,Test,Analyses	检查,试验,分析
ITAAC	Inspections, Test, Analyses, and Acceptance Criteria	检查,试验,分析和验收准则
ITAC	Integrated Test and Acceptance Criteria	整体试验和验收标准
ITB	Instructions to Bidders	招标细则
ITB	Invitation to Bid	招标邀请
ITC	Isothermal Temperature Coefficient	等温温度系数
ITER	International Thermonuclear Experimental Reactor	国际热核试验反应堆
ITT	Invitation to Tender	邀请投标,邀标书
LBB	Leak before Break	先漏后破
LC	Locked Closed	锁定在关闭状态
LD	Liquidated Damages	违约罚金
LDE	Lens Dose Equivalent (cornea)	透镜计量当量(角膜)
LHR	Linear Heat Rate	线功率密度,线发热率

英文简称	英文全称	中文名称
LHRA	Locked High Radiation Area (>1 000 mrem)	高辐照封闭区(>1 000 mrem)
LIP	Locked in Position	闭锁到位
LLRT	Local Leak Rate Test	局部泄漏率试验
LO	Locked Open	锁定于开位
LOCA	Loss of Coolant Accident	冷却剂丧失事故,失水事故(轻水堆)
LOF	Loss-of-Flow with Failure to Scram	失流未能停堆事故
LOFT	Loss of Flow Test	失流试验
LOI	Letter of Intent	意向书
LOOP	Loss of Offsite Power	失去厂外电源
LOSP	Loss of System Pressure with Degraded ECCS Operation	应急堆芯冷却系统动作失常时的系统失压
LOTI	List of Test and Inspections	试验和检验清单
LP	Low Pressure	低压
LPPT	Low Power Physics Test	低功率物理试验
LPS	Low Power and Shutdown	低功率和停堆状态
LPZ	Low Population Zone	低密度人口区
LSB	Last Stage Blade	末级叶片
LT	Level Transmitter	液位传感器
LTOP	Low Temperature Overpressure Protection	低温超压保护
LWA	Limited Work Authorization	特种工作授权(委任,认可)
LWR	Light Water Reactor	轻水反应堆

英文简称	英文全称	中文名称
M&E	Mass and Energy	质量能量
M&TE	Measuring and Test Equipment	测量与试验设备
MA	Mill Annealed	工厂退火
MAAP	Modular Accident Analysis Programs	模块化事故分析程序
MAF	Modular Application Function	模块使用功能
Max	Maximum	最大
MBTU	Million British Thermal Units	英制热单位
MCC	Motor Control Center	电机控制中心
MCL	Maximum Contaminated Level	最大被污染水平
MCR	Main Control Room	主控制室
MCRHS	Main Control Room Habitability System	主控制室可居留系统
MCS	Maintenance Cooling System	维修冷却系统
MECH	Mechanical	机械
MFC	Manufacturing	制造
MFCV	Main Feedwater Control Valve	主给水控制阀
MFIV	Main Feedwater Isolation Valve	主给水隔离阀
MG	Motor Generator	电动发电机
MIL-STD	Military Standards and Specifications	军标及规范
Min	Minimum	最小
MMC	Metal Matrix Composite	金属基体复合
MMI	Man Machine Interface	人机接口
M-MIS	Man-Machine Interface System	人机交互系统

英文简称	英文全称	中文名称
MOU	Memorandum of Understanding	谅解备忘录
MOV	Motor-operated Valve	电动阀
MPC	Maximum Permissible Concentration	最大容许浓度
MRS	Material Requisition Sheets	材料采购技术要求单
MSDS	Material Safety Data Sheet	材料安全数据单
MSIV	Main Steam Isolation Valve	主蒸汽隔离阀
MSLB	Main Steam Line Break	主蒸汽管破裂
MSSV	Main Steam Stop Valve	主蒸汽截流阀
MT	Magnetic Particle Testing	磁粉探伤
MTBE(F)	Mean Time Between Event (Failure)	失效事件间平均时间间隔
MTO	Material Take-Off	材料的替代
N/A	Not Applicable	不适用的
NAP	Nuclear Application Program	核申请程序
ND	Normally De-energized	正常降能
NI	Nuclear Island	核岛
NIMOD	Nuclear Instrumentation Module	核仪表模块
NIS	Nuclear Instrumentation System	核仪表系统
NMAC	Nuclear Maintenance Applications Center	核维修应用中心
NNS	Non Nuclear Safety-Related	非核安全相关的
NPP	Nuclear Power Plants	核电厂
NPP-BTT	Nuclear Power Plant Based on the Transferred Technology	基于技术转让的核电厂

英文简称	英文全称	中文名称
NPSH	Net Positive Suction Head	净正吸入压头
NQA	Nuclear Quality Assurance	核性能保证
Nth	Defined as the third or subsequent plant	确定为第 3 号或后续机组
O&M	Operating & Maintenance	运行与维修
OBE	Operating Basis Earthquake	运行基准地震
OBS	Organizational Breakdown Structure	组织分解结构
OD	Outside Diameter	外径
OL	Open Look	畅所欲言,坦诚直言
OMB	Outside Missile Barrier	外部飞射物屏障
OOM	Order of Magnitude	数量级
OOS	Out of Service	退役
OP	Open	打开
OPDMS	Online Power Distribution Monitoring System	在线功率分配监控系统
ORC	Outside Reactor Containment	反应堆安全壳外
ORE	Occupational Radiation Exposure	职业辐射照射
OSA	Operational Sequence Analyses	运行序列分析
OSC	Operations Support Center	运行支持中心
PABX	Private Automatic Branch Exchange	专用自动小交换机
PAC	Provisional Acceptance Certificate	临时验收证书
PAMS	Post Accident Monitoring System	事故后监测系统

英文简称	英文全称	中文名称
PAN	Pre-Award Notification	预中标通知
PAN	Preliminary Award Notice	初步的中标通知
PAR	Passive Autocatalytic Recombiner	非能动自动复合器
PBMR	Pebble Bed Modular Reactor	球床式模块反应堆
PBS	Price Breakdown Structure	价格分解结构
PC	Personal Computer	个人计算机
PCCAWS	Passive Containment Cooling Ancillary Water Storage Tank	非能动安全壳冷却水储水箱
PCCS	Passive Containment Cooling System	非能动安全壳冷却系统
PCCWS	Passive Containment Cooling Water Storage	非能动安全壳冷却水贮存
PCCWST	Passive Containment Cooling Water Storage Tank	非能动安全壳冷却水储存箱
PDA	Project Development Agreement	项目开发协议
PdMP	Predictive Maintenance Program	预定的维修大纲
PDS	Plant Design System (modeling software)	电站设计系统(模拟软件)
PFD	Process Flow Diagram	工艺流程图
P-Grid	Protective Grid	保护格架
pH	Potential of Hydrogen	氢的潜能
P&ID	Piping and Instrument Diagram	管线仪表图
PIE	Potential Initiating Event	潜在的触发事件
PIB	Project Information Book	项目资料包
PIV	Pressure Isolation Valve	压力隔离阀

英文简称	英文全称	中文名称
PLS	Precautions, Limitations, and Setpoints	预防、限制和设定值
PMS	Protection & Safety Monitoring System	保护与安全监控系统
PMSS	Probable Maximum Storm Surge	可能的最大风暴潮
PMST	Processor Module Software Testing	处理器模块软件试验
PORV	Power Operated Relief Valve	能动释放阀
PPE	Personal Protective Equipment	个人防护设备
PPG	Piping	管道
PPM	Prefabrication, Preassembly and Modularization	预制,预装和模块化
PPP	Passive Plant Projects	非能动电厂项目
PQAP	Project Quality Assurance Programme	项目质保大纲
PQAP	Project Quality Assurance Plan	项目质量保证计划
PQP	Project Quality Program	项目质量大纲
PRE-OP	Pre-Operational Testing	运行前试验
PREP	Preparation	准备
PRHR	Passive Residual Heat Removal System	非能动余热排出系统
PRHR HX	Passive Residual Heat Removal Heat Exchanger	非能动余热排出热交换器
PSAR	Preliminary Safety Analysis Report	初步安全分析报告

英文简称	英文全称	中文名称
PSI	Pounds per Square Inch	磅/英寸^2
psia	Pounds per Square Inch Absolute	磅/英寸^2(绝对压力)
PSSD	Preliminary Scheduled Shipment Date	初步计划的装船日期
PT	Penetrant liquid Testing (NDE)	渗透检验
PTLR	Pressure Temperature Limits Report	压力温度限制报告
PUAT	Preliminary upgrade Acceptance Test	初步改进验收试验
PvMP	Preventative Maintenance Program	预防性维修大纲
PWR	Pressurized Water Reactor	压水反应堆
PWSCC	Primary Water Stress Corrosion Cracking	一回路水应力腐蚀裂纹
PXS HX	Passive Core Cooling System Heat Exchanger	非能动堆芯冷却系统热交换器
PZR	Pressurizer	稳压器
QDPS	Qualified Data Processing System	合格(1E 级)数据处理系统
QMS	Quality Management System	质量管理系统
QPTR	Quadrant Power Tilt Ratio	象限功率倾斜比
QSDR	Quality Surveillance Deficiency Report	质量监督缺陷报告
QSL	Qualified Supplier List	合格供货商清单
R. G.	Regulatory Guide	管理导则

英文简称	英文全称	中文名称
RA	Radiation Area (>5 mrem, <100 mrem)	辐照区(>5 mrem,<100 mrem)
RAI	Request for Additional Information (generally an NRC request)	额外资料请求(通常指 NRC 请求)
RAM	Reliability, Availability, Maintainability	可靠性,可用性和可维护性
RAR	Request for Approval and Review	请求批准和审查
RAT	Reserve Auxiliary Transformer	备用辅助变压器
RB	Reactor Building	反应堆厂房
RCC	Reinforced Concrete Construction	钢筋混凝土建筑
RCA	Radiological Controlled Area	放射性控制区
RCCA	Rod Cluster Control Assembly	棒束控制组件
RCDT	Reactor Coolant Drain Tank	反应堆冷却剂疏水箱
RCMP	Reliability Centered Maintenance Program	以可靠性为中心的维修大纲
RCP	Reactor Coolant Pump	反应堆冷却剂泵,主泵
RCPB	Reactor Coolant Pressure Boundary	反应堆压力边界
RDMG	Rod Drive Motor Generator	控制棒驱动电动发电机
RFA	Robust Fuel Assembly	西屋公司的 Robust 型燃料组件
RFI	Request for Information	资料请求
RFI	Radio Frequency Interference	射频干扰
RFP	Request for Proposal	请求建议
RFQ	Request for Quote	报价请求
RFSAR	Revised Final Safety Analysis Report	修订后的最终安全分析报告

英文简称	英文全称	中文名称
RG	Regulatory Guide	管理导则
RGB	Red-Green-Blue (VDU color settings)	红绿蓝视觉显示单元(VDU 彩色设置)
RHR	Residual Heat Removal	余热排出
RIP	Rod Internal Pressure	燃料棒的内压
RM	Responsible Manager	责任经理
RM	Refueling Machine	换料机
RME	Reasonable Maximum Exposure	可接受的最大受照
RO	Restricting Orifice	截流孔板
RP	Radiation Protection	辐射防护
RPRS	Rapid Power Reduction System	快速降功率系统
RPV	Reactor Pressure Vessel	反应堆压力容器
RSEDs	Reusable Software Element Documents	可重复使用的软件要素文件
RSR	Remote Shutdown Room	远程停堆室
RSW	Remote Shutdown Workstation	远程停堆工作站
RT	Radiographic Test (NDE)	射线探伤
RTCOT	Reactor Trip Channel Operational Test	反应堆停堆通道运行试验
RTD	Resistance Temperature Detector	电阻温度探测器
RTDP	Revised Thermal Design Procedure	修正的热工设计程序
RTNSS	Regulatory Treatment of Non-Safety System	非安全系统的管理处理
RTP	Rated Thermal Power	额定热功率
RTS	Reactor Trip System	反应堆停堆系统

英文简称	英文全称	中文名称
RV	Reactor Vessel	反应堆压力容器
RVCH	Reactor Vessel Closure Head	反应堆压力容器上封头
RVI	Reactor Vessel Internals	堆内构件
RW	Radiation Waste Building	放射性废物厂房
RWP	Radiation Work Permit	放射性工作许可
SAT	Site Acceptance Test	现场验收试验
SAVerif	Safety Analysis Verification	安全分析验证
SC	Steel Plate Reinforced Concrete Construction	钢内衬预应力混凝土结构
SCC	Self Compacting Concrete	自紧缩混凝土
SCC	Stress Corrosion Cracking	应力腐蚀裂纹
SCF	Standard Cubic Feet	标准立方英尺
SCFM	Standard Cubic Feet per Minute	标准立方英尺/分
SCM	Supply Chain Management	供应链管理
SCV	Steel Containment Vessel	钢制安全壳
SDE	Skin Dose Equivalent	皮肤剂量当量
SDM	System Design Manual	系统设计手册
SDM	Shut Down Margin	停堆裕度,停堆裕量
SDODT	Simplified Display Object Design Tool	简化的显示目标设计工具
SDV	Safety Depressurization and Vent	安全卸压与通风
SER	Safety Evaluation Report	安全评价报告
SFP	Spent Fuel Pool	乏燃料水池
SFW	Startup Feedwater	启动给水
SG	Steam Generator	蒸汽发生器

英文简称	英文全称	中文名称
SGI	Safeguards Information	安保信息
SGTR	Steam Generator Tube Rupture	蒸汽发生器传热管破裂
SI	Safety Injection	安全注入
SIS	Safety Injection System	安全注入系统
SIT	Structural Integrity Test	结构完整性试验
SLB	Steam Line Break	蒸汽管线破裂
SOD	Scope of Document	文件范围
SOW	Scope of Work	工作范围
SPEC	Specification	规格书
SQ	Supplier Qualification	供货商资格鉴定
SRP	Standard Review Plan	标准审查计划
SRV	Safety Relief Valve	安全卸压阀,安全释放阀
SSAR	Standard Safety Analysis Report	标准安全分析报告
SSCs	Structures, System and Components	构筑物,系统和部件
SSD	Scheduled Shipment Date	计划的装船日期
SSD	System Specification Document (system description)	系统规格书文件(系统描述)
SSE	Safe Shutdown Earthquake	安全停堆地震
SSI	Soil Structure Interaction	土壤结构的相互作用
STD	Simulator Technical Documents	模拟机技术文件
STD	Standard	标准
STRM	Storm	暴风雨
SUFCV	Startup Feedwater Control Valve	启动给水控制阀
SUFIV	Startup Feedwater Isolation Valve	启动给水隔离阀

英文简称	英文全称	中文名称
SUNSI	Sensitive Unclassified Non-Safeguards Information	未分级非安保敏感信息
SUPTS	Supports	支撑,支承,支持
SUR	Startup Rate	启动速率
SUT	Startup Test	启动试验
SV	Stop Valve	截止阀
SWC	Surge Withstand Capability	抗浪涌能力
SWGR	Switch Gear	电气(开关设备)
SWR	Sewer	下水道
SYWD	Switchyard	开关站
T&M	Time & Material	时间与材料
T/T	Telegraphic Transfer	电汇
TA	Transient Analysis	瞬态分析
TAD	Transport, Aging, and Disposal (canister)	运输、老化和处置(玻璃固化体)
TADOT	Trip Actuating Device Operational Test	停堆触发设备运行试验
TC	Test Connection	试验汇总
TC	Thermocouple	热电偶
TCP/IP	Transmission Control Protocol/Internet Protocol	传输控制协议/网络协议
TCR	Technical Change Request	技术变更请求
TDC	Technical Document Control (database- system)	技术文件控制(数据库系统)
TE	Temperature Element (RTD or TC)	测温元件(电阻温度探测器或热电偶)

英文简称	英文全称	中文名称
TECH SPEC	Technical Specification	技术规格书
TEDE	Total Effective Dose Equivalent	总有效剂量当量
TG	Turbine Generator	汽轮发电机
THD	Thermo Hydraulic Design	热工水力设计
TID	Total Integrated Dose	全员总受照剂量
TMI	Three Mile Island	三哩岛
TNK	Tank	箱
TOPO	Topological/ Topographic	地质学的/地形学的
TRIR	Total Recordable Injury Rate	总的有记载的伤害率
TSC	Technical Support Center	技术支持中心
TSS	Technical Support Services	技术支持服务
TT	Technical Transfer	技术转让
TT	Thermally Treated	经热处理的
TV	Test Vent	试验通风
TV	Throttle Valve	节流阀
UAT	Unit Auxiliary Transformer	机组辅助变压器
UBC	Uniform Building Code	统一厂房代码
UL	Underwriters Laboratories	保险商实验室
UPS	Uninterruptible Power Supply	不间断电源
URD	Utility Requirements Document	电力公司要求文件
URI	Universal Resource Identifier	统一资源标识符
URL	Uniform Resource Locator	统一资源定位符
URN	Uniform Resource Name	统一资源名称

英文简称	英文全称	中文名称
USD	US Dollar	美元
USI	Unresolved Safety Issue	未解决的安全问题
USPHS	United States Public Health Service	美国公共健康服务
UT	Ultrasonic Test (NDE)	超声检验(无损检验)
V&V	Verification and Validation (software related)	软件有关的验证和确认
VAT	Value Added Taxes	增值税
VDU	Video Display Unit	视频显示器
VERT	Vertical	垂直的
VFDS	Variable Frequency Drive System	变频驱动系统
VHS	Health Physics and Hot Machine Shop Areas	保健物理及热机修车间区域
VOD	Vacuum-Oxygen Decarburization	真空-氧脱碳
VT	Visual Testing (NDE)	目视检查(无损检验)
WB	Whole Body	全身
WCAP	Westinghouse report designator, originally Westinghouse Commercial Atomic Power	西屋报告编号,西屋商用原子能的简写
WPIS	Wall Panel Information System	墙屏信息显示系统
WTR	Water	水
XFMR	Transformer	变压器

Ⅱ. 核电英语常用缩写

英文简称	英文全称	中文名称
A	Absolute/Alarm/Ammeter/Amps/Automatic	绝对的/报警(器)/电流表,安培计/安培数/自动的
AA	Adjustable Absorber/Arithmetic Average/Approximate Absolute Temperature Scale	调节棒/算术平均/近似绝对温标
AAB	Accident Analysis Branch of the USNRC	美国核管理委员会事故分析处
AAC	Alternate Alternating Current/Amplitude Absorption Coefficient/Automatic Amplitude Control	替代交流电源/振幅吸收系数/自动振幅控制
AAM	Accident Analysis Matrix	事故分析矩阵
AAW	Air-acetylene Welding	空气乙炔焊
AB	Adhesive bonding/Analysis Basis	粘接/分析基础
ABB	Asea Brown Boveri	瑞士 ABB 公司
Abbr.	Abbreviation	缩写
ABC	Automatic Boiler Control	锅炉自动控制(调节)
ABCB	Air Blast Circuit Breaker	喷射气流断路器
ABD	Aseismic Bearing Device/As-Built Data	抗地震轴承装置/竣工资料
ABM	Automatic Batch Mixing	自动化批料混合
ABS	Absolute/Air Break Switch/American Bureau of Standards	绝对的/空气断路开关/美国标准局

英文简称	英文全称	中文名称
ABSE	Absolute Error	绝对误差
ABT	About	大约
ABW	Arc braze welding	电弧钎焊
AC	Absorption Coefficient/Aerodynamic Center/Aged Condition/Alternating Current/Arc cutting/Automatic Computer/Automatic Control	吸收系数/空气动力中心/老化条件,老化状况/交流电/电弧切割/自动计算机/自动控制
AC/A/C	Access Control	出入口控制,存取控制
AC/ACCUM	Accumulator	安全注入箱,蓄压箱
A/C	Air Conditioning	空气调节,空调
ACA	Accident Consequence Assessment	事故后果评定
ACB	Air Circuit Breaker	空气断路器
ACC	Acceptable/Accident/Accuracy/Argonne Code Center/Auxiliary Core Cooling	可接受/事故/精(确)度/阿贡法规中心,阿贡程序中心/堆芯辅助冷却
ACCIS	Advanced Control Center Information System	先进控制中心信息系统
ACC/ACCRS	Accessories	附件
Acct.	Account	计算,账目,账户
ACCW/ACW	Alternating Current Continuous Waves	交流连续波
ACE	Air Conditioning Equipment	空调设备

英文简称	英文全称	中文名称
ACES	Automatic Checkout and Evaluation System	自动检查和评价系统
ACF	Average Calibration Factor	平均校准因子
ACK	Acknowledge	肯定,应答,承认收到
ACM	Alternating Cooling Mode/Association for Computing Machinery	替代冷却方式/计算机协会(美国)
ACO	Assistant Control Operator	控制室助理操纵员,助理控制操纵员
ACOM	Automatic Coding Module	自动编码模块,自动编程模块
ACOS	Automatic Coding System	自动编码系统,自动编程系统
ACP	Auxiliary Control Panel	辅助控制仪表板,辅助控制屏
ACPR	Annular Core Pulse Reactor	环状堆芯脉冲反应堆
ACR	Advanced CANDU Reactor/Action Clarification Request	先进坎杜堆/行动澄清要求
ACRDS	Automatic Control Rod Drive System	自动控制棒驱动系统
ACS	Above-Core Structure/American Chemical Society/Auxiliary Coolant System	堆芯上部构件/美国化学学会/辅助冷却系统
ACT	Acting/Active	作用的,代理的/能动的;活性的;放射性的
ACTN	Action	行动
ACTUR	Actuator	驱动器,触发器,传动装置,执行机构
ACU	Air Conditioning Unit/Air-Cooling Unit/Airlock Cooling Unit	空调机组/空冷机/空气闸门冷却装置

英文简称	英文全称	中文名称
ACW	Anti-Clockwise	反时针方向
AD	Access Door/Administrative Procedure/Advanced Development/Analysis Detail/Average Deviation	出入门/行政管理程序/升级/分析详情/平均偏差
AD/A/D	Analog to Digital/Analog/Digital	模拟数字/模拟/数字
ADA	Automatic Data Acquisition/Automatic Diagnosis of Alarm	自动数据采集/报警自动诊断
ADAPT	Adaptor	适配器
ADC	Analog-Digital Converter	模-数转换器
Add	Address	地址
ADDL	Additional	附加的,辅助的
ADDTN	Addition	加法;添加(剂),掺杂物
ADE	Automatic Design Engineering	自动设计工程
ADES	Automatic Digital Encoding System	自动数字编码系统,自动数字编程系统
ADI	Acceptable Daily Intake	日容许摄入量
ADJ	Adjustable/Adjusting	可调节的/调节
ADL	Acceptable Defect Level/Annual Dose Limit/Automatic Data Line	合格缺陷标准/年剂量限值/自动数据传输线
ADLWR	Accelerator Driven Light Water Reactor	加速器驱动的轻水堆
ADM	Administration	行政管理
ADP	Accident Diagnosis and Prognosis	事故诊断与预测
ADPC	Automatic Data Processing Center	自动数据处理中心
ADPS	Automatic Data Processing System	自动数据处理系统

英文简称	英文全称	中文名称
ADS	Automatic Depressurization System	自动卸压系统
ADV	Atmospheric Dump Valve	大气排放阀
ADX	Automatic Data Exchanger	自动数据交换
AE	Absolute Error/Assistant Engineer	绝对误差/助理工程师
AE/A/E	Architect Engineer/Architect Engineering	建筑工程师/建筑工程
AED	Acceptable Emergence Dose/Automated Engineering Design	容许应急剂量/自动工程设计
AEDS	Atomic Energy Detection System	原子能探测系统
AEN	Agence Européenne Pour L'Energie Nucléaire/Acceptable Except Noted Comments	欧洲核能局(巴黎)/有条件接受
AEOD	Analysis and Evaluation of Operation Data	运行数据分析和评价
AER	Anion Exchange Resin	阴离子交换树脂
AES	American Electromechanical Society/American Electronical Society/Atomic Emission Spectrometry	美国机电学会/美国电子学会/原子发射光谱法
AESC	American Engineering Standards Committee	美国工程标准委员会
AETR	Advanced Epithermal Thorium Reactor	先进型超热钍反应堆
AEWS	Area Emergency Warning System	地区应急报警系统
AF	Across Flats/Aerosol Filter/Amplifier/Automatic Following/Availability Factor/Axial Flow	对边宽度，扳手开口尺寸/气溶胶过滤器/放大器/自动跟踪/可利用因子/轴向流动

英文简称	英文全称	中文名称
AFA	Advanced Fuel Assembly/Audio-Frequency Amplifier	先进燃料组件/音频放大器
AFAU	Air Filtration and Analysis Unit	空气过滤与分析装置
AFBIO	Auxiliary Feed & Bleed Inlet & Outlet (Headers)	辅助给排水进出口(集管)
AFC	Automatic Frequency Control	自动频率控制,自动频率调节
AFFFS	Aqueous Film-Forming Foam System	形成水膜起泡系统
AFO	Authorized First Operator	授权的首席操作员,当班值长
AFP	Auxiliary Feedwater Pump	辅助给水泵
AFR	Acceptable Failure Rate/Away-From-Reactor	容许故障率/堆外,离堆
AFS	Available for Service Certificates	运行批准证书
AFSR	Argonne Fast Source Reactor	阿贡快中子源反应堆(美国)
AFW	Auxiliary Feedwater	辅助给水
AFWP	Auxiliary Feedwater Pump	辅助给水泵
AG	Agitator	搅拌器
AGR	Advanced Gas-Cooled Reactor	先进气冷堆
AGS	Annulus Gas System	环腔气体系统
AGSL	Auxiliary Gland Seal	辅助汽封,轴封
Agt.	Agent	代理商
AH	Ampere-Hour/Ampere Hour Meter/Available Hour	安培小时/安时表/可利用小时数
AHLCS	Auxiliary H_2O Leakage Collection System	辅助轻水泄漏收集系统
AHR	Aqueous Homogeneous Reactor	水均匀反应堆

英文简称	英文全称	中文名称
AHU	Air Handling Unit	空气控制单元
AHVC	Automatic High-Voltage Control	自动高压控制
AHW	Atomic Hydrogen Welding	原子氢焊
AI	Analytical-Internal/Artificial Intelligence/Automatic Input	内部分析的/人工智能/自动输入
AI/A/I	Analog Input	模拟量输入
AIA	American Institute of Architects/Authorized Inspection Agency	美国建筑学会/指定的检查机构
AID	Aerosol Ionization Detector/Agency for International Development	气溶胶电离探测器/国际开发署
A/M	Automatic/Manual	自动/手动
AIM	Asset and Information Management	资产与信息管理
AIMS	Accounting Information Management System	会计信息管理系统
AIW	Annunciation Interrogation Workstation	报警查询工作站,通告查询站
AL	Accident Loss/Aluminum/Assessment Letter	事故损失/铝/评估函
ALAP	As Low As Possible	尽可能低
ALARA	As Low As Reasonably Achievable	合理可行尽量低
ALARP	As Low As Reasonably Practicable	合理可行尽量低
ALEX	Argonne Liquid Metal Engineering Experiment	阿贡液态金属工程实验(装置)

英文简称	英文全称	中文名称
ALGOL	Algorithmic Language	算法语言
ALI	Annual Limit of Intake	年摄入量限值
ALR	Automatic Load Regulator	自动负载调节器
ALST	Alarm Status Table	报警状态表
ALU	Arithmetic and Logical Unit	运算逻辑单元
ALWR	Advanced Light Water Reactor	先进轻水堆
AM	Angular Momentum/Associative Memory	角动量/关联存储(器)
AMAD	Activity-Median Aerodynamic Diameter	中等活度气体动力学直径
AMMS	Automated Maintenance Management System	自动化维修管理系统
AMP	Ampere/Amperage	安培/安培数
AMS	American Material Standard/American Metallurgical Society/Administration Management System	美国材料标准/美国冶金学会/行政事务管理系统
Amt.	Amount	总额
AMTEC	Automatic Time Element Compensator	自动延时补偿器
AMU	Atomic Mass Unit	原子质量单位
AN	Air Natural (Cooled)	空气自然(冷却的)
ANEC	American Nuclear Energy Council	美国核能理事会
ANG	Angle	角(度)
ANI	Authorised Nuclear Inspector	授权的核检查员

英文简称	英文全称	中文名称
ANN/ANNU	Annunciation	报警
ANNUL	Annulus	环廊,环形(物)
ANNUN	Annunciation/Annunciator	报警/报警器
ANPSH	Available Net Positive Suction Head	可用净正吸入压头
AO	Access Opening/Auxiliary Operator/Axial Offset	人孔,检修孔,出入孔/副操纵员/轴向偏移
AO/A/O	Analog Output	模拟量输出
AOM	Abnormal Operating Manual/Assistant Operations Manager	异常运行手册/运行经理助理
AOP	Abnormal Operating Procedure/Auxiliary Oil Pump	异常运行程序/辅助油泵
AORS	Abnormal Occurrence Reporting System	异常事件报告系统
AOV	Air-Operated Valve/Automatically Operated Valve	气动阀/自动操作阀
AP	American Patent/Assembly Procedure/Automatic Programming/Auxiliary Pump	美国专利/组装程序/自动编程/辅助泵
APC	American Power Conference/Atomic Power Construction Ltd/Automatic Pressure Control/Auxiliary Pressure Control	美国电力会议/英国原子能建设有限公司/自动压力控制/辅助压力控制
APDMS	Axial Power Distribution Monitoring System	轴向功率分布监测系统

英文简称	英文全称	中文名称
APDX	Appendix	附录(件)
APL	Assembly Parts List	装配部件清单
APP	Appendix/Approved/Atomic Power Plant/Auxiliary Power Plant	附录/批准的，许可的/核电站，核电厂/辅助电源设备，辅助电厂
APPO	Assistant Power Plant Operator	电站助理操纵员
Approx.	Approximately	约计
APR	Additional Price Request	额外费用索赔申请，附加费用要求
APRM	Average Power Range Monitor	平均功率区段监测器
APRS	Automatic Production Record System	自动化生产记录系统
APS	Accessory Power Supply/Airborne Particulate Sampler/American Physical Society/Atomic Power Station/Auxiliary Power Supply	辅助电源/气溶胶粒子取样器/美国物理学会/原子能发电站/辅助电源
APTR	Advanced Pressure Tube Reactor	先进压力管式反应堆
APU	Auxiliary Power Unit	辅助电源设备，厂用电设备
APWR	Advanced Pressurized Water Reactor	先进压水堆
AQ	Any Quantity	任意量
AQCS	Analytical Quality Control Services	质量分析控制处
AQL	Acceptable Quality Level	合格质量标准，容许品质等级

英文简称	英文全称	中文名称
AR	Acceptance Requirement/Accidental Report/A nalysis Report/Argon	验收要求/事故报告/分析报告/氩
ARAVS	Auxiliary and Radwaste Area Ventilation System	辅助和放射性废物区域通风系统
ARC	Automatic Remote Control	自动遥控
ARI	Air-Conditioning and Refrigeration Institute/Alternate Rod Insertion/American Refrigeration Institute	空调和制冷学会/替代棒插入/美国制冷学会
ARIANR	Automatic Retrieval of Incidents Affecting Nuclear Reactors	核反应堆事件自动检索系统
ARIS	Automated Reactor Inspection System	反应堆自动检查系统
ARL	Acceptable Reliability Level	可接受的可靠性标准
ARM	Application Reference Manual	应用参考手册
ARMF	Advanced Reactivity Measuring Facility	先进反应性测量装置
ARRG	Arrangement	布置,配置
ARSAP	Advanced Reactor Severe Accident Program	先进反应堆严重事故处理大纲
ART	Adjusted Reference Temperature	调整的参考温度
Art.	Article	条款,物项,物品
ARU	Automatic Range Unit	自动排列单元,自动装置
ARV	Automatic Relief Valve	自动卸压阀

英文简称	英文全称	中文名称
AS	Aeroball System/American Standard/Assistant Supervisor	气动小球系统/美国标准/副主管，助理监督
ASAP	As Soon As Possible	尽快
ASCII	American Standard Code for Information Interchange	美国信息交换标准代码
ASCOT	Assessment of Safety Culture in Organization Teams	机构团队的安全文化评估
ASCS	Automatic Stabilization and Control System	自动稳定和控制系统
ASD	Anthracene Scintillation Dosimeter/Assessment Document/Atmospheric Steam Dump	蒽闪烁剂量计/评价文件/蒸汽大气排放
ASDS	Auxiliary Shutdown System	辅助停堆系统
ASDV	Atmospheric Steam Discharge Valve	大气蒸汽释放阀
ASG	Acoustic Strain Gauges	音频(振弦)应变仪
ASI	AECL Subject Index	AECL 主题索引
ASMFP	Adjusted Speed Motor Feedwater Pump	调速电动给水泵
ASMS	Alarm and Status Management System	报警与状态管理系统
ASO	Area of Safe Operation/Auxiliary Switch (breaker) Normally Open	安全作业区/辅助开关(断路器)正常断开

英文简称	英文全称	中文名称
ASP	Arc spraying	电弧喷涂
ASR	Acceptance Summary Report/Annual Safety Report/Automatic Speed Run-up/Available Supply Rate	验收总结报告/年度安全报告/自动升速/有效供给速率，可用供应率
ASSC	Auxiliary System Shield Cooling	辅助系统屏蔽冷却
ASSET	Analysis of Significant Safety Events Terms/A ssessment of Safety Significant Events Team	安全重要事项分析/安全重要事件评价组
ASSY	Assembly	组件，装配
ASV	Acceleration-Switching Valve/Automatic Self-Verification	快关阀/自动自检验
ASWG	American Standard Wire Gage	美国标准线规
AT	Acceptance Test/Air Temperature/Air Test/Ambient Temperature/Atmospheric/Auxiliary Transformer/Assistant Tester	验收试验/气温/大气试验/环境温度/大气压的/辅助变压器，(核电厂)厂用变压器/助理试验员
ATA	Atmosphere Absolute	绝对大气压
ATBM	Average Time Between Maintain	维修平均间隔时间
ATC	Amplitude-to-Time Converter/Analog Trip Comparator/Automatic Temperature Compensator/Automatic Train Control System	幅-时转换器/模拟紧急停机比较器/自动温度补偿器/自动序列控制系统
ATD	Analogue to Time to Digital	模拟-时间-数字转换
ATLAS	Advanced Test Loop and Simulator	先进试验回路及模拟器

英文简称	英文全称	中文名称
ATM	Atmosphere	大气压
ATOG	Abnormal Transient Operation Guidelines	异常瞬态运行导则
ATP	Acceptance Test Procedure/Authorization to Proceed	验收试验程序/(合同)批准执行,批准开工
ATR	Acceptance Test Report/Advanced Thermal Reactor/Authorization Training Requirements	验收试验报告/先进热中子反应堆/培训授权要求
ATRC	Advanced Test Reactor Critical	先进试验堆临界实验装置
ATS	Automatic Turbine Start-up (System)/Assistant to Test Supervisor	汽轮机自动启动(系统)/试验副主管,试验监督助理
ATWS	Anticipated Transient Without Scram	未能紧急停堆的预期瞬态
ATWT	Anticipated Transient Without Trip	未能紧急停堆的预期瞬态
AU	Atom Unit/Automatic	原子单位/自动的;自动装置
AUC	Audit Checklists	监查检查表
AUD	Audit Department	审计部
AUR	Audit Reports	监查报告
AUS	Auxiliary Switch	辅助开关
AUTOGRP	Automatic Grouping System	自动分组系统
AUX	Auxiliary	辅助的
AVF	Available Factor	可利用系数

英文简称	英文全称	中文名称
AVG	Average	平均的
AVR	Automatic Voltage Regulator	自动电压调整器
AVT	All Volatile Treatment	全挥发处理
AW	Acid Waste/Arc Welding/Atomic Weight	废酸，酸性废物/电弧焊/原子量
AWC	Absolute Worst Case	绝对最坏情况
AWG	American Wire Gauge	美国线规
AWN	Additional Work Notice	额外工作通知
AWP	Approved Work Practice	批准的工作实践(做法)
AX	Axial/Axis	轴(向)的/中心轴，轴线
AZL	Average Zone Level	平均区域液位，平均区域水平
B	Blower/Brazing/Bus	风机/铜焊，软钎焊/母线
B. O. NSSS	Balance of Nuclear Steam Supply System	核蒸汽供应系统配套设施
B/E	Both Ends	两端
B/L	Bill of Lading	提(货)单
BAC	Binary-Analog Conversion	二进制模拟转换
BAEA	British Atomic Energy Authority	英国原子能管理局
BAERE	British Atomic Energy Research Establishment	英国原子能科学研究中心
BAR	Barometric	气压的
BAT	Battery	电池组，蓄电池
BAU	British Association Unit	英国标准单位
BB	Ball Bearing	滚珠轴承
BB	Block Brazing	块钎焊

英文简称	英文全称	中文名称
BBSL	Boiler Blowdown System Line	锅炉(蒸汽发生器)排污系统管道
BC	Binary Code/Bolt Circle	二进制码/螺栓分布圆
BCA	Benefit-Cost Analysis	收益-成本分析,效费分析
BCD	Burst Can Detection/Burst Cartridge Detector/Binary Coded Decimal	包壳破损探测/燃料元件包壳破损探测器/二进制编码小数
BCI	Bechtel China, Inc.	Bechtel 中国公司
BCL	Backfilling Check List	回填土检查单
BCO	Battery Cut-off	蓄电池电路自动断路器,蓄电池断开
BCS	Blowdown Cleanup System/Bulk Capacity Storage/Burner Control System	排污净化系统/大容量存储器/燃烧器控制系统
BCT	Bushing Current Transformer	套管式电流互感器,通芯变流器
BD	Block Diagram	方块图,方框图
BDBA	Beyond Design Basis Accident	超设计基准事故
BDC	Binary Digital Counter	二进制数字计数器
BDHT	Blowdown Heat Transfer	(失水事故时)喷放(阶段)传热
BDL	Bundle	束,棒束
BDPM	Battery-Operated Digital Pressure Monitor	电池数字压力监测仪
BDV	Blowdown Valve/Breakthrough Voltage	排污阀/击穿电压
BEA	Background Equivalent Activity	本底当量放射性

英文简称	英文全称	中文名称
BECHTEL	Overseas Bechtel, Incorporated	Bechtel 海外公司
BEP	Building Emergency Procedures	厂房应急程序
BF	Back Feed/Blank Flange/Bottom Face	反馈/盲板法兰/底面
BFP	Backup Feedwater Pump/Boiler Feed Pump	备用给水泵/锅炉给水泵
BFT	Blister Formation Threshold	气(汽)泡生成阈
BFW	Boiler Feed Water	锅炉给水
BH	Brinell Hardness	布氏硬度
BH	Bulkhead	隔板
BHP	Brake Horsepower/British Horsepower	制动马力/英制马力
BIL	Basic Impulse Level/Basic Insulation Level/Bulk Items List	基本脉冲电平/绝缘基本冲击耐压水平/散装货物清单
BIS	Bid Invitation Specification	招标说明书,邀标书
BITUM	Bitumen	沥青
BK	Brake	制动器,闸
BL	Bar List/Base Line/Base Load/Black	钢筋清单/基线/基本负荷/黑色(的)
BLC	Boiler Level Control	蒸发器水位控制
BLD	Bleed	排(泄)放
BLDG	Building	厂房;建筑物
BLK	Block	块;组
BLTC	Bottom Loading Transfer Cask	底部装料运输容器

英文简称	英文全称	中文名称
BLWS	Bellows	波纹管
BM	Bending Moment/Bill of Material/Breakdown Maintenance	弯矩/材料清单/故障维修
BMAW	Bare Metal Arc Welding	无保护金属丝电弧焊
BMW	Boiler Make-up Water	锅炉(蒸汽发生器)补水
BN	Bolt and Nut	螺栓与螺母
BNDG	Binding	结合,连接
BNI/BONI	Balance of Nuclear Island	核岛配套设施
BNSP	Balance of Nuclear Steam Plant	核蒸汽厂房辅助设施
BO	Boiler/Bulk Order	锅炉/大宗采购订单
BOD	Biochemical Oxygen Demand/Biological Oxygen Demand	生化需氧量/生物耗氧量
BOF	Basic Oxygen Furnace	氧气顶吹转炉
BOI	Basis of Issue/Beginning of Life/Break of Inspection	论据/寿期初/检查中断
BOLT	Beam-of-Light-Transistor	光束晶体管
BONUS	Boiling Nuclear Superheat Power Station	沸腾超热核电站
BOP	Balance of Plant	电站辅助设施,核电厂配套设施
BOQ	Bill of Quantities	工程量清单
BOT	Build-Operate-Transfer	建造-运行-移交方式
BOTI	Balance of Turbine Island	汽轮机岛配套设施,常规岛配套设施

英文简称	英文全称	中文名称
BP	Backplane/Back Pressure/Barometric Pressure/Base Plate/Boiling Point/Booster Pump	底(后挡)板/背压/大气压力/底板/沸点/增压泵
BPC	Boiler Pressure Control	锅炉压力控制
BPD	Burst Pin Detection	破损燃料细棒探测
BPRA	Burnable Poison Rod Assembly	可燃毒物棒组件
BPV	Bypass Valve	旁路阀
BR	Brass/Break Request/Breeding Ratio/Buchholz Relay	黄铜/中断请求/增殖比/瓦斯继电器
BRC	Below-Regulatory-Concern	管理当局不关心的
BRD	Braid/Braided	编织(物)
BRET	Breeder Reprocessing Engineering Test	增殖堆后处理工程试验(装置)
BRG	Bearing	轴承
BRID	Bridge	桥架
BRK	Break	破(断)裂
BRKT	Bracket	托架,牛腿,支撑架
BRP	Building Removal Plan	厂房拆除计划
BRS	Boron Recovery System	硼回收系统
BRV	Body Relief Valve	阀体泄压阀
BRZ	Bronze	青铜
BS	Bellows Sealed/Bessemer Steel/Binary Subtract/Blocking Switch/British Standard/Button Switch	波纹管密封的/酸性转炉钢,贝斯麦钢/二进制减法/闭锁开关/英国标准/按钮开关

英文简称	英文全称	中文名称
BSD	Burst Slug Detector	嵌条破损检测器,燃料元件破损探测器
BSF	Bulk Shielding Facility	整体屏蔽装置
BSHG	Bushing	套管(筒)
BSI	Basic Subject Index/British Standards Institute	基本主题索引/英国标准学会
BT	Burst Test	爆破试验
BTG	Boiler-Turbine-Generator	锅炉-汽轮机-发电机
BTM	Bottom	底
BTN	Button	按钮
BTRFY	Butterfly	蝶形
BTS	Book of Technical Specifications	技术规格书
BTU	British Thermal Unit	英热单位
BU	Back Up	备用
BUL	Bulletin	公报,会刊
BV	Ball Valve/Breakthrough Voltage	球阀/击穿电压
BW	Butt Weld	对(接)焊
BW	Braze welding	钎焊,铜焊
BWR	Boiling Water Reactor	沸水(反应)堆
BWST	Borated Water Storage Tank	加硼水贮存箱
BWV	Back-Water Valve	回水阀
BYP	Bypass	旁路;旁通管
C	Cable/Celsius/Closing	电缆/摄氏(温度)/关闭,闭合;接通

英文简称	英文全称	中文名称
C&C	Control and Coordination/Commissioning and Coordination	控制与协调/调试与协调
C&F	Cost and Freight	成本加运费价
C&F FO	Cost and Freight Free Out	成本加运(船方不负担卸货)费用
C&I	Control and Instrumentation	控制和仪表
C&MS	Control and Monitor Subsystem	控制与监测子系统
C. O.	Commercial Operation	商业运行,投产
C/A	Conductivity Analyzer	电导率分析仪
C/A	GEC Alstom	通用电气/阿尔斯通公司
C/B	Cast Bronze	铸造青铜
C/D	Certificate of Delivery	交货证书
C/I	Certificate of Insurance	保险证明
C/M	Certificate of Manufacturer	制造厂证明书
C/O	Care of/Certificate of Origin	转交/原产地证明书
C/P	Control Panel	控制屏
C/W	Cement-Water Ratio	灰水比
C/W	Complete With	配有
C to C	Center to Center	中心(轴间)距
CA	Cancel/Compressed Air/Condition Assessment	取消/压缩空气/条件评估
CAB	Controlled Atmosphere-Brazing	保护气体钎焊
CABT	Cabinet	(小)室(柜,隔间)
CABW	Carbon Arc Braze Welding	碳弧钎焊
CAC	Carbon Arc Cutting	碳弧切割
CAC-A	Air Carbon Arc Cutting	空气碳弧切割

英文简称	英文全称	中文名称
CACS	Core Auxiliary Cooling System	堆芯辅助冷却系统
CAD	Cadmium/Cash Against Documents/Computer Aided Design/Computer Aided Diagnosis	镉/凭单付款/计算机辅助设计/计算机辅助诊断
CADD(S)	Computer Aided Design and Drawings(Drafting)	计算机辅助设计与制图
CADPL	Cadmium Plated	镀镉的
CAE	Canadian Atomic Energy/Computer Aided Engineering/Certified as Executed	加拿大原子能/计算机辅助工程/竣工验收
CAF	Cost and Freight	成本加运费
CAGS	Compressed Air and Gas System	压缩空气和气体系统
CAL	Calculated Average Life/Calibration	平均计算寿命,计算平均寿命/校准,标定
CAM	Computer Aided Manufacturing	计算机辅助制造
CAMAC	Computer Aided Measurement and Control	计算机辅助测量和控制(系统)
CAMLS	Computerized Annunciation Message List System	计算机化报警信息列表系统
CAN	Cell Average Normalization/National Standard of Canada	栅元平均归一化/加拿大国家标准
CANATOM	Canatom Inc.	加拿通公司
CANDU	Canada Deuterium Uranium	加拿大重水天然铀(反应)堆,坎杜堆

英文简称	英文全称	中文名称
CANNET	CANDU Network	坎杜网络，坎杜网
CAP	Capacity/Commodity Access Passage	容量，功率/货物出入通道
CAPES	Containment Atmosphere Purge Exhaust System	安全壳气体净化排放系统
CAPP	Computer Aided Process Planning	计算机辅助工艺规划
CAR	Corrective Action Reports/Corrective Action Request	纠正措施(行动)报告/纠正措施(行动)请求
CARB	Carbon	碳
CARR	Carriage	托架
CAS	Central Alarm Station/Compressed Air System/Concession Application to Supplier	中央警报站/压缩空气系统/对供货商的特许权申请
CAT	Control and Test/Control Attenuator Timer	控制和测试/控制衰减计时器
CAT/Catal.	Catalogue	目录
CAV	Cavitation	气蚀，气穴现象
CAW	Carbon Arc Welding	碳极电弧焊
CAW-G	Gas Carbon Arc Welding	气体保护碳弧焊
CAW-S	Shielded Carbon Arc Welding	有保护碳弧焊
CAW-T	Twin Carbon Arc Welding	双碳极间电弧焊
CB	Circuit Breaker/Common Base/Contact Breaker/Control Board/Control Bus/Control Button	断路器/共基极/接触断路器/控制板，控制盘/控制总线/控制按钮

英文简称	英文全称	中文名称
CBA	Computerized Blocking Assistance	计算机辅助隔离系统
CBI	Confederation of British Industry/Configuration Baseline Index	英国工业联合会/配置基准索引
CBM	Contract Budget Management	合同预算管理
CBP	Condensate Booster Pump	凝结水增压泵
CBR	Control Board Requirements	控制盘要求
CC	Characteristic Curve/Channel Controller/Chemical Composition/Closed Circuit/Construction Contractor/Control Console/Counter Clockwise/Critical Condition	特性曲线/通道控制器/化学成分/闭路/建造承包商/控制台/反时针方向/临界条件
CCA	Commissioning Completion Assurance/Configuration Change Approval/Contract Change Analysis	调试完工确认/配置变更批准/合同变更分析
CCB	Change Control Board/Configuration Control Board	变换控制板(委员会)/构形控制板
CCC	Cargo Condition Certificate/Common Control Center/Community Coordination Center/Console Control Circuits	货物状态证书/公用控制中心/地区协调中心/控制台控制电路
CCCO	Construction Contract Change Order	建造合同变更令

英文简称	英文全称	中文名称
CCD	Charge Coupled Device	电荷耦合器件
CCDF	Cross Compound Double Flow	双轴双排汽汽轮机
CCE	Command Control Equipment	指令控制设备
CCF	Common Cause Failure	共因故障
CCFI	Commissioning Construction Field Instruction	调试建造现场指令
CCH	Conductivity Cell Holder	电导室支撑架,传导单元支架
CCIB	China Commodity Inspection Bureau	中国商品检验局
CCIC	Core Coolant Inventory Control	堆芯冷却剂装量控制
CCKW	Counter-Clockwise	逆时针方向
CCL	Critical Crack Length/Concreting Check List	临界裂纹长度/灌砼检查单
CCMS	Committee on Challenges of Modern Science	现代科学委员会,迎接现代科学挑战委员会
CCN	Contract Change Notification	合同变更通知
CCP	Centrifugal Charging Pump/Commissioning Control Point/Construction Contract Package/Critical Channel Power/Critical Compression Pressure/Contractual Change Proposal	离心式上充泵/调试控制点/建造合同包/临界通道功率/临界压缩压力/商务变更建议,合同变更建议
CCR	Central Control Room/Commissioning Clarification Request	中央控制室/调试澄清要求

英文简称	英文全称	中文名称
CCS	Clearance Control System/Containment Cooling System/Coordinated Control System/Critical Compression Stress	间隙控制系统/安全壳冷却系统/协调控制系统/临界压缩应力
CCSF	Cross-Compound Single Flow	双轴单排汽汽轮机
CCSM	Construction Contractor's Site Manager	建造承包商现场经理
CCSP	Construction and Commissioning Spares	建造和调试备件
CCT	Cause-Consequence Tree/Cold Commissioning Test	因果树(分析)/冷(态)调试
CCTV	Closed Circuit Television System	闭路电视系统
CCU	Camera Control Unit	摄像机控制装置
CCVT	Coupling Capacitor Voltage Transformer	耦合电容变压器
CCW	Condenser Cooling Water/Counter-Clockwise	凝汽器冷却水系统/反时针方向
CCWS	Component Cooling Water System	设备冷却水系统
CD	Coastdown/Component Design/Conceptual Design/Condenser/Control Schematic Diagram	惰走/部件设计/概念设计/凝汽器/控制示意图
CDC	Control Digital Computer/Control Distribution Center	控制数字计算机/控制分配中心

英文简称	英文全称	中文名称
CDCR	Construction Design Clarification Request	建造设计澄清要求
CDD	Conceptual Design Description	概念设计描述
CDF	Central Distribution Frame/Control Distribution Frame/Core Damage Frequency	中央配线架/控制配线架/堆芯损坏频度,堆芯损坏概率
CDL	Contract Documents List	合同文件清单
CDN	Conceptual Design Note	概念设计说明
CDP	Conceptual Decommissioning Plan/Conceptual Design Package/Critical Decision Point	概念退役计划/概念设计包/临界判定点
CDR	Conceptual Design Requirement/Construction Deficiency Report	概念设计要求/建造缺陷报告
CDROM	Compact Disk Read Only Memory	光盘只读存储器
CDS	Conceptual Design Specification	概念设计规格书
CDSM	Claim Delivery Sheet Management	索赔单管理
CDSR	Condenser	凝汽器,冷凝器
CDV	CANDEV Reports	CANDEV 报告
CDWS	Chilled Water System	冷冻水系统
CE	Chemical Engineer/Chief Engineer/Civil Engineer	化学工程师/总工程师/土木工程师
CEA	Commissariat de Energie Atomique/Control Element Assembly	法国原子能委员会/控制棒组件,控制元件组件

英文简称	英文全称	中文名称
CEAR	Construction & Erection All Risks	建安一切险
CED	Contract Effective Date	合同生效日期
CEDB	Component Event Data Bank	部件事件数据库
CEDM	Control Element Drive Mechanism	控制棒驱动机构
CEDS	Control Element Drive System	控制棒驱动系统
CEG	Commissioning Execution Group	调试执行组
CEM	Comprehensive Emergency Management	综合应急管理
CEMA	Canadian Electrical Manufacturer's Association	加拿大电器制造商协会
CEMR	Center for Energy and Mineral Resources	能源和矿物质资源中心(美国)
CEN	Cell Edge Normalization	栅元边缘归一化
CEOC	County Emergency Operating Centre	地区应急操控中心
CEP	Condensate Extraction Pump	凝结水抽水泵
CERN	European Organization for Nuclear Research	欧洲核研究机构
CF	Centrifugal Force/Conversion Facility/Conversion Factor/Correct Factor/Cost and Freight	离心力/转化设备/转换系数/修正因子/成本加运费价格
CFA	Central Facilities Area/Clear for Action	中心设施区/可以应用,可以行动

英文简称	英文全称	中文名称
CFAR	Commissioning Field Action Request/Commissioning Final Acceptance Report	调试现场活动申请/调试最终验收报告
CFC	Containment Fan Cooler/Certified for Construction	安全壳风冷却器/确保施工的
CFC-A	Oxyacetylene Cutting	氧乙炔切割
CFC-H	Oxyhydrogen Cutting	氢氧切割
CFC-N	Oxynatural Gas Cutting	氧天然气切割
CFCR	Commissioning Field Change Request	调试现场变更要求
CFE	Contractor Furnished Equipment/Criteria for Evaluation	承包商供应的设备/评价准则
CFI	Crystal Frequency Indicator	晶体频率指示器
CFR	Code of Federal Regulation/Common Field Rack	联邦管理法规(美国)/公用现场机架
CFS	Common Field Stand/Core Flooding System	公用现场试验台架/堆芯淹没系统
CFSTI	Clearinghouse for Federal Scientific and Technical Information	联邦科技情报交换中心(美国)
CFT	Cold Functional Test	冷态功能试验
CFTL	Component Flow Test Loop/Core Flow Test Loop	部件流动试验回路(美国)/堆芯流量试验回路
CFV	Channel Flow Verification	通道流量验证
CGA	Compressed Gas Association	压缩气体协会

英文简称	英文全称	中文名称
CGCS	Combustible Gas Control System/Cover Gas Cleanup System	可燃气体控制系统/覆盖气体净化系统
CGE	Canadian General Electric Co.	加拿大通用电气公司
CGR	Carbon Dioxide Cooled Graphite Moderated Reactor/Ceiling Grounding Receptacle	CO_2 冷却石墨慢化(反应)堆/天棚接地插座
CGS	Carrier Grounding Switch	线盘接地开关
CGSB	Canadian General Standard Bureau	加拿大标准总局
CH	Channel/Checking	通道/检查(验)
CHC	Choke Coil	扼(阻)流圈
CHCF	Component Handling and Cleaning Facility	部件操作及清洗装置
CHD	Construction History Docket	建造历史大事记
CHF	Critical Heat Flux	临界热流密度
CHFR	Critical Heat Flux Ratio	临界热流密度比
CHFRE	Chexal-Hedges Flow Ratio Equalizer	Chexal-Hedges 流率补偿器
CHK	Check	检查
CHKG	Checking	检查
CHMBR	Chamber	室
CHOC	Civil Works Hand Over Certificate	建筑物移交证书
CHPP	Combined Heat and Power Plant	热电联供厂
CHRS	Containment Heat Removal System	安全壳排热系统

英文简称	英文全称	中文名称
CHT	Cold Hydraulic Test	冷态水压试验
CHU	Caloric Heat Unit/Centigrade Heat Unit	卡热单位/摄氏热单位
CHW	Chilled Water	冷冻水
CI	Calibration Information (Drawing)/Cast Iron/Chlorine Institute/Circuit Interrupter/Confidence Interval/Contact Input/Conventional Island/Cut-In	标定信息(图)/铸铁/氯气学会/电路中断器/置信区间/触点输入,接触输入/常规岛/接入,插入
CIAS	Containment Isolation Actuation Signal	安全壳隔离触发信号
CIC	Compensated Ionization Chamber	补偿电离室
CID	Change in Design	设计更改
CIF	Cost, Insurance and Freight	到岸价(货价、保险及运费)
CIGAR	Channel Inspection and Gauging Apparatus for Reactors	反应堆通道检查和测量装置
CIIR	Central Institute for Industrial Research	中央工业研究所
CILC	Crud Induced Localized Corrosion	积垢引起的局部腐蚀
CILRT	Containment Integrated Leak Rate Test	安全壳整体泄漏率试验
CIN	Component Intervention Notice	设备干预通知
CIP	Carriage and Insurance Pay to/Construction Implementation Plan	运费和保险付至(指定目的地)/建造实施计划

英文简称	英文全称	中文名称
CIR	Circuit/Customer Inspection Record	电(回)路/买主检查记录
CIRC	Circular	圆
CIRR	Commercial Interest Reference Rate	商业参考利率
CIS	Chinese Industrial Standards/Containment Isolation Signal/Containment Isolation System	中国工业标准/安全壳隔离信号/安全壳隔离系统
CIV	Containment Isolation Valve	安全壳隔离阀
CJ	Cold Junction/Construction Joint	冷接点/施工结合处
CK	Call Key/Check/Circuit	呼叫键/检验,核对/电路
CL	Cable List/Center (Central) Line/Class/Clutch/Critical Level	电缆清单/中心线/等级,类/离合器/临界液位
CLCS	Closed-Loop Control System	闭路控制系统
CLK	Clock	时钟
CLMP	Clamp	夹具,夹子
CLN	Clean	清洁(干净)的
CLP	China Light & Power Co. Ltd./Classification Package	中华电力有限公司/分类包
CM	Center of Mass/Centimeter/Circuit Module/Commissioning Manual/Control Motor/Corrective Maintenance	质心/厘米/电路组件/调试手册/控制马达/故障检修
CMAA	Crane Manufactures Association of America	美国吊车制造商协会

英文简称	英文全称	中文名称
CMF	Common-Mode-Failures	共模故障
CMMS	CANDU Material Management System	坎杜材料管理系统
CMP	Core-Melt Probability/Corrugated Metal Pipe	堆芯熔化概率/金属波纹管
CMR	Common Mode Rejection (Ratio)/Continuous Maximum Rating	共态抑制(比)/持续(运行)最大功率
CMRCL	Commercial	商业(用)的
CMS	Cycle Monitoring System	循环监测系统
CMT	Construction Management Team	建造管理队
CMTP	Calibration and Maintenance Test Procedure	校准与维护测试程序
CMU	Control Monitor Unit	监控装置
CN	Change Notice	变更通知
CNAA	Continuous Neutron Activation Analysis	连续中子活化分析
CND	Construction Notice of Deviation	建造偏差通知
CNEC	China Nuclear Engineering Construction Corporation	中国核工程建设总公司
CNEIC	China Nuclear Energy Industrial Company	中国原子能工业公司
CNI	Chinese Nuclear Industry	中国核工业
CNPM	Canatom NPM	加拿通 NPM
CNSC	Canadian Nuclear Safety Council	加拿大核安全委员会

英文简称	英文全称	中文名称
CNVTR	Converter	转(变)换器
CNVYR	Conveyer	传(输)送机
CO	Carbon Monoxide/Carry over/Component Specification/Control Operator/Cut-Out	一氧化碳/带水率/部件说明书/控制操纵员/切断,关闭,结束工作
Co.	Company	公司
Co. Ltd.	Company Limited	有限公司
COA	Commissioning Operations Authority	调试运行授权
COC	Change of Contract/Chemical Oxygen Consumption	合同变更/化学耗氧量
COD	Chemical Oxygen Demand/Commercial Operation Date/Crack Opening Displacement	化学需氧量/商业运行日期/裂纹开口位移
COED	Concentration on Engineering Design	专注于工程设计
COF	Cooling of Fuel	燃料冷却
COFACE	Compagnie Francaise d' Assurance Pour le Commerce Enterieur	法国出口信贷担保局
COG	CANDU Owners' Group	坎杜业主联合会,坎杜堆业主集团
COGA	Combustible Gas Analyzer	可燃气体分析装置
COGO	Coordinate Geometry	坐标几何学
COI	Communication Operating Instruction	通信操作指南

英文简称	英文全称	中文名称
COINS	Computer and Information Sciences	计算机和信息科学
COL	Collar/Construction and Operation License	环,圈/建造和运行许可证
COM	Contact	接触;触点
COMB	Combination	联(组、化、结)合
COME	Computer Output Microfilm Equipment	计算机缩微胶片输出机
COMM	Commissioning	调试
COMMUN	Communication	通信,交流,联络
COMP	Compression	压缩
COMPD	Compound	化合物
COMPR	Compressor	压缩机
CONCEN	Concentric	同心
COND	Condenser/Condition/Conditioning	凝汽器,冷凝器/条件/调节
CONDE	Condensate	冷凝水(液)
CONF	Conference	会议
CONN	Connection/Connector	连接/连接器
CONS	Construction	建造,施工
CONT	Continuous	连(继)续
Cont.	Contract/Continued	合同/续
CONTD	Contained	包含(括)
CONVEN	Conventional	普通的,常规的,通用的
CONVN	Conversion	转换

英文简称	英文全称	中文名称
COP	Coefficient of Performance/Continuous Operation Procedure	性能系数/连续操作程序
COR	Center of Rotation	转动中心
CORR	Corrosion	腐蚀
CORRU	Corrugated	波纹(形)
COS	Crack Opening Stretch	裂纹开口扩展
COSCO	China Ocean Shipping Corporation	中国远洋运输集团公司
COSMOS	Core On-Line Surveillance Monitoring and Operation System	堆芯在线监测与运行系统
COSP	Construction Spare	施工备件
COST	Contaminated Oil Settling Tank	污油沉淀槽
COT	Calandria Outlet Temperature	排管容器出口温度
COTS	Checkout Test Set	检测设备
COUP	Coupling	耦合
COV	Cover/Cross-Over Valve/Cut off Valve	转换阀/截止阀
CP	Cable Pan/Caloric Power/Closure Plug/Commissioning Procedure/Condensate Pump/Construction Permit/Control Panel/Control Point/Cooling Pond/Critical Pressure	电缆槽/发热能力 密封塞/调试程序/凝结水泵/建造许可/控制盘/控制点/冷却水池/临界压力
CP1	Level 1 Commissioning Procedure	一级调试程序
CP2	Level 2 Commissioning Procedure	二级调试程序

英文简称	英文全称	中文名称
CP3	Level 3 Commissioning Procedure	三级调试程序
CP4	Level 4 Commissioning Procedure	四级调试程序
CPA	Closest Point of Approach/Construction Permit Applications	最接近点/建造许可证申请
CPAA	Charged Particle Activation Analysis	带电粒子活化分析
CPBA	Competitive Protein Binding Assay	竞争性蛋白质结合化验
CPC	Commissioning Planning & Control	调试计划和管理
CPD	Contact Potential Difference	接触电位差
CPDC	Core Power Distribution Control	堆芯功率分布控制
CPDP	Core Power Distribution Prediction	堆芯功率分布预测
CPE	Charged Particle Equilibrium/Construction Project Evaluation Programme/Core Performance Evaluation	带电粒子平衡/建设项目评估计划,建造方案评估计划/堆芯性能评价
CPF	Coated Particle Fuel	包覆颗粒燃料
CPIS	Containment Purge Isolation Signal	安全壳吹扫隔离信号
CPL	Conformity Punch List	符合性消缺单
CPM	Counts Per Minute/Critical Path Method	每分钟计数/关键路径法
CPOS	Continuous Production Operation Sheet	流水线作业图表

英文简称	英文全称	中文名称
CPP	Card-Punching Printer/Condensate Polishing Plant	卡片穿孔打印机/凝结水净化装置
CPPF	Channel Power Peaking Factor	通道功率峰因子
CPR	Critical Power Ratio	临界功率比
CPS	Central Processing System/Condensate Polishing System/Count per Second	中央处理系统/凝结水净化系统/每秒计数
CPU	Card Punching Unit/Central Processing Unit	卡片穿孔装置/中央处理装置
CPV	Concrete Pressure Vessel	混凝土压力容器
CQAM	Commissioning Quality Assurance Manual	调试质量保证手册
CQD	Come Quick, Danger	遇难求救信号
CQOR	Commissioning Quality Observation Report	调试质量观察报告
CQS	Construction Quality Surveillance (Plan)	建造质量监督(计划)
CR	Capacitance-Resistance/Carriage Return/Cathode Ray/Change Request/Chloroprene Rubber/Clarification Request/Cold Rolled/Commissioning Records/Control Room/Controlled Rectifier/Conversion Ratio	阻容,容抗/回车/阴极射线/变更申请/氯丁橡胶/澄清要求/冷轧的/调试记录/控制室/可控整流器/转换比
CRA	Control Rod Assembly	控制棒组件

英文简称	英文全称	中文名称
CRAM	Card Random Access Memory	卡片随机存取存储器
CRAS	Coder and Random Access Switch	编码和随机存取开关
CRAVS	Control Room Area Ventilation System	控制室区域通风系统
CRB	Control Rod Bank	控制棒组
CRBR	Controlled Recirculation Boiling Water Reactor	受控再循环沸水堆
CRC	Control and Reporting Centre	控制和报告中心
CRCE	Cavity Reactor Critical Experiment	腔式反应堆临界实验(装置)
CRD	Client Requirements Document/ Control Rod Drive	客户要求文件/控制棒驱动
CRDCS	Control Rod Drive Control System	控制棒驱动控制系统
CRDHS	Control Rod Drive Hydraulic System	控制棒水力驱动系统
CRDM	Control Rod Drive Mechanism	控制棒驱动机构
CRES	Corrosion Resistant Steel	耐蚀钢
CRF	Carryout Rate Fraction/Clarification Request Form	夹带率份额/澄清申请表
CRGT	Control Rod Guide Tube	控制棒导向管
CRL	Chalk River Laboratory	乔克河实验室

英文简称	英文全称	中文名称
CRM	Confinement of Radioactive Material/Counting Rate Meter	放射性物质的密封/计数率仪
CRO	Cathode-Ray Oscillograph/Computer Request to Order/Control Room Operator	阴极射线示波器/采购申请计算机管理/控制室操纵员
CRP	Control Room Panel/Control of Reactor Power	控制室控制台屏/反应堆功率控制
CRQ	Current Requirements	现行要求
CRS	Client Requirement Statement/Control Room Supervisor/Cross-Section	客户要求报告书/控制室值长/横截面，截面
CRT	Cathode Ray Tube/Classroom Training	阴极射线管，显示屏/课堂培训
CRTC	China Road Transportation Company	中国汽车运输总公司
CRV	Combined Reheat Valve	组合再热阀
CS	Carbon Steel/Component Specification/Concession Application/Control Signal/Control Switch/Controlled Shutdown/Core Spray/Cross Section/Current Strength	碳钢/设备规格书/特许权申请/控制信号/控制开关/受控停堆，计划停堆/堆芯喷淋/截面，横截面/电流强度
CSA	Canadian Standards Association/Core Support Assembly	加拿大标准协会/堆芯支承组件
CSAL	Construction Allowance	建造经费限额

英文简称	英文全称	中文名称
CSAR	Conceptual Safety Analysis Report	概念安全分析报告
CSAS	Containment Spray Actuation Signal	安全壳喷淋触发信号
CSB	Core Support Barrel	堆芯支承吊篮
CSC	Change Schedule Chart	进度(变化)表
CSCS	Core Spray Cooling System/Core Standby Cooling System	堆芯喷淋系统/堆芯备用冷却系统
CSD	Catenaary System Drive/Computer System Design/Criticality Safety Document	悬链系统传动/计算机系统设计/临界安全文件
CSDV	Condenser Steam Discharge Valve	凝汽器蒸汽排放阀
CSE	Containment System Experiment	安全壳系统实验
CSEC	Cross-Section	截面,横截面
CSF	Concentric Saturated Finishing/Condensate Storage Facility/Core Support Floor	同心饱和精加工/冷凝液贮存设施/堆芯支承底板
CSG	Commissioning Support Group	调试支持组
CSIS	Containment Spray Injection System	安全壳喷淋注入系统
CSISRS	Cross Section Information Storage and Retrieval System	截面数据储存与检索系统
CSL	Current Switch Logic	电流开关逻辑
CSMOL	Control Station Manual Operating Level	控制站手控工作水平
CSMPS	Computerized Scientific Management Planning System	计算机化的科学管理计划系统

英文简称	英文全称	中文名称
CSO	Commissioning Specification & Objective	调试规格书与目标
CSP	Containment Spray Pump/Critical Safety Parameter	安全壳喷淋泵/关键安全参数
CSPMS	Critical Safety Parameter Monitoring System	关键安全参数监测系统
CSR	Comprehensive Study Report	综合研究报告
CSRS	Containment Spray Recirculation System/Core Spray Recirculation System	安全壳喷淋再循环系统/堆芯喷淋再循环系统
CSS	Containment Spray System/Component Specification Sheet	安全壳喷淋系统/设备说明书
CST	Condensate Storage Tank/Control System Test/Countdown Sequence Timer	冷凝液贮存箱/控制系统试验/倒计时器
CSTS	Calandria Side Tube-Sheet Sub-Assembly/Combined System Test Stand	(重水堆)排管容器端管板子组件/组合系统试验台
CSUM	Consumable Supplies	消耗品
CTC	Centralized Control	集中控制
CTE	Cable Termination Equipment	电缆终端设备
CTF	Component Test Facility	部件试验设施,设备试验设施
CTG	Coating	涂(镀)层,包覆
CTI	Comparative Tracking Index/Conductivity-Temperature Indicator	比较径迹指数/导热率-温度显示器

英文简称	英文全称	中文名称
CTL	Certified Tool List	检定合格的工具清单
CTM	Channel Temperature Monitoring (System)/Component Test Memo	通道温度监测(系统)/部件测试备忘录,设备测试备忘录
CTMOD	Commissioning Temporary Modification	调试临时变更
CTMS	Channel Temperature Monitoring System	(燃料)通道温度监测系统
CTOA	Crack Tip Opening Angle	裂纹顶尖张开角度
CTOP	Commissioning Temporary Operation Procedure	调试临时操作程序
CTP	Contractual Training Program	合同培训大纲
CTR	Center/Contract Technical Requirements/Critical Temperature Resistor	中心/合同技术要求/临界温度电阻器
CTRG	Centering	定(中)心
CTRL	Control	控制
CTS	Cartographic Test Standard/Compact Tension Specimen/Component Test Set	制图测试标准/紧凑拉伸试样/部件测试组,设备测试装置
CTU	Centigrade Thermal Unit	摄氏热量单位
CU	Carryunder/Copper	水带汽,带汽率/铜
CUF	Cumulative Usage Factor	累计利用因子
CUFT	Cubic Feet	立方英尺
CUM	Cubic Meter	立方米
CUW	Call upon Warranty	要求(供货商)履行保证条款

英文简称	英文全称	中文名称
CV	Check Valve/Constant Velocity/Containment Vessel/Control Valve	止回阀/恒速/安全壳(壳体)/控制阀
CVCS	Chemical and Volume Control System	化学和容积控制系统
CVIS	Containment Ventilation Isolation System	安全壳通风隔离系统
CVPETS	Condensate Vacuum Pump Effluent Treatment System	凝结水真空泵排水处理系统
CVR	Component Verification Report	设备/部件验证报告
CVS	Component Verification Specification	设备/部件验证规格书
CVSS	Control Valve Specification Sheet	控制阀规格表
CW	Cladding Waste/Clockwise/Constant Wave/Cooling Water/Cosine Wave	(燃料)包壳废物/顺时针/恒波/冷却水/余弦波
CW	Cold Welding	冷压焊
CWD	Capacitance Water-Level Detector	电容式水位检测器
CWP	Construction Work Package	土建工作包
CWS	Central Wireless Station/Chilled Water Supply/Circulating Water System/Cooling Water System	无线电中心台/冷冻水供应/循环水系统/冷却水系统
CX	Coaxial	同轴的
CY	Calendar Year	日历年
CYAN	Cyanide	氰化物
CYL	Cylinder	圆柱(筒);汽缸
CYN	Cyanide	氰化物

英文简称	英文全称	中文名称
D	Depth	深度
D&A	Decision and Action	决策与行动
D & D	Decontamination and Decommissioning	去污和退役
D&D	Decontamination and Disposal	去污和处置
D. t.	Delivery Time	交货时间
D/A	Digital to Analog	数模转换
D/C	Degasser Cooler	除气冷却器
D/F	Defective Fuel	破损燃料
D/I	Digital Input	数字输入
D/O	Delivery Order/Digital Output	提货单,出栈凭单/数字输出
DA	Data Acquisition/Delayed Action	数据采集(获取)/延期作用,延迟动作
DAA	Data Access Arrangement	数据存取安排
DAC	Data Acquisition Computer/Derived Air Concentration/Digital-Analog Converter	数据采集计算机/导出的空气浓度/数字-模拟转换器
DADS	Data Acquisition and Display System	数据获取与显示系统
DAISY	Data Acquisition and Interpretation System	数据收集和整理系统
DALD	D_2O in Air Leak Detection	空气中重水检漏
DAPS	Data Acquisition and Processing System/Direct Access Programming System	数据获取和处理系统/随机存取程序设计系统,直接存取程序设计系统
DAS	Data Acquisition System	数据采集系统

英文简称	英文全称	中文名称
DASTAR	Data Storage and Retrieval System	数据储存与检索系统
DAVDS	Data Acquisition and Visual Display System	数据采集与直观显示系统
DAW	Dry Active Waste	干放射性废物
DB	Data Base/Data Bus/Design Baseline/Distribution Box	数据库/数据总线/设计基准/配电箱
DB	Dip Brazing	沉浸钎焊
DBE	Design Basis Earthquake	设计基准地震
DBF	Design Basis Failure/Design Basis Fire	设计基准故障/设计基准火灾
DBFL	Design Basis Flooding Level	设计基准淹没水位,设计基准洪水水位
DBLOCA	Design-Basis Loss-of-Coolant Accident	设计基准冷却剂丧失事故
DBM	Design Bill of Material	设计材料清单
DBMS	Database Management System	数据库管理系统
DBR	Design Basis Review	设计基础审查
DBS	Design Basis Storm	设计基准风暴
DBTT	Ductile-Brittle Transition Temperature	延性-脆性转变温度
DC	Density Controller/Design Change/Design Criteria/Digital Computer/Direct Current/Distribution Coefficient	密度控制器/设计变更/设计标准/数字计算机/直流/分配系数
DCA	Design Completion Assurance/Deuterium Critical Assembly/Digital Computers Association	设计完工保证/重水临界(实验)装置/数字计算机协会

英文简称	英文全称	中文名称
DCC	Degasser Cooling Condenser/Design Change Committee/Digital Control Computer/Distribution Control Center	除气冷却冷凝器/设计变更委员会/数字控制计算机/分配控制中心
DCD	Design Concept Decision	设计概念决策
DCH	Drain Collection Header	疏水收集总管
DCI	Design Change Instruction/Double Cycle Inverse/Drawing Change List/Ductile Cast Iron	设计变更令/反相双循环/图纸更改一览表/球墨铸铁
DCM	Design Change Manual/Design Clarification Memo	设计变更手册/设计澄清备忘录
DCN	Design Change Notice	设计变更通知
DCP	Design Change Package	设计变更包
DCR	Design Change Request/Device Control Requirement/Direct Conversion Reactor/Document Correction Request	设计变更请求/设备控制要求/直接转换反应堆/文件修正申请
DCRV	Degasser Condenser Relief Valve	除气冷凝器卸压阀
DCS	Data Communication System/Data Control System/Design and Calculation Sheet/D_2O Collection System	数据传输系统，数据通信系统/数据控制系统/设计和计算表/重水收集系统
DCUS	D_2O Clean-Up System	重水净化系统
DD	Design Description/Design Deviation/Detail Design/Detail Drawing/Digital Display/Direct Drive/Double Diode	设计描述/设计偏差/详细设计/详图/数字显示/直接驱动(装置)/双二极管

英文简称	英文全称	中文名称
DDA	Digital Differential Analyzer	数字微分分析器
DDAPS	Digital Data Acquisition and Processing System	数字数据获取与处理系统
DDC	Defense Documentation Center for Scientific and Technical Information/Direct Digital Control	国防科技情报资料中心(美国)/直接数字控制
DDD	Deadline Delivery Date/Dilute, Disperse and Decontaminate	交货截止日期/稀释、弥散和去污
DDF	Digital Distribution Frame	数字配线架
DDG	Digital Display Generator	数字显示发生器
DDH & DS	Digital Data Handling and Display System	数据处理和显示系统
DDI	Depth Deviation Indicator	深度偏差指示器
DDL	Data Drawing List	资料图纸清单
DDP	Delivered Duty Paid/Detailed Decommissioning Plan	完税交货(价)/详细退役计划
DDR	Design Development Record/Deviation Disposition Request	设计进展记录/偏差处理要求
DDU	Delivered Duty Unpaid	未完税交货(价)
DDW	Detail Design Drawing	详细设计图
DE	Digital Element/Display Equipment/Dose Equivalent/Double End	数字元件/显示设备/剂量当量/双端
DECON	Decontamination	去污
DEGB	Double-Ended Guillotine Break	双端切断破裂
DEH	Digital Electrohydraulic	数字电动液压

英文简称	英文全称	中文名称
DEHG	Digital Electro-Hydraulic Governor	数字电液调节器
DEI	Design Engineering Inspection	设计工程检查
DEIB	Design Engineering Inspection Board	设计工程检查委员会
DEL	Delete Character/Derived Emission Limit	删除符/导出的释放限值
DEN	Design Evaluation Notice	设计评估通知
DEP	Deflection Error Probable	可几偏转误差,预期偏离误差
DESTN	Destination	目的地
DET	Detector	探(检)测器
DETP	Design Evaluation Test Program	设计评估试验大纲
DEUT	Deuteration	氘化
DF	Decontamination Facility/Decontamination Factor/Diffuser	去污设施/去污因子/扩散器
DFA	Damage Function Analysis	损伤功能分析
DFB	Departure from Film Boiling	偏离膜态沸腾
DFB	Diffusion Brazing	扩散钎焊
DFCS	Digital Feedwater Control System	数字给水控制系统
DFR	Dounreay Fast Reactor	唐瑞快(中子)反应堆(英国)
DFW	Diffusion Welding	扩散焊
DG	Design Guide/Diesel Generator	设计导则/柴油发电机
DGCAIES	Diesel Generator Combustion Air Intake and Exhaust System	柴油发电机燃烧、进气与排气系统
DGCR	Direct Cycle Gas Cooled Reactor	直接循环气冷反应堆
DGFOSTS	Diesel Generator Fuel Oil Storage and Transfer System	柴油发电机燃料油贮存和输送系统

英文简称	英文全称	中文名称
DGS	Design Guide Supplement	设计导则补充
DH	Design Handbook/Difference in Height	设计手册/高度差
DHCV	Delayed Hydride Cracking Velocity	延迟氢化断裂速度
DHE	Data Handling Equipment	数据处理设备
DHLW	Defense High-Level Waste	国防高放射性废物
DHR	Decay Heat Rejection	衰变热排出
DHRIL	Decay Heat Removal Intermediate Loop	衰变热排出中间回路
DHRS	Decay Heat Removal System	衰变热排出系统
DHX	Dump Heat Exchanger	事故排放热交换器
DI	Decommissioning Instruction/Decontamination Index/Digital Input/Ductile Iron	退役说明,退役指南/去污指数/数字输入/球墨铸铁
DIA	Diameter	直径
DIAPH	Diaphragm	隔膜(板)
DIB	Department Information Bulletin/Design Information Bulletin	部门信息公告/设计信息公报
DIC	Digital Integrated Circuit	数字集成电路
DID	Digital Information Display/Direct inward Dialing	数字信息显示/直接接入拨号
DIE	Design Industrial Engineering	设计工业工程学
DIFF	Differential	微分,差分(动)
DIG RO	Digital Readout	数字读出
DIL	Derived Intervention Level	导出的干预水平

英文简称	英文全称	中文名称
DIR	Direct/Direction/Directional	直接/方向/定向的，方向的
DIS	Doppler Inertial System/Draft International Standard	多普勒惯性系统/国际标准草案
DIV	Division	部(分)，处
DK	Deck	台面
DL	Datum Level/Dead Load/Drawing List	基准面/静载荷/图纸清单
DLD	D_2O Leak Detection	重水检漏，重水泄漏检测
DLEC	Dose Limit Equivalent Concentration	剂量限值等效浓度
DLF	Dynamic Load Factor	动态载荷因子
DLL	Design Limit Load	设计极限载荷
DM	Damage Monitor/Design Manual/Doctor of Mathematics	损伤监测仪/设计手册/数学博士
DMA	Design Manual Addendum	设计手册补遗
DMDS	D_2O Management-D_2O Supply	重水管理-重水供应
DME	Distance Measuring Equipment	测距仪
DMF	Dose Modifying Factor	剂量修正因子
DMR	Defective Material Report	损坏材料报告
DMS	Document Management System	文件资料管理系统
DN	Design Note/Delayed Neutron	设计说明/缓发中子
DNB	Departure From Nucleate Boiling	偏离泡核沸腾
DNBR	Departure From Nucleate Boiling Ratio	偏离泡核沸腾比
DND	Delayed Neutron Detector	缓发中子探测器

英文简称	英文全称	中文名称
DNM	Delayed Neutron Monitor	缓发中子监测器
DNS	Directorate of Nuclear Safety	原子核工作安全委员会,核安全委员会
DO	Digital Output/Drawing Office	数字输出/制图室
DOC	Documentation	文件,文件编制
DOD	Direct outward Dialing	直接外拨
DOE	Department of Energy	(美国)能源部
DOL	Derived Operational Limit	导出的运行限值
DOM	Deputy Operations Manager/Drawing Office Manual/Dryout Margin	电站运行副经理/绘图室手册/干涸裕度
DOR	Division of Responsibility	责任分工
DOS	Disk Operating System	磁盘操作系统
DOW	Division of Work	业务部
DP	Data Processing/Dead Point/Design Plan/Design Pressure/Dew Point/Diametral Pitch/Differential Pressure/Dye Printing	数据处理/死点/设计方案/设计压力/露点/径向节距/差压/着色探伤
DPC	Data Processing Center	数据处理中心
DPDT	Double-Pole Double-Throw	双刀双掷
DPE	Data Processing Equipment	数据处理设备
DPLTD	Depleted	贫化的,耗尽的
DPN	Diamond Pyramid Number	威氏硬度值
DPR	Demanded Power Routine/Demonstration Power Reactor	通常所需功率/示范动力堆
DPS	Data Processing System	数据处理系统

英文简称	英文全称	中文名称
DPST	Double-Pole Single-Throw	双刀单掷
DPV	Drop Pressure Valve	降压阀
DR	Design Requirement/Design Review/Deviation Ratio	设计要求/设计审查/偏移系数，偏离比
Dr.	Drum	桶；卷(盘)
DRAP	Design Reliability Assurance Plan	设计可靠性保证计划
DRD	Deep Rock Disposal	深岩层处置
DREE	Direction Des Relations Economiques Exterieures	法国对外经济关系总局
DRF	Dose Reduction Factor	剂量减低系数，剂量降低因子
DRL	Derived Release Limits	导出的释放限值
DRN	Drain	疏(排)水
DRP	Design Report/Double Rotatable Plug	设计报告/双旋塞
DRR	Design Review Report	设计审查报告
DRS	Data Reduction System/Design Registration Submission	数据简化系统/设计注册提交文件
DS	Decontamination System/Design Specification/Disconnecting Switch/Dousing System/Drywell Spray	去污系统/设计规格书/隔离开关/干井喷淋
DS	Dip Soldering	沉浸钎焊
DSA	Differential Strain Analysis	微分应变分析
DSC	Document Service Center	资料服务中心
DSDD	Dousing System Detail Drawing	喷淋系统详图

英文简称	英文全称	中文名称
DSHE	Downstream Heat Exchanger	顺流热交换器
DSIR	Department of Scientific and Industrial Research	科学和工业研究局(英国)
DSK	Disk	磁盘
DSL	Design Safety Limit	设计安全限值
DSLD	Dousing System Line Detail	喷淋系统管道详图
DSM	Design Standards Manual	设计标准手册
DSP	Dynamic Speaker	电动式扬声器
DSR	Data Survey Report/Design Submission Report	资料述评报告,资料调研报告/设计提交报告
DSS	Decision Support System	决策支持系统
DSW	Dry Solid Waste	干固体废物
DT	Destructive Test/Dimensional Check/Down Time/Dwell Time	破坏性试验/尺寸检查/停工期/保压时间,停留时间
D-T	Deuterium-Tritium/Deuterium-Tritium Reactor	氘-氚/氘-氚反应堆
DTCS	Data Transmission and Control System	数据传递与控制系统
DTG	Date-Time Group	日期分组
DTHYL	Ethylene	乙烯
DTL	Detail/Detroit Testing Laboratory	详图/底特律测试实验室
DTMF	Dual-Tone Multifrequency	双音多频
DTP	Design Transfer Package/Dynamic Test Plan	设计转让包/动态试验计划
DTS	Design Technical Specification/Detailed Test Specification	设计技术规格书/试验详细说明书

英文简称	英文全称	中文名称
DTT	Design Transition Temperature	设计转变温度
DUGS	D_2O Upgrading System	重水升级系统
Dupl	Duplicate	副本
DUT	D_2O Upgrading Tower	重水升级塔
DVM	Digital Voltmeter	数字式电压表
DVP	Design Verification Plan	设计验证计划
DVPS	Degraded Voltage Protection System	电压下跌保护系统
DVR	Discharge Voltage Regulator	放电调压器
DVSP	Digital Valve Sequence Programmer	数字阀门顺序编程器
DVST	Direct View Storage Tube	直观存储显像管
DW	Dead Weight/Demineralized Water/Distilled Water/Drinking Water	重载(荷)/除盐水/蒸馏水/饮用水
DWC	Deadweight Capacity/Drain Water Circuit	载重能力/泄水管路
DWG	Drawing	图纸
DWL	Derived Working Limit	导出的工作限值
DWP	Daily Work Program/Decommissioning Work Plan (Package)	每日工作计划/退役工作计划(包)
DWS	Demineralized Water System	除盐水系统
DWST	Demineralized Water Storage Tank	除盐水贮存箱
DWT	Deadweight/Deep Well Thermometer/Drop-Weight Test	重载/深井温度计/落锤试验
DWTT	Drop Weight Tear Test	落锤扯裂试验

英文简称	英文全称	中文名称
DYNO	Dynamometer	功率计
E&O. E., EOE	Errors and Omissions Excepted	错漏不在此限
E&SP	Equipment and Spare Parts	设备与备件
E/C	Eddy-Current	涡流
E/M	Earthquake Monitor	地震监测器
EA	Energy Absorption/Energy-Analysis/Environment Assessment	能量吸收/能量分析/环境评价
EAA	Engineers and Architects Association	工程师与建筑师协会
EAB	Exclusion Area Boundary	隔离区边界,禁区边界
EACS	Emergency Air Cleaning System	应急空气净化系统
EAES	European Atomic Energy Society	欧洲原子能学会
EAF	Effective Attenuation Factor/Equivalent Availability Factor	有效衰减因子/等效可利用因子
EAM	Engineering Administration Manual	工程管理手册
EAP	Equivalent Air Pressure	当量空气压力,等效空气压力
EAR	Engineering Analysis Report	工程分析报告
EARI	Equipment Acceptance Requirement and Inspection	设备验收的要求与检查
EAS	European Atomic Society	欧洲原子能学会
EBBW	Electron Beam Braze Welding	电子束钎焊
EBC	Electron Beam Cutting	电子束切割
EBIC	Electron-Beam-Induced Conductivity	电子束感应电导率
EBOP	Emergency Bearing Oil Pump	应急轴承油泵

英文简称	英文全称	中文名称
EBP	Exhaust Back Pressure	排气背压
EBR	Electronic Beam Recorder/Experimental Breeder Reactor	电子束记录仪/实验性增殖反应堆
EBS	Electron Beam Scanning System	电子束扫描系统
EBW	Electron Beam Welding	电子束焊
EBW-HV	High Vacuum Electron Beam Welding	高真空电子束焊
EBW-MV	Medium Vacuum Electron Beam Welding	中真空电子束焊
EBW-NV	Non Vacuum Electron Beam Welding	非真空电子束焊
EC	Earth Current/Elasticity Coefficient/Electronic Computer/Enamel-Covered Wire/Environment Control/Error Correcting	接地电流/弹性系数/电子计算机/漆包线/环境控制/误差校正
ECASS	Electronically Controlled Automatic Switching System	电子控制自动交换系统
ECC	Eccentricity/Emergency Core Cooling/Equipment Condition Certification	偏心，偏心率/应急堆芯冷却/设备状态证书
ECCS	Emergency Core Cooling System	应急堆芯冷却系统
ECD	Environmental Conditions Determination	环境条件确定
ECEPDI	East China Electrical Power Design Institute	华东电力设计院
ECG	Electrical Cable Gallery	电气电缆走廊

英文简称	英文全称	中文名称
ECGD	Export Credits Guarantee Department	出口信贷担保局(英国)
ECI	Emergency Coolant Injection	冷却剂应急注入
ECL	Electronics Components Laboratory/Equipment Component List	电子元器件实验室/设备部件表
ECM	European Common Market	欧洲共同市场
ECN	Engineering Change Notice	工程设计变更通知
ECNR	European Council for Nuclear Research	欧洲原子核研究理事会
ECPL	Engineering Component Parts List	工程部件清单
ECR	Emergency Control Room/Engineering Closeout Report	应急控制室/工程设计关闭报告
ECROS	Electrically Controllable Read-Only Storage	电子可控只读存储器
ECS	Engine Control System/Engineering Change Notice Subpackage	发动机调节系统/工程设计变更通知子项包
ECT	Eddy Current Test	涡流探伤检验
ECU	European Currency Unit/Exhaust Cleanup Unit	欧洲货币单位/排风净化机组
ED	Elementary Diagram/Electrical Diagram	原理图/电气图
EDC	Engineering Data Center/Engineering Drawing Change/Export Development Corporation	工程资料中心/工程图纸变更/出口开发公司
EDCC	Environmental Detection Control Center	环境探测控制中心

英文简称	英文全称	中文名称
EDD	End Delivery Date	最后交货日期
EDD	Engineering Design Data	工程设计数据
EDECAIES	Emergency Diesel Engine Combustion Air Intake and Exhaust System	应急柴油机燃烧空气吸入和排出系统
EDECWS	Emergency Diesel Engine Cooling Water System	应急柴油机冷却水系统
EDELS	Emergency Diesel Engine Lubrication System	应急柴油机润滑系统
EDESS	Emergency Diesel Engine Starting System	应急柴油机启动系统
EDF	Effective Decontamination Factor/Evaporator Decontamination Factor	有效去污因子/蒸发器去污因子
EDG	Emergency Diesel Generator	应急柴油发电机
EDITP	Engineering Development Integration Test Program	工程进展综合试验大纲
EDL	Engineering Drawing List	工程图纸清单
EDM	Electric Dipole Moment/Electro-Discharge Machining/Engineering Design Modification	电偶极矩/放电加工/工程设计修改
EDMS	Electronic Document Management System	电子文件管理系统
EDO	Executive Director for Operations	运营执行主任
EDPE	Electronic Data Processing Equipment	电子数据处理设备

英文简称	英文全称	中文名称
EDR	Engineering Design Review	工程设计审查
EDT	Equipment Drain Tank	设备疏水箱
EE	Electrical Engineer/Errors Excepted	电气工程师/允许误差
EEC	Electronic Equipment Committee	电子设备委员会
EEC	European Economic Community	欧洲经济共同体
EEMAC	Electrical and Electronic Manufacturer Association of Canada	加拿大电气和电子制造商协会
EES	Emergency Exhaust System	应急排放系统
EESR	End of Erection Status Report	(安装)竣工报告
EF	Exhaust Fan/External Floods	排气风扇/外界洪水
EFDS	Equipment and Floor Drainage System	设备和地面疏水系统
EFF	Effect/Efficiency	效应,效果/效率
EFI	Engineering Field Instruction	工程现场指令
EFIC	Emergency Feedwater Initiation and Control	应急给水触发与控制
EFPD	Effective Full Power Days/Equivalent Full Power Days	有效满功率天数/等效满功率日
EFPH	Effective Full Power Hours	有效满功率小时
EFPY	Effective Full Power Years	有效满功率年
EFR	Event and Failure Record	事件与故障记录
EFRUG	European Fast Reactor Utilities Group	欧洲快堆电力集团公司
EFS	Emergency Feedwater System	应急给水系统
EFTA	European Free Trade Association	欧洲自由贸易协会

英文简称	英文全称	中文名称
EFU	Emergency Filter Unit	应急过滤装置
EFWP(M)	Emergency Feedwater Pump (Motor)	电动应急给水泵
EFWP(T)	Emergency Feedwater Pump (Turbine)	汽动应急给水泵
EFWS	Emergency Feedwater System	应急给水系统
EFWST	Emergency Feedwater Storage Tank	应急给水储存箱
EG	Equipment Guarantee	设备担保
EGA	Evolved Gas Analysis	离析气体分析
EGD	Evolved Gas Detection/Evolved Gas Detector	析出气体检测/析出气体检测器
EGT	Exhaust Gas Temperature	排气温度
EGW	Electrogas welding	气电立焊
EH	Electric Heater	电炉,电加热器
EHC	Electro-Hydraulic Controller	电-液控制器
EHF	Extremely High-Frequency	极高频
EHG	Electro-Hydraulic Governor	电液调节器
EHP	Effective Horsepower	有效马力
EHPH	Electric Horsepower Hour	电动马力-小时
EHRS	Emergency Heat Removal System	应急排热系统
EHV	Extra-High Voltage	超高压
EHWL	Extreme High Water Level	极高水位,高-高水位
EI	External Insulation	外部绝缘
EIA	Electronic Industries Association	电子工业协会

英文简称	英文全称	中文名称
EIB	Enforcement and Inspection Bulletin	强制执行与检查公告
EICA	Electronics Corporation of America	美国电子有限公司
EID	Engineering Item Description	工程项目说明
EIL	Error in Label	标号错误
EIN	Equipment Installation Notice	设备安装通知
EIO	Error in Operation	操作错误
EIR	Environmental Impact Report	环境影响报告
EIRP	Equivalent Isotropic Radiated Power	等效各向同性辐射功率
EISR	Environmental Impact Statement Report	环境影响陈述报告
EIT	Electrical Integrated Team	综合电气队
EJ	Expansion Joint	膨胀节
EJA	Engineering Job Analysis	工程分析
EJMA	Expansion Joint Manufacturers Association	膨胀节制造商协会
EL	Elastic Limit/Elevation/Equipment List	弹性极限/海拔标高/设备清单
ELA	Electron Linear Accelerator	电子线性加速器
ELC	Elastic Limit Under Compression	抗压弹性极限
ELEC	Electrical	电(气)的
ELEM	Element	元素(件)
ELEV	Elevation	标高;正视图
ELF	Extremely Low Frequency	极低频

英文简称	英文全称	中文名称
ELL	Electrical Load List	电气负荷清单
ELPHR	Experimental Low-Temperature Process Heat Reactor	低温供热试验(反应)堆
EM	Engineering Manual/Evaluation Model/Expanded Metal	工程手册/评价模型/膨胀金属
EMAR	Electron Microscope Autoradiography	电子显微镜自动射线照相
EMB	Embedded	预埋的,埋置的
EMC	End Manufacturing Certificate	制造完工合格证
EMD	Electric Motor Driven	电动机驱动的
EMDP	Electromotive Difference of Potential	电动势
EMER	Emergency	应急
EMI	Electric-Mechanical Indicator/Electromagnetic Indicator/Electromagnetic Interference	机电式指示仪/电磁指示器/电磁干扰
EMP	Effective Mean Pressure/Electro Mechanical Package	有效平均压力/机电(工程)包
EMR	End of Manufacturing Report/Engineering Material Research	设备出厂报告/工程材料研究
EMS	Effluent Management Software/Electrical Management System/Emergency Switch/Environment Monitor System	流出流管理软件/电气管理系统/应急开关/环境监测系统
EMT	Engineering Management Team	工程管理队
ENA	Environmental Assessment	环境评估

英文简称	英文全称	中文名称
Encl.	Enclosure	附件
ENEA	European Nuclear Energy Agency	欧洲核能机构
ENG	Engineering/Engineering Procedure Manual	工程;工程设计/技术实施细则,工程规程手册
ENPEP	Energy and Power Evaluation Programme	能源和动力评价计划
ENS	Emergency Notice System	应急通知系统
ENSM	Engineering Nuclear Safety Manager	工程核安全经理
EO	Engine Oil/Equipment Operator/Executive Order	发动机滑油/设备操纵员/执行指令
EOC	Emergency Operation Center/End of Contract/End of Cycle	应急操作中心/合同终止/燃料循环末期
EOF	Emergency Operating Feature	应急操作特性
EOL	End of Life	寿期末
EOM	End of Month	月底
EOMM	Equipment Operation and Maintenance Manual	设备运行维修手册
EOMR	End of Manufacturing Report	设备出厂报告
EONR	European Organization for Nuclear Research	欧洲原子核研究组织
EOP	Emergency Oil Pump/Emergency Operating Procedure/Equipment Operational Procedure	应急油泵/应急运行规程/设备操作程序
EOS	End of Season	季度末

英文简称	英文全称	中文名称
EOT	Electronic Overspeed Trip/End of Test	电子超速停机/试验结束
EOV	Electrically Operated Valve	电动阀
EP	Earth Plate/Edge Preparation/Embedded Parts/Embedded Plate/Emergency Preparedness/Emergency Procedure/Engineering Project/Explosion-Proof/Extreme Pressure	接地板/坡口制备/预埋件/预埋板/应急准备/应急程序/工程项目/防爆的/极端压力
EPA	Electron Probe Analysis/Environment Protection Agency	电子探针分析/环境保护局(美国)
EPCS	Emergency Plant Cooldown System	电站应急冷却系统
EPD	Earliest Possible Date	最早可能的日期
EPDCC	Electric Power Development Coordinating Council	电力发展协调委员会
EPFM	Elasto-Plastic Fracture Mechanics	弹性-塑性断裂力学
EPG	Emergency Procedure Guideline	应急程序导则
EPI	Electronic Position Indicator	电子位置指示器
EPMA	Electron Probe Microanalyzer	电子探针微量分析仪
EPOF	Equivalent Planned Outage Factor	等效计划停运因子
EPR	Emergency Procurement Requisition/Ethylene-Propylene Rubber	应急采购申请/乙烯-丙烯橡胶
EPROM	Electrically Programmable Read Only Memory/Erasable Programmable Read Only Memory	可电可编程序只读存储器/可擦可编程只读存储器
EPS	Electric Power Storage/Emergency Power Supply System	电力存储/应急供电系统

英文简称	英文全称	中文名称
EPSM	Engineering Purchasing Specification Manual	工程采购规格书手册
EPSU	Emergency Power Supply Unit	应急发电机组
EPT	Excitation Power Transformer	励磁电源变压器
EPV	Electropneumatic Valve	电力气动阀,电空阀
EQ	Environmental Qualification/Engineering Quotation	环境评定/工程报价
EQA	Environmental Qualification Assessment	环境定级评估
EQAV	Equivalent Average	当量平均
EQC	Environmentally Qualified Components List	环保合格设备清单
EQCL	Environmental Qualification Component List	需环保评定的设备清单
EQDB	Equipment Qualification Data Bank	设备鉴定数据库
EQL	Environmental Qualification List Development Package	环保评定清单开发包
EQR	Engineering Quotation Request	工程报价要求
EQTB	Environmental Qualification Technical Basis	环境评定技术基础
EQUIP	Equipment	设备
EQUIV	Equivalent	等价(效,量)的
ERA	Europe Representative Agency	驻欧办事处

英文简称	英文全称	中文名称
ERAB	Energy Research Advisory Board	能源研究咨询委员会(美国)
ERC	Electronic Research Center/Emergency Response Center	电子研究中心/应急响应中心
EREA	Equivalent Radon Emanating Area	当量氡析出面积
ERER	Equivalent Radon Emanation Rate	当量氡析出率
ERF	Emergency Response Facility	应急响应设施
ERFS	EESR Reservation Follow-up Sheet	安装竣工报告(EESR)遗留后续行动单
ERI	Engineering Research Institute	工程研究所
ERIS	Emergency Response Information System	应急响应信息系统
ERL	Emergency Reference Level	应急参照水平
ERP	Effective Radiated Power/Error Recovery Procedure	有效辐射能力/纠错程序
ERQC	Engineering Reliability and Quality Control	工程可靠性和质量控制
ERS	Equipment Requirement Specification	设备要求规格书
ERW	Electrical Resistance Weld	电阻焊
ES	Earth Switch/Electrochemical Society/Engineering Standard	接地开关/电化学学会/工程设计标准
ESAR	Electronically Steerable Array Radar	电子可控阵列雷达
ESCA	Electron Spectroscopy for Chemical Analysis	用于化学分析的电子能谱学

英文简称	英文全称	中文名称
ESD	Emergency Shutdown/Emergency Shutdown Device	事故停机;应急停堆/应急停堆装置
ESDA	Emergency Services and Disaster Agency	应急服务与救灾机构(美国)
ESDR	End-State Decommissioning Report	退役终态报告
ESFVS	Engineered Safety Feature Ventilation System	专设安全设施通风系统
ESG	Engineering Service Group	工程维修组
ESHP	Equivalent Shaft Horse Power	等效轴马力
ESI	Emergency Stop Indicator	紧急停止指示器
ESM	Engineering Standards Manual	工程设计标准手册
ESP	EESR Subsystem Package	安装竣工报告(EESR)子系统包
ESR	Environmental Screening Report	环境筛选报告
ESRS	Engineering Safety Review Service	工程安全评审服务,专设安全评审服务
ESS	Engineered Safety System	专设安全系统
ESSM	Essential System Status Monitor	主要系统状态监测器
ESTE	Engineering Special Test Equipment	工程特殊试验设备
ESU	Electro Static Unit	静电单位
ESV	Emergency Stop Valve	危急遮断阀,主汽门,紧急止流阀
ESW	Electroslag Welding	电渣焊
ESW	Emergency Service Water System	应急厂用水系统
ESW-CG	Consumable Guide Eletroslag Welding	熔嘴电渣焊

英文简称	英文全称	中文名称
ET	Eddy-Current Techniques	涡流技术
ETA	Estimated Time of Arrival	估计的到达时间
ETCC	Environmental Test Control Center	环境试验管理中心
ETD	Estimated Time of Delivery/Estimated Time of Departure	估计的送达时间/估计的出发时间
ETDR	Equipment Technical Document Request	设备技术文件要求
ETF	Engine Test Facility/Engineering Test Facility/Environmental Test Facility/Execution Tracing File	发动机试验设施/工程试验设施/环境试验设施/施工跟踪档案
ETHO	Ethylene Oxide	环氧乙烷
ETL	Engineering Test Laboratory	工程试验实验室
ETO	Engineering Test Order	工程试验指令
ETP	Engineering Test Procedure/Evaluation Test Procedure	工程试验程序/评估试验程序
ETR	Engineering Test Reactor/Estimated Time of Return	工程试验反应堆/估计的返回时间
ETS	Encapsulated Test Specimen/Evaluation Test Specification	封装试验样品/评价试验说明书
EU	Energy Unit/Entropy Unit/Equivalent Unit	能量单位/熵的单位/等效单位
EUOF	Equivalent Unplanned Outage Factor	等效非计划停运因子
EUR	Equalization Unavailable Ratio	均等不可利用比

英文简称	英文全称	中文名称
EURATOM	European Atomic Energy Community	欧洲原子能联营
EURO-CHEMIC	European Company for the Chemical Processing of Irradiated Fuels	欧洲辐照燃料化学处理公司
EUV	Extreme Ultraviolet Radiation	超紫外辐射
EVHM	Ex-Vessel Handling Machine	容器外燃料装卸机
EW	Early Warning/Electrical Welding	早期警告/电焊
EWR	Emergency Water Reservoir	应急水池
EWS	Emergency Water System	应急水系统
EX	Extra	额外的;超,极,特
EXB	Exothermic Brazing	铝热钎焊
EXBW	Exothermic Braze Welding	铝热反应钎焊
EXHST	Exhaust	排气,抽空,耗尽
EXIM	Export-Import Bank	进出口银行
EXP	Exciter Panel/Expansion/Experiment	励磁机台屏/膨胀/实验
EXPD	Experimental Data	实验数据
EXS	Excitation System	励磁系统
EXT	Exterior/External	外(面,部)(的)
EXTEN	Extension	伸长,扩张,延伸
EXTND	Extended	伸长的,扩张的,延伸的
EXW	Explosion Welding	爆炸焊
EZB	Exclusion Zone Boundary	隔离区边界
F	Fahrenheit/Fan/Fault/Flow Meter/Fuel	华氏(温度)/风机/故障/流量计/燃料

英文简称	英文全称	中文名称
F/C	Fuel Channel	燃料通道
F/CH	Fuel Channel	燃料通道
F/C,FC	Fail Closed/Fast Closing/Flow Control	故障关/快关/流量控制
F/D VLV	Fill and Drain Valve	注入和排放阀
F/H	Fuel Handling	燃料操作
F/HT	Free Height	自由高度
F/M	Fueling Machine	装卸料机
F/O	Fail Open	故障开
F/S	Framatome-Spie Batignolles	法马通/斯比公司
F/T	Full Thread	全螺纹
FA	Failure Analysis/Final Acceptance/Flame Arrester/Fuel Assembly	故障分析/最终验收/灭火器/燃料组件
FAC	Flow-Accelerated Corrosion/Final Acceptance Certificate/Final Acceptance Criteria/Force Air Cooling System	流动加速腐蚀/最终验收证书/最终验收准则/强迫空气冷却系统
FACIL	Facility	设施
FACT	Fully Automatic Compileing Technique	全自动编辑技术
FAF	Fueling Against Flow/Flow Assisted Fueling	逆流装料/流动辅助装料
FAI	Fail as Is	照故障时原样
FARE	Flow Assisted Ram Extension	流动辅助冲杆延伸

英文简称	英文全称	中文名称
FAS	Federation of Atomic Scientists/Free Alongside Ship	原子科学家联合会/船边交货价
FAT	Factory Acceptance Test	工厂验收试验
FATT	Fracture Appearance Transition Temperature	出现断裂的过渡温度
FB	Feedback	反馈
FB	Furnace Brazing	炉中钎焊
FBFM	Fractional Batch Fuel Management	小批量燃料管理
FBG	Fractional Breeding Gain	小增殖增益
FBI	Federation of British Industries	英国工业联合会
FBPC	Fast Boiler Pressure Control	锅炉压力快速控制
FBR	Fast Breeder Reactor	快中子增殖堆
FBRNPP	Fast Breeder Nuclear Power Plant	快增殖堆核电厂
FC	Facility Control/Fail Close/Ferrite Check/Forced Convection/Frequency Changer/For Construction	设施控制/故障关/铁素体检查/强迫对流/变频器/作施工准备
FC	Field Change	现场变更
FCAW	Flux Cored Arc Welding	药芯焊丝电弧焊,包芯焊丝电弧焊
FCB	Fast Cut Back	甩负荷,快速切负荷
FCD	First Concrete Date	第一罐混凝土浇灌日
FCI	Fuel Channel Inspection/Fuel-Coolant Interaction	燃料通道检查/燃料-冷却剂相互作用

英文简称	英文全称	中文名称
FCIT	Fuel Channel Installation Tool	燃料通道安装工具
FCN	Field Change Notice	现场变更通知
FCO	Field Change Order/Field Checkout/Fuel Change Order	现场更改令/现场检验/换料令
FCR	Field Change Request/Full Core Reserve	现场变更要求/一炉料储备
FCS	Feedwater Control System/Forged Carbon Steel	给水控制系统/锻压碳钢
FCST	Federal Council for Science and Technology	联邦科学技术委员会(美国)
FCV	Flow Control Valve	流量控制阀
FCW-G	Gas-shielded Flux Cored Arc Welding	气体保护药芯焊丝电弧焊
FCW-S	Self-shielded Flux Cored Arc Welding	自保护药芯焊丝电弧焊
FD	Floor Drain/Flow Diagram/Forced Draft/Functional Diagram	地漏/流程图/强制通风/功能图
FDCD	Facility Design Criteria Document	设备设计标准文件
FDCR	Field Design Clarification Request	现场设计澄清要求
FDE	Failure Detection and Estimation	故障探测与评估
FDF	Forced-Draft-Fan	送风机
FDR	Feeder	供(给)水装置
FDS	Floor Drain System	地面疏水系统

英文简称	英文全称	中文名称
FDT	Floor Drain Tank	地面疏水箱
FDWS	Feedwater System	给水系统
FE	Finite Element/Flow Element	有限元/流量元件
FEDAL	Failed Element Detection and Location	破损元件的探测和定位
FELIX	Fusion Electromagnetic Induction Experiment	聚变电磁感应实验(装置)(美国)
FEM	Female/Finite Element Method	阴性的,凹形/有限元法
FER	Field Engineering Request	现场工程要求
FERC	Federal Energy Regulatory Commission	联邦能源管理委员会(美国)
FERD	Fuel Element Rupture Detector	燃料元件破损探测器
FERMILA	Fermi Laboratory	费密实验室
FESW	Fuel Element Storage Well	燃料(元件)存放井
FET	Field Effect Transistor	场效应晶体管
FF	Flat Face	平面
F-F	Flip-Flop	触发器,触发电路,双稳态多谐触发器
FFC	Fissile Fuel Content	易裂变燃料含量
FFD	Failed Fuel Detection/Failed Fuel Detector/Functional Flow Diagram	破损燃料(元件)探测/破损燃料探测器/功能流程图
FFDL	Failed Fuel Detection and Location	破损燃料探测和定位
FFEL	Failed Fuel Element Locator/Fresh Fuel Element Loading	破损燃料元件定位器/新燃料元件装堆

英文简称	英文全称	中文名称
FFF	Fusion Fuel Facility/Fusion Fuel Factory	聚变燃料设施/聚变燃料厂
FFFF	Fussion-Fission Fuel Factory	聚变-裂变燃料厂
FFMS	Fuel Failure Monitoring System	燃料破损监测系统
FFP	Fissile Fuel Producer	易裂变燃料生产堆
FFSG	Fitness for Service Guideline	适用于维修的导则
FFT	Fast Fourier Transform	快速傅里叶变换
FFWT	Final Feedwater Temperature	最终给水温度
FGM	Fission-Gas Monitor	裂变-气体监测器
FH	Fire Hose/Fuel Handling	消防软管/燃料操作
FHAHVS	Fuel Handling Area Heating and Ventilation System	燃料操作区暖通系统
FHB	Fuel Handling Building	燃料操作厂房
FHC	Fuel Handling Control	燃料操作控制
FHEB	Fourth Harbour Engineering Bureau	(交通部)第四航务工程局
FHIH	Fuel Handling Interconnection Hardware	燃料装卸互连硬件
FHM	Fuel Handling Machine	装卸料机
FHPO	Fuel Handling Panel Operator	燃料操作台操纵员
FHS	Fuel Handling System	燃料操作系统
FHY	Fire Hydrant	消防栓
FI	Fail Indeterminate/Flow Indicator	故障时不确定/流量指示器
FIC	Film Integrated Circuit	膜集成电路
FIFA	Fissions per Initial Fissionable Atoms	菲法(燃耗表达方式之一,为平均每个初始装载入堆的易裂变原子已引起的裂变数)

英文简称	英文全称	中文名称
FIFO	First-in First-Out/Floating Input Floating Output	先进先出/浮点输入浮点输出
FIG	Figure	图
FIL	Fillister	凹槽,槽口
FILT	Filter	滤(波)器,过滤器
FIMA	Fissions per Initial Heavy Metal Atoms	菲 马(燃耗表达方式之一,为平均每个初始装载入堆的重金属原子已引起的裂变数)
FIN	Finish	结束;精加工;光洁度
FINA	Following Items Not Available	下列各项现缺,下列各项不适用
FIR	Facility Installation Review/Fissile Inventory Ratio	设备安装检查/易裂变核素存量比
FIREX	Fire Extinguisher	灭火器
FIST	Full Integral Simulation Test	整体全模拟试验
FIX	Fixture	夹具;固定
FL	Fail Locked (Lockout)/Foundation Level/Full Load	故障锁定(闭锁)的/基准线/满负载,满载
FL/MTR	Flow Meter	流量计
FLB	Flow Brazing	熔钎焊
FLECHT	Full Length Emergency Cooling Heat Transfer	全长度应急冷却传热
FLEC-HTSET	Full Length Emergency Cooling Heat Transfer System Effects Test	全长度应急冷却传热系统效应试验
FLEX	Flexible	软的,柔性的
FLG	Flange	法兰,凸缘

英文简称	英文全称	中文名称
FLLS	Fuel Low Level Sensor	燃料低液位传感器
FLOW	Flow Welding	熔焊
FLR	Filler/Flow Rate	填料，填充物/流量；流速
FLSP	Flame Spraying	火焰喷涂
FLSP-W	Wire flame spraying	丝材火焰喷涂
FLT	Full-Load Torque	满载转矩
FLU	Fault Location Unit	故障定位设备
FM	Field Maintenance/Flow Meter/Frequency Modulated/Frequency Modulation/Fueling Machine	现场维修/流量计/（已）调频的/调频/装卸料机
FMA	Free Mineral Acidity	游离无机酸
FMBE	Fuelling Machine-Bridge Electrical	装卸料机-桥架电气（器件）
FMC	Fuelling Machine-Carriage	装卸料机-滑车
FMCOH	Fuelling Machine-Carriage Oil Hydraulic	装卸料机-滑车油压
FMCRD	Fine Motion Control Rod Drive	微调控制棒传动，控制棒微调
FMDP	Fuel Management Design Program	燃料管理设计大纲
FMDS	Fuelling Machine-D_2O Supply	装卸料机-重水供应
FMEF	Fuels and Material Examination Facility	燃料与材料检验装置（美国）
FMF	Fuel Material Facility	燃料材料装置
FMMI	Fuelling Machine Miscellaneous Items	装卸料机-杂项
FMML	Fuelling Machine-Maintenance Lock	装卸料机-维修闭锁

英文简称	英文全称	中文名称
FMP	Fuel Maintenance Panel	燃料检修操作台
FMR	Field Material Requisition/Financial Management Review	现场领料单/财务管理检查
FMRP	Field Manufacturing Request Purchase Order	现场制造申请订货单
FMRQ	Field Manufacturing Request Quotation	现场制造申请报价
FMS	Fueling Machine & Storage	装卸料机和存放
FMSDP	Fueling Machine-Shielding Door Panel	装卸料机-屏蔽门控制屏
FMTD	Fueling Machine-Transfer Door	装卸料机-转运门
FNAA	Fast Neutron Activation Analysis	快中子活化分析
FNG	Fast Neutron Generator	快中子发生器
FNP	Floating Nuclear Power Plant	(海上)浮动核电厂
FNPT	Female-National Pipe Thread	国标管内螺纹
FO	Fail Open/Field Operator/Firm Offer/Fuel Oil	故障开/现场操纵员/实盘/燃油
FOB	Free on Board	离岸价,船上交货
FODT	Fuel Oil Day Tank	燃油日用量储存箱
FOH	Forced Outage Hours	强迫停运小时数
FOR	Free on Rail	火车上交货价
FOS	Factor of Safety	安全因子
FOSAR	Foreign Object Search and Retrieval	异物搜寻与回取
FOST	Fuel Oil Storage Tank	燃油储存箱
FOT	Free on Truck	敞车交货价

英文简称	英文全称	中文名称
FOTP	Fuel Oil Transfer Pump	燃油输送泵
FOW	Forge Welding	锻焊
FP	Feed Pump/Feedback Positive/Fire-Proof/Fire Protection/Fire Pump/Fission Product/Flash Point/Freezing Point/Full Power	给水泵/正反馈/耐火的，防火的/防火/消防水泵/裂变产物/燃点，着火点/冰点/满功率
FPA	Failure Probability Analysis	故障概率分析
FPCCS	Fuel Pit Cooling and Cleanup System	燃料池冷却与净化系统
FPD	Fission-Product Detection/Flame Photometric Detector/Full Power Days	裂变产物探测/火焰光度探测器/满功率天
FPDS	Fire Protection Data System	防火数据库
FPLCS	Fission Product Leakage Control System	裂变产物泄漏控制系统
FPM	Facility Power Monitor	设施功率监测器
FPMS	Fueled Prototype Mockup System	装燃料的原型模拟体系统
FPO	Facilities Purchase Order	设备采购订单
FPP	Fire Protection Program	防火大纲，消防大纲
FPRT	Fission Product Release and Transport	裂变产物释放与转移
FPSR	Full Power Steaming Rate	满功率蒸汽流量
FPT	Feed Pump Turbine	给水泵汽轮机
FPT	Full Power Trial	满功率试验
FPU	Fuel Purification Unit	燃料纯化装置
FR	Failure Rate	故障率，失效率

英文简称	英文全称	中文名称
FRA	Framatome	法国法马通公司
FRAMEX	Framatome	法马通海外检修公司
FRCTF	Fast Reactor Core Test Facility	快堆堆芯试验装置,快堆零功率装置
FREQ	Frequency	频率
FRM	Frequency Meter	频率计
FROG	Framatome Owners Group	法马通业主协会
FRP	Fiber Glass Reinforced Plastics/Fuel-Reprocessing Plant	玻璃钢/燃料后处理工厂
FRS	Floor Response Spectra/Fuel Recording System	楼面响应谱/燃料记录系统
FRSF	Fuel Receiving and Storage Facility	燃料接收与贮存设施
FRSS	Fuel Receiving and Storage Station	燃料接收与贮存站
FRT	Fortnight	两星期
Frt	Freight	货运,货物
FRW	Friction Welding	摩擦焊
FRW-DD	Direct Drive Friction Welding	径向摩擦焊
FRW-I	Inertia Friction Welding	惯性摩擦焊
FS	Factor of Safety/Ferrous Steel/Flow Switch/Flowsheet	安全系数/铁素钢材/流量开关/流程图
FS	Furnace Soldering	炉中钎焊
FSBC	Free Standing Breaker Check	独立式断路器检查
FSDF	Full-Scale Demonstration Facility	全尺寸示范装置
FSI	Fluid-Structure Interaction	流体-结构相互作用,流固耦合
FSM	Flowsheets Modification	流程图修改

英文简称	英文全称	中文名称
FSPPO	Field Senior Power Plant Operator	电站现场高级操纵员
FSR	Flowable Solids Reactor/Fuel-Sodium Reaction	可流动的固体燃料反应堆，熔盐堆/燃料-钠反应
FSS	Filter Specification Sheet/Fuel Safety System/Full Scope Simulator	过滤器规格表/燃料安全系统/全范围模拟机
FSTS	Fueling Side Tubesheet Sub-Assembly	装料侧管板子组件
FSU	Full Scale Unit	全尺寸装置
FSW	Friction Stir Welding	搅拌摩擦焊
FT	Field Test/Firing Point/Flow Transmitter/Functional Test	现场试验/着火点/流量变送器/功能试验
FTA	Fault Tree Analyses	故障树分析
FTC	Facility Terminal Cabinet	设备接线盒，设备终端柜
FTCCG	Fuel Transfer Canal Containment Gate	燃料传送通道安全壳闸门
FTE	Factory Test Equipment	工厂试验设备
FTG	Fitting	装配，配件
FTP	Field Test Procedure/File Transfer Protocol/Functional Test Procedure	现场试验规程/文件传输协议/功能试验规程
FTS	Fuel Transfer System/Functional Test Specification	燃料输送系统/功能试验规范书
FTSO	Fuel Transfer & Storage Operation	燃料传送和存放操作
FU	For Use/Fuse	供使/熔断器，保险丝
FUM	Fuel Unloading Machine	卸料机

英文简称	英文全称	中文名称
FUP	Fusion Point	熔点;核聚变温度
FV	Flush Valve/Folio Verso/Fuel Valve	冲洗阀/见本页背面/燃料阀
FW	Feedwater/Finishing Work/Fresh Water/Full Wave	给水/完工/淡水/全波
FW	Flash Welding	闪光焊
FWCV	Feedwater Control Valve	给水调节阀
FWF	Fueling-With-Flow	顺流装料
FWHM	Full-Width at Half-Maximum	半高全宽
FWIV	Feedwater Isolation Valve	给水隔离阀
FWL	Feedwater Line	给水管道
FY	Fiscal Year	财政年度
G	Generator/Ground	发生器;发电机/接地;地线
G/P	General Purpose	通用的
G/S	Garter Spring	环状螺旋弹簧,夹紧弹簧
GA	Gas/Gauge/General Arrangement/General Assembly	气体/量规/总体布置,总图/总装配(图)
GAL	Gallon	加仑
GALV	Galvanized	电镀的,镀锌的
GAM	Gross Activity Monitoring	总放射性监测
GAS	Gas Analysis System	气体分析系统
GASL	General Applied Science Laboratories	通用科学实验室
GC	Gas Chromatography/Gas Compressor	气相色谱法/气体压缩机

英文简称	英文全称	中文名称
GCB	Gas-Cooled Breeder	气冷增殖(反应)堆
GCLM	Grass Containment Leakage Monitoring	安全壳总泄漏监测
GCM	Greatest Common Measure	最大公约数
GCR	Gas-Cooled Reactor	气冷(反应)堆
GCS	Generator Control System	发电机控制系统
GDCS	Gravity Driven Cooling System	重力驱动冷却系统
GDDS	Gamma Dose Detection System/ Generic Design Decision Summary	γ剂量探测系统/通用设计决策概述
GE	Gas Injection/General Electric Company/Generator End/Ground Equipment	气体喷射/通用电气公司(美国)/发电机端/地面设备
GECA	General Electrical-Alsthom Corp.	通用电气-阿尔斯通公司(英、法)
GED	Gas Evolution Detection/Gasoline Engine Driven	释出气体检测法/汽油发动机驱动的
GEM	Gaseous Effluent Monitor	气态流出物监测器
GEN	Generator	发电机
GEPB	Guangdong Environmental Protection Bureau	广东省环保局
GER	General Engineering Research	通用工程研究
GEV	General Exhaust Ventilation	整体排气通风
GFBR	Gas-Cooled Fast Breeder Reactor	气冷快中子增殖(反应)堆
GFFD	Gross Failed Fuel Detector	燃料破损总探测器
GFP	Gas Fission Product	气态裂变产物
GFPM	Gas Fission Product Monitor	气态裂变产物监测器

英文简称	英文全称	中文名称
GFRP	Glass Fiber Reinforced Plastic	玻璃钢
GG	Gas Generator	气体发生器
GHR	Gross Heat Rate	总热率,毛热率
GIB	Gas Insulated Bus	气体绝缘母线
GIC	Gas Insulated Conductor	气体绝缘导线
GIS	Gas Inlet System/Gas Insulation Substation/Gas Insulated Switchgear/General Information System	进气系统/气体绝缘变电站/气体绝缘开关/通用信息系统
GKT	Gasket	垫圈(片),衬垫
GL	Ground Low Tension Side	地线低拉力端
GLM	Gross Leakage Monitoring	总泄漏监测
GLS	Glass	玻璃
GLV	Globe Valve	球阀
GM	Geiger-Müller/General Management/General Manager/Governor Motor	盖革-弥勒计数管/总经理部/总经理/调速马达
GMAC	Gas Metal Arc Cutting	熔化极气体保护电弧切割,气体保护金属极电弧切割
GMAW	Gas Metal Arc Welding	熔化极气体保护电弧焊,气体保护金属极电弧焊
GMAW-P	Pulsed Arc	熔化极气体保护脉冲电弧焊,气体保护金属极脉冲电弧焊
GMAW-S	Short Circuiting Arc	熔化极气体保护短路电弧焊,气体保护金属极短路电弧焊
GMG	Gross Maximum Generation	毛最大发电量

英文简称	英文全称	中文名称
GMM	General Maintenance Manuals	通用维修手册
GMP	Generator Monitoring Panel	发电机监控屏
GMR	Graphite-Moderated Reactor	石墨慢化(反应)堆
GMS	Gravitational Mass Sensor	引力质量传感器
GND	Ground	地;接地;地线
GNIC	Guangdong Nuclear Power Investment Co., Ltd.	广东核电投资有限公司
GNPJVC	Guangdong Nuclear Power Joint Venture Co. ,Ltd.	广东核电合营有限公司
GNPS	Guangdong Nuclear Power Station	广东(大亚湾)核电站
GNR	Generator	发生器;发电机
GNRB	General Nuclear Review Board	通用核评审委员会
GNSI	General Nuclear Services Inc.	通用核服务公司
GOP	General Operating Program	总运行大纲
GOR	General Operating Rules	运行总则
GOS	General Operating Specification	总体运行性能规格书,一般操作规程
GOV	Governor	调速器
GP	Gauge Pressure/Generalized Programming/Geometric Progression	表压力/通用程序设计/几何级数,等比级数
GPHC	Guangdong Electric Power Holding Co.	广东省电力集团公司
GPS	Gas Purification System	气体纯化系统
GPT	Grounding Potential Transformer	接地电压互感器
GR	Gear Ratio/Grey Rod	齿速比/灰棒

英文简称	英文全称	中文名称
GRND	Ground	地面
GRO	Guangdong Regional Office (NNSA)	(国家核安全局)广东监督站
GRPDB	Generic Reliability Parameter Data Bank	通用可靠性参数数据库
GRS	Ground Response Spectra	地面响应谱
GRWPS	Gaseous Radioactive Waste Processing System	放射性废气处理系统
GRWS	Gaseous Radioactive Waste System	放射性废气系统
GS	Generating Station/Guide Sleeve	发电站/导向套筒
GSC	Gland Steam Condenser	汽封蒸汽凝汽器,轴封蒸汽凝汽器
GSE	Ground Support Equipment	地面支持设备
GSH	Gland Steam Heater	汽封加热器
GSI	General Subject Index	一般主题索引
GSKT	Gasket	衬垫,填料
GSRS	Generic Seismic Response Spectrum	通用地震响应谱
GSS	Gas Sampling System/Gland Seal System/Guaranteed Shutdown State	气体采样系统/轴封系统/保证停堆状态
GSTR	Geological Survey Triga Reactor	地质普查铀氢锆(反应)堆
GSV	Globe Stop Valve	球心节流阀,截止球阀
GT	Gas Turbine/Ground Tertiary/Guide Tube	燃气透平(轮机)/第三级接地/导向管
GTA	Gate Tower Accessories	门塔附件
GTAC	Gas Tungsten Arc Cutting	气体保护钨极电弧切割

英文简称	英文全称	中文名称
GTAW	Gas Tungsten Arc Welding	气体保护钨极电弧焊
GTAW-P	Pulsed Arc	气体保护钨极脉冲电弧焊
GTC	Gas Turbine Compressor/General Terms and Conditions	燃气轮机压缩机/一般条款和条件
GTCC	Greater-Than-Class-C	活性大于 C 等级的(放射性废物)
GTE	Ground Trip Element	接地跳闸元件
GTV	Gate Valve	闸阀
GV	Governor Valve	调节阀
GW	Gross Weight	总重,毛重
GWMS	Gaseous Waste Management System	气体废物管理系统
GWPS	Gaseous Waste Processing System	气体废物处理系统,废气处理系统
H	Height/High	高度/高
H&G	Harden and Grind	硬化与研磨
H&T	Handling and Transportation	装卸与运输
H. O.	Head Office	总公司,总部
H/D	Heavy Duty	重载,重负荷
H_2CO_2S	H_2 and CO_2 System	H_2 和 CO_2 系统
H_2CS	Hydrogen Cooling System	氢冷系统
H_2SOS	Hydrogen Seal Oil System	氢密封油系统
HA	High Amplitude/Hydraulics Accumulator/Hypothetic Accident	高振幅/液压蓄压箱/假想事故
HAF	Chinese Nuclear Safety Codes and Guides	中国核安全法规和导则

英文简称	英文全称	中文名称
HAMMLAB	Halden Man-Machine System Laboratory	哈顿人-机系统实验室
HANJUNG	Korea Heavy Industries & Construction Co., Ltd.	韩国重工业株式会社
HAW	High-Active Waste	高放(射性)废物
HAZ	Heat Affected Zone	热影响区
HAZEL	Homogeneous Assembly Zero Energy Level Reactor	均匀化零功率(反应)堆(英国)
HB	Horizontal Beam	水平射束
HBAL	Heat Balance	热平衡
HBWR	Heavy Boiling Water Reactor	沸腾重水(反应)堆
HC	Hand Control/Handling Capacity/Heating Coil/High Capacity/High Conductivity/Hot Conditioning/Hydrogen Concentration	手控/处理能力/加热线圈/高容量/高电导率/热处理/氢浓度
HCCM	HCCM Nuclear Power Construction Joint Venture Co.	HCCM 核电建设合营公司
HCDA	Hydrodynamics of Core Disruptive Accident/Hypothetical Core Disruptive Accident	堆芯损坏事故流体动力学/假想堆芯损坏事故
HCF	Highest Common Factor	最大公因子
HCL	Horizontal Center Line	水平中线
HCLWR	High Conversion Light Water Reactor	高转换比轻水(反应)堆
HCN	Hydrocyanic Acid	氰化氢
HCSHT	High Carbon Steel Heat Treated	热处理过的高碳钢

英文简称	英文全称	中文名称
HCU	Hydraulic Control Unit	液压控制装置
HCV	Hand Control Valve/Hydraulic Control Valve	手动调节阀/液压控制阀
HD	Head/Header/Heat Distortion/Heavy Duty/High Density/History Docket	封头,顶盖/集管,联箱/热变形/重载,重负荷/高密度/历史大事记
HDD	Hardware Design Description	硬件设计描述
HDG	Hot Dip Galvanized	热浸镀
HDLG	Handling	装卸;处理,操作
HDN	Harden	硬化
HDR	Header	集(总,母)管,联箱
HE	Head End/Heat Engine/Heat Exchanger/High Efficiency/Hydrogen Embrittlement	首端/热机/热交换器/高效率/氢脆
HEAC	High Energy Advisory Committee	高能咨询委员会
Heat No.	Heat Number	钢锭炉号
HECC	Hyundai Engineering & Construction Co., Ltd.	现代建设株式会社
HEI	Heat Exchanger Institute	热交换器研究所
HEL	Helical	螺旋
HENRE	High Energy Neutron Reaction Experiment	高能中子反应实验装置(美国)
HER	Extreme High Reliability	极高的可靠性
HESIR	Hybrid-Encapsulated Straight-Individually-Replaceable	混合封装可直接单独更换的
HEU	High (Highly) Enriched Uranium	高浓铀

英文简称	英文全称	中文名称
HEX	Hexagon	六角(边)形
HF	Hard Facing/High Frequency/History File	硬表面/高频/历史文件
HFA	High-Frequency Amplifier	高频放大器
HFBR	High Flux Beam Reactor	高通量(反应)堆,高中子束流反应堆
HFC	High-Frequency Choke/High-Frequency Current	高频扼流圈/高频电流
HFD	Horizontal Flux Detector	水平(中子)注量率探测器
HFEF	Hot Fuel Examination Facility	热燃料检查设备
HFM	Homogeneous Equilibrium Critical Flow Model	均匀平衡态临界流模型
HFO	High-Frequency Oscillator	高频振荡器
HFP	Hydraulic Fluid Pump	液压泵
HFR	High Flux Reactor	高通量(材料试验)堆
HFRG	Human Factors Review Group	人为因素评审组
HGR	Hanger	吊架
HHP	High Head Pump	高压水泵,高扬程泵
HHSI	High-Head Safety Injection	高压安全注入
HHSIP	High-Head Safety Injection Pump	高压安全注入泵
HHSIS	High-Head Safety Injection System	高压安全注入系统
HI	Half-in	一半插入
HIC	High Integrity Container/Hydrogen Induced Cracking	高完整性安全壳/氢致开裂

英文简称	英文全称	中文名称
HIF	Heavy Ion Fusion/Human Initiated Failure/Hybrid Interface	重离子聚变/人为故障/模拟-数字量混合接口
HIPW	Hot Isostatic Pressure Diffusion Welding	热等静压扩散焊
HIS	Hardware Interrupt System/Health Physics Information System/Hydraulic Institute Standards	硬件中断系统/保健物理信息系统/水力学会标准
HKNIC	Hongkong Nuclear Power Investment Co., Ltd.	香港核电投资有限公司
HL	Hole and Embedded Parts List/Hot Leg/Hot Line	预留孔和预埋件清单/热管段/热管线
HLLW	High Level Liquid Waste	高放液体废物,高放射性废液
HLSW	High Level Solid Waste	高放固体废物
HLW	High Level Waste	高放(射性)废物
HM	Heavy Metal/Humidifier	重金属/空气增湿器,加湿器
HMI	Human-Machine Interface	人-机接口
HNMC	Huainan Nuclear Maintenance Company	淮南核电检修公司
HO	Half-Out	一半抽出
HOE	Hydraulically Operated Equipment	液压(动)设备
HORZ	Horizontal	水平的,卧式
HOS	Hose	软管
HOTCE	Hot Critical Experiment	热临界实验(装置)(美国)

英文简称	英文全称	中文名称
HP	Hand Pump/Health Physics/High Pressure/Hold Point/Horizontal Plane/Horsepower/Hot Particle/Hydrostatic Pressure	手动泵/保健物理/高压/停工待检点/水平面/马力/强放射性粒子/静(水)压
HPAB	High Pressure Accumulator Building	高压安全注入箱厂房
HPBFP	High Pressure Boiler-Feed Pump	高压锅炉给水泵
HP-BP	High Pressure By-Pass	一级减温减压装置,高压(再热)旁路
HPC	High Pressure Ccompressor/High Pressure Cylinder/Hydrogen Pressure Control	高压压气机/高压缸/氢压控制
HPCI	High Pressure Coolant Injection	高压冷却剂注入
HPCIP	High Pressure Coolant Injection Pump	高压冷却剂注入泵
HPCIS	High Pressure Coolant Injection System	高压冷却剂注入系统
HPCR	High Power Channel Type Reactor	大功率列管式(反应)堆(前苏联RBMK型堆)
HPCS	High Pressure Core Spray System	高压堆芯喷淋系统
HPD	High Power Discharge	大功率放电
HPE	High Pressure Extraction	高压抽汽
HPECC	High Pressure Emergency Core Cooling	高压应急堆芯冷却
HPES	Human Performance Evaluation System/Human Power Evaluation System	人员行为素质评价系统/人力资源评价系统

英文简称	英文全称	中文名称
HP-ESV	High Pressure Emergency Steam Valve	高压主汽阀,高压蒸汽危急遮断阀
HP-EV	High Pressure Extraction Valve	高压抽汽阀
HPF	High Pass Filter/Highest Possible Frequency	高通滤波器/最高可能频率
HPH	High Pressure Heater	高压加热器
HPIS	High Pressure Injection System	高压注入系统
HPIX	High Pressure Ion Exchanger	高压离子交换器
HPLC	High Pressure Liquid Chromatography	高压液相色谱法
HPN	Health Physics Net/Horse-Power Nominal	保健物理网/公称马力
HPPL	High Pressure Packing Leakage	高压汽封漏汽
HPRM	High Power Range Monitor	高功率区段监测器
HPRR	Health Physics Research Reactor	保健物理研究(反应)堆
HPRS	High Pressure Recirculation System	高压再循环系统
HPS	High Pressure Steam/Hydraulic Power Supply	高压蒸汽/水(液)压供应
HPSH	Highest Possible Suction Head	最大可能吸入高度
HPSI	High Pressure Safety Injection	高压安全注入
HPSIS	High Pressure Safety Injection System	高压安全注入系统
HPSWS	High Pressure Service Water System	厂用高压水系统

英文简称	英文全称	中文名称
HPT	High Pressure Test/High Pressure Turbine/Hot Performance Test	高压试验/高压汽轮机/热态性能试验
HPTR	Horizontal Pressure Tube Reactor	卧式压力管式反应堆
HPW	Hot Pressure Welding	热压焊
HQR	Hardware Qualification Report	硬件鉴定报告
HR	Heat Rate/Heater/Homogeneous Reactor/Hot Rolled	热耗率/加热器/均匀反应堆/热轧
HS	Heat Shield/Heat Shrink/Heating Surface/High Speed/High-Speed Steel/Hollow Spindle/Hydraulic System	热屏蔽/热收缩/加热面,受热面/高速/高速钢/空心轴/液压系统
HSD	Housed	封装的,内藏的
HSF	High Steam Flow	高蒸汽流量
HSG	Housing	箱,壳
HSI	Horizontal Situation Indicator	水平位置指示器
HSP	Handswitch Position	手动开关位置
HSS	High Speed Stop/Hot Shutdown System	高速停机/热停堆系统
HSV	Hydraulic Servomotor	液压伺服电机
HT	Heat Treatment/High Temperature/Hydrostatic Test	热处理/高温/水压试验
HTB	High-Temperature Blanket/Hydraulic Test Bench	高温再生区,高温转换区/液压试验台

英文简称	英文全称	中文名称
HTC	Head Transport Cart/Heat Transfer Coefficient/Heat Transport Control	顶盖转运车/传热系数/热传输控制
HTDC	High Tension Direct Current	强直流电
HTF	Hydraulic Test Facility	水力学试验装置
HTGR	High-Temperature Gas-Cooled Reactor	高温气冷反应堆
HTHP	High Temperature and High Pressure	高温高压
HTM	High-Temperature Heat-Transfer Medium/High-Temperature Material	高温传热介质/高温材料
HTMDB	High-Temperature Material Data Bank	高温材料数据库
HTMSR	High Temperature Molten Salt Reactor	高温熔盐(反应)堆
HTPR	Horizontal Pressure Tube Reactor	卧式压力管反应堆
HTR	High Temperature Reactor/Homogeneous Thorium Reactor	高温(气冷)(反应)堆/均匀钍(反应)堆
HTRE	Heat-Transfer Reactor Experiment	传热实验(反应)堆
HTS	Heat Transport System/High-Tensile Steel/High-Tension Supply	热传输系统/高强度钢/强流电源
HTSG	High-Temperature Strain Gauge	高温应变计
HTT	Heat Transport Trip/Heat Treatment Temperature	热传输脱扣(停堆)/热处理温度

英文简称	英文全称	中文名称
HTW	High Temperature Water	高温水
HUCR	Highest Useful Compression Ratio	最高有效压缩比
HV	Heating and Ventilation/High Vacuum/High Velocity/High Voltage	供暖与通风，暖通/高真空/高速/高电压
HVD	Half-Value Depth	半值层深度，半值厚度
HVDC	High Voltage Direct Current Transmission	高压直流输电
HVL	Half-Value Layer	半值层，半吸收层
HVOF	High Velocity Oxyfuel Spraying	高速氧燃喷涂
HVPS	High Voltage Power Supply	高压电源
HVT	Half-Value Thickness	半值层厚度
HVY	Heavy	重(型)的
HW	Heavy Water/Hot Water	重水/热水
HWARE	Hardware	硬件
HWCTR	Heavy Water Components Test Reactor	重水部件试验反应堆
HWGCR	Heavy Water Moderated Gas Cooled Reactor	重水慢化气冷(反应)堆
HWL	Hot Water Line	热水管路
HWLWR	Heavy Water Moderated Boiling Light Water Cooled Reactor	重水慢化沸腾轻水冷却(反应)堆
HWOCR	Heavy Water Organic Cooled Reactor	重水(慢化)有机冷却(反应)堆
HWR	Heavy Water Reactor/Hot Water Return	重水(反应)堆/热水回流

英文简称	英文全称	中文名称
HWSR	Heavy Water Suspension Reactor	重水悬浮液(反应)堆
HX/HD	Hex Head	六角头(顶盖)
HXCC	Huaxing Construction Company	华兴建设公司
HXSS	Heat Exchanger Specification Sheet	热交换器规格表
HYDRZ	Hydrazine	肼,联氨
I	Indicator	指示器,指针
I&M	Inspection and Maintenance	检查与维修
I/D	Iron Ductile	铁的延性
I/F	Irradiated Fuel	辐照过的燃料
IABSE	International Association for Bridge and Structural Engineering	国际桥梁和结构工程协会
IAC	Incompleted Test but Acceptance/Interim Acceptance Criteria	试验未完成但可接收/临时验收准则
IAD	Internal Assessment Document	内部评估文件
IAE	Institute of Advance Engineering	先进工程技术研究所
IAHR	International Association for Hydraulic Research	国际水力研究协会
IALL	Instrument Air Load List	仪用空气负载表
IAM	International Association of Machinists	国际机械师协会
IAMAP	International Association of Meteorology and Atmospheric Physics	国际气象与大气物理协会
IAR	Institute for Atomic Research	原子能研究所(美国)

英文简称	英文全称	中文名称
IAS	Instrument Application Sheet/Instrument Application Specification	仪表申请单/仪器操作规范
IASCC	Irradiation Assisted Stress Corrosion Cracking	辐照促成的应力腐蚀破裂
IATM	International Association for Testing Material	国际材料试验协会
IB	Induction Brazing	感应钎焊
IB	Isophase Bus	同相母线
IBC	Interbuilding Coffin	厂内(屏蔽运输)包装容器
IBM	International Business Machines Corporation	国际商用机器公司(美国)
IBPA	Integrated Burnable Poison Absorber	一体化可燃毒物吸收体
IC	Indicating Controller/Input Circuit/Integrated Circuit/Intercooler/Intermediate Cooling/Ion Chamber/Ion Chromatograph	指示控制器/输入电路/集成电路/中间冷却器/中间冷却/电离室/离子色谱
ICA	Instrument Compressed Air	仪表压缩空气
ICC	Ignition Control Compound/Instrumentation Control Center/Integrated Component Circuit/International Chamber of Commerce/International Computation Center	点火器控制部件/仪表控制中心/集成元件电路/国际商会/国际计算中心
ICCS	In-Core Control Structure	堆内控制构件
ICE	Institution of Civil Engineers	土木工程师学会(英国)

英文简称	英文全称	中文名称
ICEM	Instrumentation Control and Electrical Maintainer	仪控和电气维修人员
ICF	Inertial Confinement Fusion	惯性约束聚变
ICFD	Ion Chamber & Flux Detector	电离室和注量率探测器
ICG	Internal Conversion Gain/Internal Coordination Guide	内转换增益/内部协调指南
ICI	Imperial Chemical Industries	英国化学工业公司
ICL	Item Classification List	物项分类清单
ICONTT	International Conference on Nuclear Technology Transfer	国际核技术转让会议
ICPS	Integrated Control and Protection System/International Conference on the Properties of Steam	集成控制与保护系统/国际蒸汽品质会议
ICR	Initial Conversion Ratio/Instrumentation Control Racks/Inventory Change Report	初始转换比/仪表控制架/存量变动报告
ICRP	International Commission on Radiological Protection/International Committee of Radiation Protection	国际放射防护委员会
ICRU	International Commission on Radiological Units and Measurements	国际辐射单位与测量委员会
ICS	Instrumentation Checkout Station/Integrated Control System	仪表检测站/一体化控制系统

英文简称	英文全称	中文名称
ICTD	Integrated Circuit Temperature Device	集成电路温度控制
ICV	Intercept Valve	截止阀(再热调节阀)
ID	Indicating Device/Inside Diameter/Item Description	指示器/内径/项目说明,物项描述
IDA	International Development Association	国际开发协会
IDC	Interest During Construction	建造期利息
IDF	Induced Draft Fan	引风机
IDLH	Immediately Dangerous to Life or Health	对生命或健康有即刻危险
IDP	Integrated Data Processing/Internal Department Procedure	综合数据处理/部门内部程序
IDS	Instrument Device Specification/Interim Decay Storage	仪表设备规格书/临时衰变贮存
IE	Initiating Event/(Office of) Inspection and Enforcement	始发事件/检查与强制执行办公室(美国核管会)
IEA	International Energy Agency	国际能源机构
IEC	International Electric Corporation/International Electrotechnical Commission	国际电气公司/国际电工委员会
IEE	Institution of Electrical Engineers/Institute of Environmental Engineers	电气工程师学会/环境工程师学会
IEI	Industrial Engineering Institute	工业工程学院

英文简称	英文全称	中文名称
IEMC	International Electronic Manufacturing Company	国际电子仪器制造公司
IEN	Incompleted Test Except Noted	试验未完成仍可接受,但注明者除外
IERE	Institute of Electronic and Radio Engineers	电子与无线电工程师学会
IES	Inductive Energy Storage/Interim End State (Report)	感应贮能/临时完工状态(报告)
IF	Information Flowsheet/Institute of Fuel	参考流程图拉虚席流程/英国燃料学会
IFB	Inlet Flow Blockage/Invitation for Bid	进口流阻塞/招标,邀标书
IFC	International Finance Corporation/International Formulation Committee	国际金融公司/国际公式化委员会
IFCC	Incremental Fuel Cycle Cost	逐渐增长的燃料循环成本
IFF	Inert Fluid Fill	惰性流体充填
IFIP	International Federation of Information Processing	国际信息处理联合会
IFM	Intermediate Flow Mixer	中间流量混合器
IFORS	International Federation of Operational Research Societies	国际运筹学学会联合会
IFP	Intermediate-Frequency Preamplifier	中频前置放大器
IFR	Integral Fast Reactor	一体化快中子(反应)堆

英文简称	英文全称	中文名称
IFRB	International Frequency Registration Board	国际频率注册委员会
IFSMTF	International Fusion Superconducting Magnet Test Facility	国际聚变超导磁铁试验设施(美国)
IG	Implementation Guide/Inertial Guidance	实施导则/惯性制导
IGA	Intergranular Attack	晶间腐蚀
IGC	Intergranular Corrosion	晶间腐蚀
Igpm	Imperial Gallon Per Minute	每分钟英制加仑数(4.546L)
IGS	Implementation Guide Supplement/Inertial Guidance System	实施导则补充/惯性制导系统
IGU	International Gas Union	国际气体工业联合会
IGV	Inlet Guide Vane	进口导叶
IH	Indirect Heating/Instrument Head/Inverted Hour	间接加热/表头/倒时数
IHD	Inlet Header	进口集管
IHF	Inhibit Halt Flip-Flop	禁止停机触发器
IHP	Indicated Horse-Power	指示马力
IHS	In-House Standard	内部标准,厂内标准
IHSI	Induction Heating Stress Improvement	感应加热应力改善(技术)
IHTS	Intermediate Heat Transport System	中间热传输系统
IHX	Intermediate Heat Exchanger	中间热交换器
IIB	International Investment Bank	国际投资银行
IIP	Incident Investigation Program	事件调查大纲

英文简称	英文全称	中文名称
IISI	International Iron and Steel Institute	国际钢铁学会
IIT	Incident Investigation Team	事件调查组
IIW	International Institute of Welding	国际焊接学会
IL	Improved L Mode/Indicating Light/Instrument List	改进的 L 模式/指示灯/仪表清单
ILC	International Law Commission	国际法委员会
IM	Index Manual/Induction Motor	索引手册/感应(异步)电动机
IMA	International Mineralogical Association	国际矿物学会
IMC	Institute of Measurement and Control	测量与控制研究学会
IME	Installation & Maintenance Equipment/Institute of Mechanical Engineers	安装和维修设备/机械工程师学会
IMFR	Intec Modification Field Request	Intec 现场变更要求
IMM	Institute of Mining and Metallurgy	矿冶学会
IMP	Industrial Management Procedure/Integrated Monitor Panel	工业管理程序/综合监控台屏
IMS	Information Management System/International Metallographic Society	信息管理系统/国际金相学会
IN	Inlet/Inlet Valve	进口/进汽阀
INAA	Instrumental Neutron Activation Analysis	仪表化中子活化分析

英文简称	英文全称	中文名称
INCC	International Nuclear Cooperation Center	国际核合作中心(日本)
INCOT	In-Core Instrument Test Facility	堆芯仪表试验装置
INCOT-ERMS	International Rules for the Interpretation of Trade Terms	(国际商会编写的)国际贸易条件解释通则
INDC	International Nuclear Data Committee	国际核数据委员会
INDIV	Individual	个别,单独
INES	International Nuclear Event Scale	国际核事件分级表
INFCE	International Nuclear Fuel Cycle Evaluation	国际核燃料循环评价
ING	Intense Neutron Generator	强中子发生器
INIS	International Nuclear Information System	国际核信息系统
INLCK	Interlock	联锁
INPO	Institute of Nuclear Power Operations	核电运行研究院(美国)
INS	Insulation/Insulated	绝缘,绝热,保温
INS	Iron Soldering	烙铁钎焊
INSAG	International Nuclear Safety Advisory Group	国际核安全顾问组
INSARR	Integrated Safety Assessment of Research Reactors	研究堆整体安全评估
INSER	Insertion	插入(物)

英文简称	英文全称	中文名称
INST	Installation	安装
INSTR	Instrument	仪器(表)
Intec	Integrated Electrical and Control Database (Developed By AECL)	(AECL 开发的)集成电气和控制数据库
INTL	Internal	内(部)
INTLK	Interlock	联锁
Inv. No	Invoice Number	发票号
Io	Inoperability	不可用
IOCS	Input-Output Control System	输入-输出控制系统
IOL	Input/Output List	输入/输出清单
IP	Identification of Position/Identification Point/Index of Performance/Implementation Procedure/Industrial Production/Initial Point/Installation Procedure/Instrument Panel/Intermediate Pressure/Internet Protocol/Intersecting Point	位置识别/识别点/性能指标/执行程序/工业生产/起点/安装程序/仪表盘/中压/网络协议/交点横切点
IPA	Intermediate Power Amplifier	中间功率放大器
IPCEA	Insulated Power Cable Engineers Association	绝缘电力电缆工程师协会
IPE	Individual Plant Examination	单个电厂考核
IPERS	International Peer Review Service	国际同行评审(服务)
IPL	Information Processing Language	信息处理语言
IPM	Instrument Procedure Maintenance	仪表程序化维护

英文简称	英文全称	中文名称
IPNS	Intense Pulsed Neutron Source	强脉冲中子源
IPR	Instrument Procurement Release	仪表采购释放
IPRM	Intermediate Power Range Monitor	中等功率区段监测器
IPS	Instrument Power Supply/International Pipe Standard/International Plutonium Storage	仪表供电/国际管材标准/国际钚贮存(量)
IPSAR	Integrated Plant Safety Assessment Report	综合电厂安全评价报告
IPWR	Integrated Pressurized Water Reactor	一体化压水堆
IQI	Image Quality Indicator	影像质量指示器
IR	Induction Regulator/Infrared/Inside Radius/Installation Requirements/Instrument Reading/Insulation Resistance/Internal Resistance	电感式调压器/红外线/内半径/安装要求/仪表读数/绝缘电阻/内阻
IRB	Infrared brazing	红外钎焊
IRCP	In-Core Processing Code Package	堆芯测量数据处理程序包
IRDS	Integrated Reliability Data System	综合的可靠性数据系统
IRE	Institute of Radio Engineers	无线电工程师学会
IREP	Interim Reliability Evaluation Program	临时可靠性评估大纲
IRF	Irradiation Research Facility	辐照研究设施
IRM	Intermediate Range Monitor	中间(功率)区段监测器
IRRT	International Regulatory Review Teams	国际管理评审组

英文简称	英文全称	中文名称
IRS	Incident Reporting System/Interface Requirements Specification(Section)	事件报告系统/接口要求规格书(部分)
IRS	Infrared soldering	红外钎焊
IRSE	International Reactor Safety Evaluation	国际反应堆安全评价
IS	Individual Shield/Industrial Safety/Industry Issues/Instrumentation Schematics/Internal Shield/Internal Surface/Isometric(Drawings)	个人(单独)屏蔽/工业安全/工业议题/仪表示意图/内部屏蔽/内表面/等轴图
IS	Induction soldering	感应钎焊
ISA	Instrument Society of America/International Standard Atmosphere/International Standardization Association	美国仪表学会/国际标准大气压/国际标准化协会
ISAR	Integrated Safety Assessment Report/Intermediate Safety Analysis Report	综合安全评价报告/中期安全分析报告
ISC	Industrial Security Committee	工业安全委员会
ISD	Instrument Device Specification	仪表设备规格书
ISDA	International Swap Dealers Association	国际交易商协会
ISDN	Integrated Service Digital Network	综合服务数字网络

英文简称	英文全称	中文名称
ISFM	Instrument Stand-Floor Mounted/International Spent Fuel Management	仪表台架-楼面安装/国际乏燃料管理
ISFS	Independent Spent Fuel Storage/International Spent Fuel Storage Facility	单独设置的乏燃料贮存/国际乏燃料贮存设施
ISI	In-Situ Inspection/In-Service Inspection	就地检查/在役检查
ISM	Industrial Security Manual	工业安全手册
ISO	Individual System Operation/International Standardization Organization/International Standards Organization/Isometric	单系统运行/国际标准化组织/国际标准协会/等轴图
ISOL	Isolate/Isolation	隔离
ISP	Interim System Procedures/Internally Stored Program/Internet Service Provider	临时系统运行程序/内设程序/因特网服务提供者
ISRU	International Scientific Radio Union	国际无线电联合会
ISS	Industrial Standard Specifications/Integrated Safeguards System	工业标准规范/一体化保障体系
ISTF	Instrument Senser Test Facility	仪表传感器试验装置
IT	Instrumentation Tubing	仪表管
ITE	Instantaneous Trip Element	瞬时跳闸元件,瞬时脱扣元件
ITO	International Trade Organization	国际贸易组织

英文简称	英文全称	中文名称
ITOCHU	Itochu Corporation	伊滕忠株式会社
ITP	Individual Training Programme/Inspection Test Plan	个人培训计划/检查试验计划
ITR	In-Core Thermionic Reactor/Instrument Task Report	中心束热离子反应堆/仪表功能报告
ITRL	Instrument Test Repair Laboratory	仪表检修实验室
ITS	Information Technology Service/In-Tank-Solidification	信息技术服务/槽中固化
ITT	Incident Investigation Team	事件调查组
ITV	Inspection of Television	电视检查,视频检查
IU	International Unit	国际单位
IUA	International Union of Architects	国际建筑师联合会
IUC	International Union of Chemistry	国际化学联合会
IV	Independent Variable/Initial Velocity/Intercept Valve	自变量,独立变量/初速/节流阀(再热调节阀)
IVR	Inspection and Verification Report	检查和验证报告
IW	Induction Welding	感应焊
IW	Isotopic Weight	同位素重量
IWP	Installation Work Package	安装工作包
IX	Ion Exchange	离子交换
JB	Junction Box/Junction Box List	端子箱,接线盒/接线箱清单
JB-PB	Junction Box with Push Button	带按钮的端子箱
JCAE	Joint Committee on Atomic Energy	原子能联合委员会(美国)
JCCCNRS	Joint Coordinating Committee for Civilian Nuclear Reactor Safety	民用核反应堆安全联合协调委员会(美俄)

英文简称	英文全称	中文名称
JEC	Japanese Electrotechnical Committee	日本电工技术委员会
JEPIC	Japan Electric Power Information Center	日本电力信息中心
J-EXIM	Export-Import Bank of Japan	日本进出口银行
JIC	Joint Intelligence Committee	联合情报委员会
JIS	Japanese Industrial Standards	日本工业标准
JKT	Jacket	套管(筒)
JKTD	Jacketed	装套的
JMP	Jumper	搭接线,跨接线
JNACC	Joint Nuclear Accident Coordination Center	核事故联合协调中心
JNFI	Japan Nuclear Fuel Industries Co.	日本核燃料工业公司
JNFS	Japan Nuclear Fuel Service Co.	日本核燃料服务公司
JOM	Job Operation Manual	操作手册
JOP	Jacking Oil Pump	顶轴油泵
JPA	Job Pack Area	作业装配区
JPRS	Jumpers	跳线,搭接线,跨接线
JRDB	Joint Research and Development Board	联合研究与发展委员会
JRTR	Job Responsibility Training Requirement	岗位责任培训要求
JSME	Japanese Society of Mechanical Engineers	日本机械工程师学会
JUNCT	Junction	连接,接头

英文简称	英文全称	中文名称
JW	Jacket Water	水套冷却水
K	Kelvin (Absolute Scale)	开氏(绝对温标)
KE	Kinetic Energy	动能
KEEI	Korea Energy Economics Institute	韩国能源经济研究所
KEPCO	Korean Electric Power Company	韩国电力公司
KING	Kinetic Intense Neutron Generator	动态强中子发生器
KMP	Key Measurement Point	关键测量点
KSNR	Korea Standard Nuclear Reactor	韩国标准核反应堆
L	Length/Level	长度/水平(位);电平
L/C	Letter of Credit	信用证
L/D	Length-to-Diameter	长/径比
L/G	Letter of Guarantee	信用担保书
L/M	List of Material	材料单
L/T	Liquid Tight	液体密封
L/W	Lock Washer	锁紧垫圈
LA	Life Assessment/Lightning Arrestor/Limited Applicability	寿期评估/避雷器/有限可应用度
LAB	Laboratory	实验室
LAC	Levelized Annual Cost/Local Air Cooler	平准化年成本/就地空气冷却器
LAFR	Linear Accelerator Fuel Regenerator	直线加速器燃料再生装置
LAN	Local Area Network	局域网
LANPC	Ling Ao Nuclear Power Company Ltd	岭澳核电有限公司

英文简称	英文全称	中文名称
LANS	Guangdong Ling Ao Nuclear Power Station	岭澳核电站
LAS	Light Activated Switch/Low-Alloy Steel	光敏开关/低合金钢
LBBW	Laser Beam Braze Welding	激光钎焊
LBC	Laser Beam Cutting	激光切割
LBC-EV	Evaporative Laser Beam Cutting	蒸汽激光切割
LBC-IG	Inert Gas Laser Beam Cutting	惰性气体激光切割
LBC-O	Oxygen Laser Beam Cutting	氧气激光切割
LBD	Licensing Basis Document	执照申领基础文件
LBLOCA	Large-Break LOCA	大破口失水事故
LBP	Lumped Burnable Poison	块状可燃毒物
LBW	Laser Beam Welding	激光焊
LC	Level Control/Load Center/Local Control/Locked Closed/Low Carbon	水(液)位控制/负荷中心/就地控制/锁定在关闭状态,闭位锁定/低碳
LC IV	Loss of Class Ⅳ Power	失去Ⅳ级电源
LCC	Local Control Cubicle	就地控制开关柜
LCCW	Loss of Component Cooling Water	设备冷却水丧失
LCD	Late Core Disassembly/Liquid Crystal Display/Local Control and Display	晚期堆芯解体/液晶显示/就地控制与显示
LCI	Load Convertible Inverter	负荷逆变装置
LCK	Lock/Locking	闭锁/锁定
LCKR	Lock Ring	锁紧环

英文简称	英文全称	中文名称
LCM	Least Common Multiple	最小公倍数
LCO	Limited Condition for Operation	极限运行工况
LCR	Low Cost Reactor	低成本(反应)堆
LCSS	Lower Core Support Structure	堆芯下部支承构件
LCTL	Large Component Test Loop	大部件试验回路
LCV	Level Control Valve/Liquid Control Valve/Logic Control Variable/Low Calorific Value	液位控制阀/液体控制阀/逻辑控制变量/低热值
LD	Leak Detector/Logic Diagram/Loop Diagram (Instrument)	泄漏检测/逻辑图/回路图(仪表测量)
LDB	Licensing Design Basis	取照设计依据,执照申领设计依据
LDR	License Delivery Reports/Light-Dependent Resistors	执照递交报告,执照申领报告/光敏电阻
LDV	Laser-Doppler-Velocimeter	激光多普勒速度计
LE	Lead Engineer/Lifting Eye/Low Efficiency	首席工程师/吊耳/低效
LED	Light Emitting Diode	发光二极管
LEM	Liquid Effluent Monitor	液态流出物监测
LEOC	Local Emergency Operating Center	地区应急指挥中心
LEP	Lowest Effective Power	最低有效功率
LER	Licensee Event Report	许可证持有者的运行事件报告
LESS	Least Cost Estimating and Scheduling	最低成本估算与分解
LET	Linear Energy Transfer	传能线密度
LEU	Low Enriched Uranium	低浓铀,低富集铀

英文简称	英文全称	中文名称
LF	Levelizing Factor/Load Factor/Load Fuel/Low Frequency	平准化因子/负荷因子/装料/低频
LFA	Low-Frequency Amplifier	低频放大器
LFC	Low-Frequency Choke/Low-Frequency Current	低频扼流圈/低频电流
LFO	Low Frequency Oscillator	低频振荡器
LFR	Low Flux Reactor	低通量(反应)堆,低注量率反应堆
LFTG	Lifting	提升,起吊
LFV	Low Frequency Vibration	低频振动
LH	Latent Heat/Left Handed/Liquid Hydrogen	潜热/左手(侧,边)的/液态氢
LHC	Liquid Hydrogen Container	液氢容器
LHD	Load-Haul-Dump Unit	装运卸货(料)机
LHGR	Linear Heat Generation Rate	线功率,线释热率
LHSI	Low Head Safety Injection	低压安全注射
LHSP	Low Head Safety Injection Pump	低压安全注射泵
LHST	Laundry and Hot Shower Tank	洗涤与热淋浴水箱
LI	Level Indicator/Long Delay & Instantaneous Elements	液位指示器/长延时和瞬时元件
LIBOR	London Interbank Offered Rate	伦敦银行同业拆借利率
LIC	Level Indicator/Controller	液位指示/控制器
LILW	Low and Intermediate Level Radioactive Waste	低中放废物
LIM	Limit	极限,限值
LIMDAT	Limiting Date	限制日期

英文简称	英文全称	中文名称
LIN	Linear/Liquid Injection Nozzle/ Liquid Nitrogen	直线,线性/液体(毒物)注入管嘴/液态氮
LIN-N	Linear Ion Chamber Power Measurement	线性电离室功率测量
LIQ	Liquid	液体
LISS	Liquid Injection Shutdown System	液体毒物注入停堆系统
LITR	Low Intensity Test Reactor	低(中子)腔度试验(性反应)堆
LIVC	Low Input Voltage Converter	低输入电压变换器
LKI	Leak Indicator	泄漏指示器
LL	Label List/Line List/Liquid Level/Live Load/Lower Limit	标签清单/管线清单/液位/工作负载/下限
LLD	Lower Level Discriminator/ Lower Limit of Detection	较低阈甄别器/探测下限
LLFM	Low Level Flux Monitor	低注量率监测仪
LLHS	Light Load Handling System	轻载装卸系统
LLI	Liquid Level Indicator	液位指示器
LLLP	Low Leakage Loading Pattern	低泄漏装料模式
LLW	Low Level Waste	低放(射性)废物
LM	Level Meter/Light Metals/List of Material/Liquid Metals	电平计/轻金属/材料清单/液态金属
LMFBR	Liquid-Metal Fast Breeder Reactor	液体金属快增殖(反应)堆
LMFR	Liquid-Metal-Fueled Reactor	液体金属燃料(反应)堆
LMTD	Logarithmic Mean Temperature Difference	对数平均温差
LN	Liquid Nitrogen	液(态)氮
LND	Lined	有衬里的,内衬的

英文简称	英文全称	中文名称
LNG	Liquefied Natural Gas	液化天然气
LO	Liquid Oxygen/Locked Open/Lot/Low Alarm/Lubricating Oil	液态氧/锁定在开位/批,套,组/低限报警/润滑油
LOAC	Low Accuracy	低精确度
LOC	Letter of Credit/Location/Locator	信用证/定位/定位器
LOCACS	Loss of Core Auxiliary Cooling System Accident	堆芯辅助冷却系统丧失事故
LOCE	Loss of Coolant Experiment	冷却剂流失实验,失水实验
LOCT	Loss of Coolant Test	冷却剂丧失试验,失水试验
LOCV	Loss of Condenser Vacuum	凝结器真空丧失
LOE	Licensing Operational Event	与核安全有关的运行事件
LOECC	Loss of Emergency Core Cooling	应急堆芯冷却系统丧失
LOFA	Loss of Flow Accident	失流事故
LOFW	Loss of Boiler Feedwater	蒸汽发生器给水丧失
LOG-N	Logarithmic Ion Chamber Power Measurement	电离室对数功率测量
LOI	Letter of Intent/Low Operation Interval (RRA)	意向书/RRA 短运行间隔时间
LOLP	Loss of Load Probability	甩负荷概率
LOM	List of Modifications	更改清单
LOMC	Loss of Moderator Circulation	慢化剂循环失效
LOOP	Loss of Off-Site Power/Loss of Operating Power	失去厂外电源/丧失运行功率

英文简称	英文全称	中文名称
LOP	Line of Position/Local Operational Panel/Logic Processor/Loss of Offside Power/Lubricating Oil Pump	定位线/就地操作屏/逻辑处理机/失去厂外电源/滑油泵
LOPI	Loss of Piping Integrity	失去管系完整性
LOPO	Low Power Output Reactor/Low Power Water Boiler	低功率(反应)堆/低功率水锅炉
LOR	Loss of Regulation	失调
LORC	Loss of Reactivity Control	反应性控制失效
LOS	Lube Oil System	润滑油系统
LOSW	Loss of Service Water	丧失工业(厂)用水
LOU	Letter of Understanding	协议书,谅解函
LP	Lattice Pitch/Licensing Package/Low Pressure/Lower Plenum	栅格间距,栅距/取照包/低压(力)/下腔室
LPCIS	Low Pressure Coolant Injection System	低压冷却剂注射系统
LPCRS	Low Pressure Coolant Recirculation System	低压冷却剂再循环系统
LPCSS	Low Pressure Core Spray System	低压堆芯喷淋系统
LPD	Least Perceptible Chromaticity Difference	最低可分辨色差
LPFCS	Low Power Feedwater Control System	低功率给水控制系统
LPFS	Low Pressure Flooding System	低压淹没系统
LPG	Liquefied Petroleum Gas	液化石油气
LPH	Low Pressure Heater	低压加热器

英文简称	英文全称	中文名称
LPIS	Low Pressure Injection System	低压注射系统
LPP	Low Pressure Pump	低压泵
LPR	Low Pressure Reactor	低压(反应)堆
LPRM	Local Power Range Monitor	局部功率区段监测器
LPRR	Low Power Research Reactor	低功率研究(反应)堆(美国)
LPRS	Low Pressure Recirculation System	低压再循环系统
LPS	Logic Protection System	逻辑保护系统
LPSIS	Low Pressure Safety Injection System	低压安全注射系统
LPT	Liquid Penetration Test/Low Power Test/Low Pressure Turbine	着色检查,液体渗透检查/低功率试验/低压汽轮机,低压透平
LPTF	Low Power Test Facility	低功率试验装置(美国)
LPWR	Large Pressurized Water Reactor/Light Water PWR	大型压水堆/轻水压水堆
LR	Laboratory Reactor/Level Recorder/Lightning Rod/Load Ratio/Long Radius	实验室(反应)堆/液位(电平)记录仪/避雷针/载荷比/长半径
LRB	Licensing Review Basis	执照审批基础,执照评审依据
LRC	Linear Responsibility Chart	条形责任卡
LRHSC	Large Radioisotope Heat Source Capsule	大型放射性同位素热源盒
LRO	Low Resistance Ohmmeter	低电阻欧姆表
LRP	Logic Relay Panel	逻辑继电器盘
LRRP	Lowest Required Radiated Power	要求的最低辐照能力
LRTS	Liquid Radwaste Treatment System	放射性废液处理系统

英文简称	英文全称	中文名称
LRU	Least Recently Used	最近最少使用的
LRV	Liquid Relief Valve	液体释放阀
LRWC	Liquid Radioactive Waste Concentration	放射性废液浓缩
LRWPS	Liquid Radioactive Waste Processing System	放射性废液处理系统
LS	Laboratory System/Level Switch/Limit Switch/Licensing Submission/Long & Short Delay Elements/Low Speed	实验室系统/液位开关/限位开关/提交执照申领文件/长与短延时元件/低速
LSA	Local Stress Analysis	局部应力分析
LSC	Large Scale Computer	大型计算机
LSD	Large Screen Display	大屏幕显示
LSDS	Large Screen Display System/Liquid Shutdown System	大屏幕显示系统/液体停堆系统
LSFCR	Large Scale Fuel Channel Replacement	大规模燃料通道更换
LSPB	Large Scale Prototype Breeder	大型原型增殖(反应)堆
LSS	LOCA Serveillance System	失水事故监督系统
LST	Light Sensitive Tube/Liquid Storage Tank/Lowest Service Temperature	光敏管/液体贮存箱/最低使用温度
LSTG	Large Steam Turbine Generator	大型汽轮发电机
LT	Leak Test/Low Temperature/Low Torque	泄漏试验/低温/低转矩
LTC	Low Temperature Coefficient	低的温度系数

英文简称	英文全称	中文名称
LTPD	Lot Tolerance Percent Defective	批内允许次品率
LTR	Letter	信函,字母
LTSF	Lid Tank Shielding Facility	遮盖(水)箱的屏蔽装置
LUB	Lubrication	润滑
LV	Limit Value/Linear Velocity/Low Voltage	极限值/线速度/低(电)压
LVAS	Large Volume Air Sampler	大容量空气取样器
LVDT	Linear Variable Differential Transformer	线性可变差动变压器
LVF	Linear Vector Function	线性矢量函数
LVPS	Low Voltage Power Supply	低压电源
LWB	Light Water Breeder	轻水增殖(反应) 堆
LWBR	Light-Water Breeder Reactor	轻水增殖(反应)堆
LWGR	Light-Water Cooled Graphite Moderated Reactor	轻水冷却石墨慢化(反应)堆
LWHCR	Light Water High Converter Reactor	高转换型轻水(反应)堆
LWL	Low Water Level	低水位
LWOR	Light Water Organic Reactor	轻水(慢化)有机(物冷却)(反应)堆
LWPS	Liquid Waste Processing System	废液处理系统
LZC	Liquid Zone Controller	液体区域控制器
LZCU	Liquid Zone Control Unit	液体区域控制装置
M	Main/Module	总管/模数(量)
M&IR	Manufacturing and Inspection Record	制造与检查记录

英文简称	英文全称	中文名称
M&O	Maintenance and Operation	检修与运行，维护与操作
M/A	Maintenance Analysis	维护分析，检修分析
M/C	Manual Control	手动控制
M/D	Man Day	人日
M/H	Man-Hour	工时
M/I	Malleable Iron	可锻铁
M/R	Maintenance and Repair	维护与修理
M/S	Mild Steel	低碳钢
MA	Main Alarm	主报警
MAC	Maximum Admissible Concentration/Maximum Air Concentration	最大允许浓度/最大空气浓度
MACH	Machine	机器(械)
MAD	Machine Analysis Display/Main Assembly Drawing/Maintenance, Assembly and Disassembly/Material Analysis Data/Multiply and Divide	机械分析显示/主装配图，总成图/维护、装配与拆卸/材料分析数据/乘与除
MADAR	Malfunction Analysis Detection and Recording	故障分析探查与记录
MADE	Multichannel Analog-to-Digital Data Encoder	多路模拟-数字数据编码器
MAE	Mean Absolute Error	平均绝对误差
MAERA	Mutual Atomic Energy Reinsurance Association	原子能共同再保险协会
MAF	Maximum Amplitude Filter	最大振幅滤波器

英文简称	英文全称	中文名称
MAG	Magnetic/Maximum Available Gain	磁(性)的/最大可用增益
MAGZ	Magazine	料箱(仓)
MAINT	Maintenance	维修
MALL	Malleable	可锻的,展性
MAN	Manual	手册;手动(控制)
MANO	Manometer	(流体)压力表
MAP	Mean Assembly Power/Minimum Allowable Performance Standard	反应堆组件平均功率/最低允许性能标准
MAPL	Manufacturing Assembly Parts List	制造装配零件表
MAR	Minimum Acceptability Requirement	最低可接受要求
MARS	Memory-Address Register Storage	存储-地址寄存器存储
MATL	Material	材料
MATL-CODE	Material Code Number	材料编码号
MATNO	Material (Requested) Is Not Available	(所需)材料现在缺货
MATSO	Material Requested Being Supplied	所需材料现正供应
MAW	Manual Arc Welding	手工电弧焊
MAX	Maximum	极大,最大
MB	Memory Buffer	存储缓冲器
M-B	Make-Break	闭合-断开
MBB	Make-Before-Break	先合后开
MBD	Million Barrels Per Day	每天百万桶

英文简称	英文全称	中文名称
MBH	Thousands of BTU Per Hour	千英国热量单位/时
MC	Magnet Contactor/Magnetic Core/Manual Control/Material Control/Megacycle/Metal Containment/Moisture Content/Momentary Contact	磁铁开关/磁芯/人工控制,手动控制/材料控制/兆周/金属安全壳/湿度/瞬时接触
MCA	Manufacturing Chemists Association/Maximum Credible Accident	制造化学师协会(美国)/最大可信事故
MCC	Main Control Console/Motor Control Center/Milestone Completion Certificate	主控制台/电机控制中心/里程碑完工证明
MCCR	Ministry of Consumer and Commercial Relation	消费和商业关系部
MCES	Main Condenser Evacuation System	主冷凝器抽真空系统
MCF	Million Cubic Feet	兆立方英尺
MCHFR	Minimum Critical Heat Flux Ratio	最小临界热流密度比
MCI	Malleable Cast Iron	韧性铸铁
MCL	Masonry Check List	砌体工程检查单
MCM	Magnetic Core Memory/Minimum Critical Mass/Monte Carlo Method	磁芯存储器/最小临界质量/蒙特卡罗法
MCO	Moisture Carryover	(蒸汽中)夹带水分

英文简称	英文全称	中文名称
MCP	Main Circulation Pump/Main Control Panel/Measurement and Calibration of Power	主循环泵/主控制盘(屏)/功率测量与标定
MCR	Main Control Room/Maintenance Control Report/Master Change Record/Maximum Continuous Rating/Maximum Continuous Revolution/Mean Conversion Ratio	主控制室/维修管理报告/主要变更记录/最大持续功率/最大连续转速/平均转换比
MCS	Main Control System/Master Calibration Sheet	主控系统/主要标定单
MCTI	Metal Cutting Tool Institute	金属切割工具研究所
MCV	Minimum Critical Volume	最小临界体积
MD	Magnetic Drum/Man-Day/Manufacturers' Drawing/Mean Deviation/Motor Drive	磁鼓/人-日/制造商图纸/平均偏差/电机驱动
MDA	Malfunction Detection Analyser/Minimum Detectable Activity	故障检测分析仪/最低可测(放射性)活度
MDC	Main Display Console/Maximum Dependable Capacity	主显示台/最大可靠容量
MDCC	Master Data Control Console	主要数据控制台
MDCF	Maximum Dependable Capacity Factor	最大可靠容量因子
MDCS	Moderator-D_2O Collection System	慢化剂-重水收集系统
MDDSA	Magazine D_2O Drain Stack Assembly	料仓重水疏水叠式组件

英文简称	英文全称	中文名称
MDF	Main Distribution Frame	主配线架
MDNBR	Minimum Departure From Nucleate Boiling Ratio	最小偏离泡核沸腾比
MDR	Memory-Data Register	存储-数据寄存器
MDS	Maintenance Data System/Malfunction Detection System	维修数据系统/故障检测系统
MDSS	Moderator D_2O Sampling System	慢化剂重水取样系统
MDT	Mean Down Time	平均停机(停用)时间
ME	Magnetic-Electric/Master of Engineering/Mechanical Efficiency/Mechanical Engineer/Milligram Equivalent	磁-电/工程硕士/机械效率/机械工程师/毫克当量
MEA	Maintenance Engineering Analysis	维修工程分析
MECH	Mechanical	机械的
MECHA	Mechanism	机制(理)
MED	Medium	介质,媒体;中间的
MEGARAM	Mega Random Access Memory	大容量随机存取存储器
MEH	Miscellaneous Electrical Hardware	其他电气硬件
MEL	Master Equipment List	主要设备清单
MEMS	Mechanical Equipment Management System	机械设备管理系统
MEP	Mean Effective Pressure	平均有效压力
MER	Material Evaluation Report	材料评定报告
MESA	Miniature Electrostatic Accelerometer	微型静电加速计

英文简称	英文全称	中文名称
MESS	Message	信(消)息,电(报)文
MET	Multi-Emitter Transistor	多发射极晶体管
METR	Material and Engineering Test Reactor	材料和工程试验(反应)堆
MEU	Moderately Enriched Uranium	中等富集的铀中浓铀
MF	Master Flowsheet/Microfarad/Microfilm/Multi-Frequency/Multiplying Factor	主流程图/微法拉/微型胶卷/多频/乘率,增加量
MFC	Microfunctional Circuit/Multi-functional Control	微功能电路/多功能控制
MFCS	Main Feedwater Control System	主给水控制系统
MFDM	Multivariable Frequency Domain Method	多变量频域法
MFIS	Main Feedwater Isolation System	主给水隔离系统
MFP	Main Feed Pump/Mean Free Path/Mixed Fission Products	主给水泵/平均自由程/混合裂变产物
MFR	Manufacturer	制造商(者)
MFSP	Manufacturing Spare	制造备件
MFT	Master Fuel Trip	主燃料跳闸
MFTF	Molten Fuel Test Facility	熔化燃料试验装置
MFWLB	Main Feedwater Line Break	主给水管线破裂
MFWP	Main Feedwater Pump/Motor Feedwater Pump	主给水泵/电动给水泵
MG	Main Generator/Motor Generator	主发电机/电动发电机
M-G	Motor-Generator	电动机-发电机
MGC	Maximum Guaranteed Capcity	最大保证功率

英文简称	英文全称	中文名称
MGCR	Marine Gas Cooling Reactor	船用气冷反应堆(美国)
MGD	Magnetogasdynamics/Million Gallons per Day	磁气体动力学/百万加仑/日
Mgr	Manager	经理
MGS	Main Generator System	主发电机系统
MHA	Maximum Hypothetical Accident	最大假想事故
MHC	Mechanical Hydraulic Control	机械液压控制
MHD	Manufacturer History Docket/Magnetohydrodynamics/Master History Docket	制造商历史记录/磁流体力学/主要历史记录
MHE	Material Handling Equipment	材料处理设备;材料运载设备
MHF	Master History File	主要历史文件
MHI	Mitsubishi Heavy Industries, Ltd.	日本三菱重工业公司
MHPE	Maximum Historically Probable Earthquake	历史最大可能地震
MHTS	Moderator Heat Transport System	慢化剂热传输系统
MHX	Moderator Heat Exchanger	慢化剂热交换器
MI	Maintenance Instruction/Malleable Iron/Manufacturing Inspector/Material Inspection	维修说明书,维修指南/可锻铁/制造检验员/材料检查
MIAC	Material Identification and Accounting Code	材料识别与核算程序
MIAW	Magnetically Impelled Arc Welding	磁推力电弧焊
MIC	Microbiological Induced Corrosion	微生物引起的腐蚀

英文简称	英文全称	中文名称
MIDAS	Master Information and Data Acquisition System	主要信息与数据获取系统
MIG	Metal Inert Gas (Arc) Welding	惰性气体保护金属极(电弧)焊
MIL	Military	军用的
MIN	Minimum/Monitoring Information Notice	极小值,最小值/监督信息通知
MINI	Miniature	小(微)型
MIP	Maintenance Improvement Program	维修改进大纲
MIR	Material Inspection Report/ Memory Information Register	材料检查报告/存储信息寄存器
MIRR	Material Inspection and Receiving Report	材料检验与接收报告
MIS	Management Information System/Metal-Insulator-Semiconductor/Miscellaneous	管理信息系统/金属绝缘半导体/其他
MISC	Miscellaneous	其他,杂项
MIST	Multi-Loop Integral System Test	多环路综合系统试验
MIT	Material Issue Ticket	发料单
MITP	Manufacturing, Inspection & Test Plan	制造、检查和试验计划
Mks	Marks	标记
ML	Material List/Mean Level/Mechanical Loss	材料单/平均水平/机械损失
MLC	Main Loop Cooling	主回路冷却

英文简称	英文全称	中文名称
MLD	Master Lesson Directive/Minimum Lethal Dose	主要教训/最小致死剂量
MLDDP	Maintenance Lock D_2O Discharge Port	维修闭锁重水排放口
MLHGR	Maximum Linear Heat Generation Rate	最大线热功率(密度),最大线发热率
MLPS	Moderator-Liquid Poison System	慢化剂液体毒物系统
MLW	Medium Level Waste	中放(射性)废物
MM	Maintenance Manual/Man-Months/Manufacturers' Manual/Megameter	维修手册/人-月/制造商手册/兆欧表
MMCS	Man-Machine Control System/Moderator Main Circulation Pump	人-机控制系统/慢化剂主循环泵
MMI	Man-Machine Interface/Modified Mercalli Intensity	人机接口/修正的麦加利地震烈度
MMM	Modified Monel Metal	改进型蒙乃尔合金
MMP	Motor-Mount Pump	电动提升泵
MMS	Main Moderator System/Maintenance Management System/Measure Management System	主慢化剂系统/维修管理系统/测量管理系统
MMT	Material Management Team	材料管理组
MNPT	Male-National Pipe Thread	国家标准管外螺纹
MO	Machine Operation/Manually Operated/Manual Output/Molybdenum/Motor Operated	机器操作/人工操作的,手动的/人工输出/钼/电动的

英文简称	英文全称	中文名称
MOC	Master Operational Controller/ Middle of Cycle	主操作控制器/循环中期
MOD	Moderator/Modification/Modulus	慢化剂/修改/模数,模量
MOE	Measure of Effectiveness	有效性测量
MOFFP	Mixed Oxide Fuel Fabrication Plant	混合氧化物燃料制造厂
MOH	Maintenance Outage Hours	维修停役小时数
MOIST	Moisture	湿度,水分
MOL	Middle of Life	寿命中期
MOM	Minutes of Meeting	会议纪要
MON	Monitor	监控(测)器
MONAL	Mobile Nondestructive Assay Laboratory	移动式无损检验实验室(美国)
MONAS	Multichannel On-Line Noise Analysis System	多通道在线噪声分析系统
MOPD	Maximum Operating Pressure Differential	最大运行压差
MOST	Ministry of Science and Technology	科学技术部
MOT	Main Output Transformer/Mechanical Overspeed Trip	主变压器/机械超速停机
MOTA	Material Open Test Assembly	材料开放试验组件
MOTIE	Ministry of Trade, Industry and Energy	贸易、工业和能源部

英文简称	英文全称	中文名称
MOTS	Module Test Set	模块试验装置
MP	Main Pump/Maintenance Period/Maintenance Point/Maintenance Prevention/Maintenance Procedure (Program)/Mechanical Part/Medium Pressure/Melting Point	主泵/维修周期/维修点/预防性维修/维修程序(大纲)/机械零件/中压/熔点
MPBB	Maximum Permissible Body Burden	体内最大容许承受量
MPC	Material Preparation Center/Maximum Permissible Concentration	材料制备中心/最大容许浓度
MPCU	Maximum Permissible Concentration of Unidentified Radionuclides	未知放射性核素最大容许浓度
MPD	Maximum Permissible Dose	最大容许剂量
MPDE	Maximum Permissible Dose Equivalent	最大容许剂量当量
MPE	Maximum Permissible Exposure	最大容许照射
MPFE	Maximum Possible Flooding Elevation	最大可能洪水标高
MPI	Magnetic Particle Inspection/Maximum Permissible Intake	磁粉检验/最大容许摄入量
MPL	Maximum Permissible Limit (Level)	最大允许限值(水平)
MPLHGR	Maximum Planar Linear Heat Generation Rate	堆芯某水平断面的最大线热功率(密度)

英文简称	英文全称	中文名称
MPO	Maximum Power Output	最大功率输出
MPP	Most Probable Position	最可能位置
MPRE	Medium Power Reactor Experiment	中等动力反应堆试验
MPS	Manufacturing Plan Sheet/Moderator Purification System	生产计划图表/慢化剂净化系统
MPSH	Mean Pressure Suction Head	平均压力下抽提高度
MPT	Magnetic Particle Testing	磁粉探伤
MPTP	Material Physical Testing Program	材料物理试验大纲计划
MPWM	Material Plan and Warehouse Management	材料计划与仓库管理
MQ	Material Query	材料质疑
MQC	Manufacturing Quality Control	制造质量控制
MR	Material Requisition/Memory Register/Modification Request/Monitor Recorder/Monorail/Monthly Report	材料申请/存储寄存器/修改申请/监测记录仪/单轨/月报
MRB	Material Review Board	材料审查委员会
MRD	Minimum Reacting Dose	最小反应剂量
MRF	Multipath Reduction Factor	多路降低因数
MRO	Manual Request to Order	人工采购请求
MRP	Main Reactor Pump/Monthly Report of Progress	反应堆主泵/进展月报

英文简称	英文全称	中文名称
MRR	Maintenance, Repairs and Replacements/Material Rejection Report/Material Reliability Report	维护、修理和更换/材料拒收报告/材料可靠性报告
MRT	Material Return Ticket	退料单
MS	Main Steam/Margin of Safety/Master Switch/Maximum Stress/Mean Square/Medium Steel/Metallurgical Society/Mild Steel/Millisievert/Moderator System/Moisture Separator	主蒸汽/安全裕量/主控开关/最大应力/均方/中碳钢/冶金学会/低碳钢/毫希沃特/慢化剂系统/汽水分离器
MSAF	Master Supervisory and Alarm Frame	主要的监视和报警架构
MSBR	Molten-Salt Breeder Reactor	熔盐增殖(反应)堆
MSDDA	Moderator System-Deuterium & De-deut. Assembly	慢化剂系统-氘和除氘装置
MSEE	Mean Square Error Efficiency	均方误差效率
MSF	Maintenance & Service Facilities	维修和服务设施
MSFRF	Maintenance Service Facilities-Rehearsal Facility	维修和辅助设施-操练装置
MSG	Modular Steam Generator	组合式蒸汽发生器,模块化蒸汽发生器
MSIVLCS	Main Steam Isolation Valve Leakage Control System	主蒸汽隔离阀泄漏控制系统
MSL	Main Steam Line/Maximum Service Life/Maximum Service Limit/Mean Sea Level	主蒸汽管/最长使用寿命/最大使用限度/平均海拔

英文简称	英文全称	中文名称
MSP	Motor Suction Pump	电动吸入泵,电动抽吸泵
MSR	Material Substitution Request/Main Steam Rate/Maximum Steam Rate/Moisture Separator Reheater	材料替代请求/主蒸汽产率/最大蒸汽产率/汽水分离再热器
MSRV	Main Steam Relief Valve	主蒸汽释放阀
MSS	Main Steam System/Manufacturers' Standardization Society/Material Submital Sheet	主蒸汽系统/制造商标准化学会/材料提交单
MSSV	Main Steam Safety Valve/Main Steam Stop Valve	主蒸汽安全阀/主蒸汽截流阀,主汽门
MST	Maintenance Surveillance Test	维修监督试验
MSTBR	Molten-Salt Thorium Breeder Reactor	熔盐钍增殖反应堆(美国)
MSV	Main Stop Valve	主汽门
MSV-BV	Main Stop Valve Bypass Valve	主汽门旁通阀
MSW	Microswitch	微动开关
MT	Magnetic Test/Main Step-up Transformer	磁粉探伤/主升压变压器
MTBA	Machine Tool Builders' Association	机床制造业联合会
MTBF	Mean Time Between Failures	平均无故障时间,故障间隔平均时间
MTBFO	Mean Time Between Forced Outage	故障停堆平均间隔时间

英文简称	英文全称	中文名称
MTBI	Mean Time Between Incidents	平均异常现象间隔时间
MTBM	Mean Time Between Maintenance	维修平均间隔时间
MTBMA	Mean Time Between Maintenance Actions	维修活动平均间隔时间
MTBR	Mean Time Between Repairs	平均维修间隔时间
MTBS	Mean Time Between Scram	紧急停堆平均间隔时间
MTBSF	Mean Time Between Significant Failures/Mean Time Between System Failures	严重故障平均间隔时间/系统故障平均间隔时间
MTC	Moderator Temperature Coefficient/Moderator Temperature Control	慢化剂温度系数/慢化剂温度控制
MTD	Mean Temperature Difference/Mounted	平均温差/被安装的
MTI	Metal Treating Institute	金属处理学会
MTL	Material/Material Testing Laboratory/Metal	原料,材料/材料试验室/金属
MTO	Modification Task Outline	改进工作大纲,整治工作概要
MTR	Magnetic Tape Recorder/Material Testing Reactor/Material Testing Report/Mean Time to Repair/Meter	磁带记录仪/材料试验反应堆(美国)/材料试验报告/正常运行平均时间/米,表(计)
MTRD	Metered	计(测)量的
MTRG	Metering	测(计)量
MTS	Main Turbine System	主汽轮机系统

英文简称	英文全称	中文名称
MTTF	Mean Time to Failure	无故障运行平均时间
MTTFF	Mean Time to First Failure	出现首次故障的平均时间
MTTFO	Mean Time to Forced Outage	平均强迫停运间隔时间
MTTOO	Mean Time to Planned Overhaul Outage	平均计划大修停运间隔时间
MTTPO	Mean Time to Planned Outage	平均计划停运间隔时间
MTTR	Mean Time to Repair	平均维修时间
MTU	Metric Tons of Uranium/Metric Units	公吨铀/公制单位
MU	Mass Unit/Measurement Unit/Multiple Unit	质量单位/测量单元;测量单位/多机组,多单元
MUF	Maximum Usable Frequency	最高可用频率;最大可利用频度
MUFW	Make-up Feed Water	补充水
MUR	Mockup Reactor	模式反应堆
MV	Mean Value/Measured Value/Medium Voltage/Motorized (Motor) Valve/Multivibrator	平均值/实测值/中压/电动阀/多频谐振动器
MWL	Mean Water Level	平均水位
MWP	Maximum Working Pressure/Membrane Waterproofing	最大工作压力/隔膜防水
MWS	Makeup Water System	补水系统
MWST	Makeup Water Storage Tank	补水贮存箱,补水箱
MWV	Maximum Working Voltage	最大工作电压
MXQ	Modular X-Ray Quantometer	标准型X射线剂量计,模块化X射线剂量仪
MY	Million Years	百万年

英文简称	英文全称	中文名称
N	Neutral	中性的
N/A	Not Affected/Not Applicable/Not Available	未受影响的/不适用的/不可用,没有
N/F	New Fuel	新燃料
N/G	Nuclear Grade	核级
N/S	Non-Standard	非标准
NA	Neutral Axis/Normal Alarm/Not Assigned	中性轴/正常报警/未指定,不赋值的
NAA	Neutron Activation Analysis	中子活化分析
NAH	Noise Analysis Hardware	噪声分析设备
NAL	National Accelerator Laboratory	国家加速器实验室(美国)
NAM	National Association of Manufacturers	全国制造商协会(美国)
NAPSD	Normalized Auto-Power Spectral Density	归一化自功率谱密度
NAT	Native/Natural	本国的/自然的
NATU	Natural Uranium	天然铀
NB	Narrow Band	窄频带
NBCC	National Building Code of Canada	加拿大国家建筑法规
NBR	Nuclear Boiler Rate	核锅炉出力
NC	Natural Convection/Noise Criterion/Non-Condensing/Normally Closed/Numerical Control	自然对流/噪声标准/不冷凝的/常关(闭)/数控
NCAI	National Committee of Atomic Information	全国核信息委员会
NCC	Nuclear Circuits Cleaning	核回路清洗

英文简称	英文全称	中文名称
NCCTM	New Core Component Transfer Machine	新堆芯部件转运机
NCGA	Neutron-Capture Gamma-Ray Analysis	中子俘获γ射线分析
NCM	Noncorrosive Metal	不受腐蚀金属,耐蚀金属
NCPSD	Normalized Cross-Power Spectral Density	归一化互功率谱密度
NCR	Non-Conformance Report	不符合项报告
NCRL	Non-Conformance Report List	不符合项报告清单
NCRP	National Committee on Radiological Protection	全国辐射防护委员会
NCS	Net Control Station/Numerical Control Society	网控站/数字控制学会
ND	Natural Draught/Not Detected/Noise Detector/Nuclear Device	自然通风/未被探(测)出/噪声探测器/核装置
NDE	Non-Destructive Examination	无损探伤
NDHX	Natural Draft Heat Exchanger	自然通风热交换器
NDI	Non-Destructive Inspection	无损探伤,无损检验
NDL	Non-RFC Document List	非建造许可证(RFC)文件清单
NDMN	Nuclear Design Modification Notice	核设计修改通知
NDT	Nil Ductility Transition/Non-Destructive Testing	无延性转变/无损检验
NDTT	Nil Ductility Transition Temperature	无延性转变温度
NEA	Nuclear Energy Agency	核能机构(巴黎)
NEANDC	NEA Nuclear Data Committee	核能机构核数据委员会

英文简称	英文全称	中文名称
NEC	National Electrical Code/Nippon Electric Company	全国电气法规(美国)/日本电气公司
NEF	National Extra Fine Thread/ Noise Equivalent Flux	国家标准极细牙螺纹/噪声等效注量率
NELPIA	Nuclear Energy Liability and Property Insurance Association	核损害责任与资产保险协会
NELT	NEA Study Group on the Long-Term Role of Nuclear Energy	欧洲核能机构(NEA)核能长期作用研究组
NEMA	National Electrical Manufacturers Association/National Emergency Management Association	全国电气制造商协会(美国)/国家应急管理联合会(美国)
NEO	Nuclear Equipment Operator	核设备操纵员
NEP	Nuclear-Electric Propulsion/Nuclear Emulsion Plate	核电推进/核乳胶板
NEPA	National Environment Policy Act/National Environment Protection Administration	国家环境政策法令/国家环境保护总局
NEPC-NP	Northeast Electric Power Construction Corporation	东北电力建设公司
NEPIA	Nuclear Energy Property Insurance Association	核能财产保险协会
NERO	New Experimental Low Energy Reactor	新型低功率实验性反应堆
NES	Nuclear Energy System	核能系统
NESCWS	Non-Essential Service Chilled Water System	非安全级厂用冷冻水系统

英文简称	英文全称	中文名称
NETAC	Nuclear Energy Trade Associations' Conference	核能贸易协会会议
NETF	Nuclear Engineering Test Facility	核工程试验装置
NETR	Nuclear Engineering Test Reactor	核工程试验反应堆
NF	Nano Farads/New Fuel/Nuclear Fission	毫微法拉,纳法/新燃料/核裂变
NFB	Negative Feedback	负反馈
NFI	Nuclear Fuel Industries Co.	核燃料工业公司(日本)
NFLCP	New Fuel-Local Control Panel	新燃料-就地控制盘
NFM	Neutrino Flux Monitoring	中微子注量率监测(装置)
NFPA	National Fire Protection Association/National Fluid Power Association	全国消防协会/全国流体动力协会(美国)
NFRE	New Fuel-Room Equipment	新燃料-室设备
NFTA	New Fuel-Transfer Auxiliary	新燃料-转运辅助设施
NFTM	New Fuel-Transfer Mechanism	新燃料-转运机械
NFTMA	New Fuel-Transfer Mechanism Air Panel	新燃料-转运机械压缩空气盘
NFU	Notified for Use	宣布可用
NG	Neutral Grounding	中性接地,中线接地
NGL	Natural Gas Liquid	液态天然气
NHP	Nominal Horse Power	额定马力
NHSD	Nuclear Health and Safety Department	核保健和安全部
NI	Nickel/Non-Inductive/Normally Indeterminate/Nuclear Island	镍/无感应的/正常时不定/核岛

英文简称	英文全称	中文名称
NIC	Nuclear Information Centre	核情报中心
NIM	Nuclear Instrument Module	尼姆(国际通用的一种核仪器标准模块)
NINF	Nuclear Information Working Group	核信息工作组
NIST	National Institute of Standards and Technology	国家标准与技术局(美国)
NITC	Not in This Contract	不在此合同中
NL	Net Loss/Nozzle List	净损失/管嘴清单,接管清单
NLGI	National Lubricating Grease Institute	国立润滑脂研究所
NLS	Non-Linear System	非线性系统
NMAC	Nuclear Maintenance Applications Center/Nuclear Maintenance Assistance Center	核维修应用中心/核维修援助中心
NMC	National Meteorological Center	国家气象中心
NMCS	Nuclear Material Control System	核材料控制中心
NMR	Normal Mode Rejection/Nuclear Magnetic Resonance	常态抑制/核磁共振
NMTBA	National Machine Tool Builders Association	全国机床制造商协会
NNDC	National Nuclear Data Centre	国家核数据中心(美国)
NNI	Non-Nuclear Instrumentation	非核测量仪表,常规测量仪表
NNSA	National Nuclear Safety Administration	国家核安全局(中国)
NO	Normally Open	常开的,静开的

英文简称	英文全称	中文名称
NOAC	Nuclear Operation Analysis Center	核运行分析中心
NOC	Notice of Change/Nuclear Oversight Committee	变更通知单/核监视委员会(美国)
NOE	Normal Operation Earthquake	正常运行地震
NOI	Notice of Intent	意向书
NOL	Normal Operating Loss	正常运行损失
NOM	Nominal	额定的,标定(称)的
NOR	Normal	正常的
NP	Name Plate/Neutral Point/Normal Pitch/Normal Pressure	铭牌/中性点/标准间距(节距)/额定压力,常压
NPAR	Nuclear Plant Aging Research	核电厂老化研究
NPCIL	Nuclear Power Co. Of India Ltd.	印度核电有限公司
NPD	Nuclear Power Demonstration	示范性核电厂
NPGD	Nuclear Power Generating Department	核发电部
NPHS	Normal Power Heat Sink	正常热阱
NPI	Nuclear Power International	核动力国际公司
NPL	National Physical Laboratory/Noise Pollution Level	国家物理实验室(英国)/噪声污染级,噪声污染水平
NPLA	Nuclear Plant Life Assurance	核电厂寿命保证
NPM	Nuclear Project Management/Nuclear Project Manager	核项目管理/核项目经理
NPO	Nuclear Plant Operator	核电厂操纵员
NPOC	Nuclear Power Oversight Committee	核电监督委员会

英文简称	英文全称	中文名称
NPPTS	Nuclear Power Plant Training Simulator	核电厂培训模拟机
NPQJVC	Nuclear Power Qinshan Joint Venture Company, Ltd.	核电秦山联营有限公司
NPR	New Production Reactor	新生产堆
NPRDS	Nuclear Plant Reliability Data System	核电厂可靠性数据系统
NPS	Neutrons per Second/Non-Interruptible Power Supply/Nuclear Power Station	每秒中子数/不间断电源/核电站
NPSH	Net Pressure Suction Head	净正吸入压头
NPSHA	NPSH Available	有效净正吸入压头
NPSHR	NPSH Required	要求的净正吸入压头
NPT	Normal Pressure and Temperature	常压常温
NQR	Non Quality Related	与质量无关的
NRC	Nuclear Regulatory Commission/Nuclear Research Centre	核管理委员会(美国)/核研究中心
NRCP	Non-Reinforced Concrete Pipe	无钢筋混凝土管
NRM	Non-Routine Maintenance	非日常维修
NRMS	Normalized Root-Mean-Square	归一化均方根
NRP	Normal Rated Power	额定功率
NRR	Office of Nuclear Regulatory Research	核管理研究局(美国)
NRT	Normal Rated Thrust	额定推力
NRV	Non-Return Valve	止回阀,逆止阀
NRZ	Non-Return-to Zero	不归零(制)

英文简称	英文全称	中文名称
NS	Not Specified/Nuclear Safety	未指定的，不规定的/核安全
NSA	National Security Agency/National Standards Association	国家安全局(美国)/国家标准协会(美国)
NSB	Nuclear Standards Board	核标准委员会
NSC	National Security Council/No Significant Counts/Nuclear Safety Center	国家安全委员会(美国)/无显著计数/核安全中心(美国)
NSEA	Nuclear Safety Environmental Analysis	核安全环境分析
NSF	National Science Foundation	国家科学基金会(美国)
NSN	Nuclear Safety Note	核安全函件，核安全短笺
NSP	Nuclear Steam Plant/Nuclear Supply Plant	核蒸汽厂房/核动力站
NSR	Non Safety Related	非安全相关的
NSRI	Nuclear Safety Research Index	核安全研究索引
NSSC	Nuclear Safety Standards Commission	核安全标准委员会(德国)
NSSS	Nuclear Steam Supply System	核蒸汽供应系统
NT	Non-Tight	不密封的
NTC	Negative Temperature Coefficient	负温度系数
NTL	Non-Threshold Logic	无阈值逻辑(电路)
NTP	Normal Temperature and Pressure	常温常压
NTS	Not to Scale	不按比例
NU	Natural Uranium	天然铀
NUC	Nuclear	核的

英文简称	英文全称	中文名称
NUCOMPS	Nuclear Core Monitoring and Prediction System	核堆芯监测与预报系统
NUMARC	Nuclear Management and Resources Council	核管理与资源理事会(美国)
NUPER	Nuclear Plant Event Report	核电厂事件报告
NUPLEX	Nuclear Plant Life Extension	核电厂寿命延长,核电厂延寿
NUSS	Nuclear Safeguards System/Nuclear Safety Standard/Nuclear Safety Standards Program	核保障体系/核安全标准/核安全标准程序
NWL	Normal Water Level	正常水位
NWP	Non Waterproof	不防水的未经耐水处理的
NWS	National Weather Service	国家气象局
NX	Nonexpendable Item	非消耗品
O&R	Overhaul and Repair	检修与修理
O/JKT	Outer Jacket	外套
O/M	Operation and Maintenance	运行与维修
O/R	On Request	根据请求,函索(即寄)
OA	Office Automatic	办公自动化
OAC	Oxygen Arc Cutting	氧气电弧切割
OAI	On-Line Analog Input	在线模拟量输入
OAO	On-Line Analog Output	在线模拟量输出
OAT	Operating Ambient Temperature/Outside Air Temperature	工作环境温度/室外温度
OAW	Operations Analysis Worksheet	运行分析工作表
OAW	Oxy-acetylene Welding	氧气乙炔焊

英文简称	英文全称	中文名称
OBI	Overseas Bechtel Incorporated	Bechtel 海外公司
OBN	Observation Note	观察通知单
OC	Oxygen Cutting	气割
OCA	Overcooling Accident	过冷事故
OCB	Oil Circuit Breaker/Overload Circuit Breaker	油开关/过载断路器
OCDR	Organic-Cooled Deuterium-Moderated Reactor	有机冷却重水慢化(反应)堆
OC-F	Flux Cutting	熔剂切割
OC-P	Metal Powder Cutting	金属粉剂切割
OCS	Overspeed Control System	超速控制系统
OCU	Operational Control Unit	操作控制设备,操作控制单元
OD	Outside Diameter/Outside Dimension/Over Drive	外径/外部尺寸/过载驱动
ODI	On-Line Digital Input	在线数字量输入
ODO	On-Line Digital Output	在线数字量输出
OECD	Organization for Economic Cooperation and Development	经济合作和发展组织
OEM	Original Equipment Manufacturer	原设备制造厂家
OF	Oil Fuel/Operation Flow-Sheet/Outside Face	燃油/运行流程图/外表面
OFA	Optimized Fuel Assembly	优化的燃料组件
OFC	Oxyfuel Gas Cutting	氧燃气切割
OFOR	Operational Forced Outage Rate	运行故障停用率
OFPDC	Operational Flowsheet Preparation During Commissioning	调试期间运行流程图的编制

英文简称	英文全称	中文名称
OFS	Operating Flowsheet	运行流程图
OFW	Oxyfuel Gas Welding	气焊
OG	Oxygen Gouging	氧气刨削
OGMS	Off-Gas Management System	废气管理系统
OGPP	Owner Group Progress Plan	业主集团进度计划
OGS	Off-Gas System	废气系统
OH	Oil Heater/Ontario Hydro/Overhaul	油加热器/安大略水电公司(加拿大)/大修
OHD	Outlet Header	出口集管
OHLHS	Overhead Heavy Load Handling System	高空重载装卸系统
OHM	Ohmmeter	欧姆表
OHT	Ontario Hydro Technologies	安大略水电技术公司(加拿大)
OHW	Oxy-Hydrogen Welding	氢氧焊
OI	Operating Instructions	操作说明
OID	Onset of Intermittent Dryout	间歇风干起始点,开始时断时续的风干过程
OIF	Operator Induced Failure	操纵员引起的故障
OIL	Operation Inspection Log	操作检查记录
OIU	Operator Interface Unit	人机接口单元
OJT	On (-the)-Job Training	岗位培训,在岗培训
OJTA	On Job Training Abnormal	异常工况的岗位培训
OJTE	On Job Training Emergency	应急工况的岗位培训
OJTN	On Job Training Normal	正常工况的岗位培训
OL	Oil Level/Open Loop/Operation Licence/Overall Length/Overlap/Overload	油位/开环,回路敞口/运行许可证,运行执照/全长/重叠,重叠度/过载

英文简称	英文全称	中文名称
OLC	Operating Limits and Conditions/Outgoing Line Circuit	运行限值和工况/输出回线
OLC	Oxygen Lance Cutting	氧矛切割
OM	Operating Manual/Operation Maintenance/Operator Manual Mode	运行手册/运行维护/运行人员手控方式
OMI	Operational Maintenance Instruction	运行维修说明书
OMR	Organic Moderated Reactor	有机物慢化反应堆
OMS	Overpressure Mitigation System	超压缓解系统
OMT	Operation, Maintenance and Test	运行、维护与试验
ON	Observation Notice	视察通知
ONB	Onset of Nucleate Boiling	泡核沸腾起始点
ONRR	Office of Nuclear Reactor Regulation	核反应堆管理局
OO	Operating Order	操作票
OOC	Out of Control	失控
OOO	Out of Order	有故障的，失常
OP	Oil Pressure/Oil Pump/Operating Procedure/Operation Permit/Operations Equipment/Over Pressure	油压/油泵/运行程序，操作规程/运行许可/运行设备/超压
OP&P, OPP	Operating Polices and Principles	运行策略和原则
OPC	Overspeed Protection Control	超速保护控制

英文简称	英文全称	中文名称
OPCOM	Operator-Process Communication System	操作过程通信联络系统
OPE	Operating Plant Experience	核电厂运行经验
OPEC	Organization of Petroleum Exporting Countries	石油输出国组织
OPG	Outage Planning Group	大修计划组
OPI	Open Package Inspection	设备开箱检查
OPP	Operating Polices and Procedures	运行策略和程序
OPR	Overpressure Protection Report	超压保护报告
OPS	Operational Protection System	运行保护系统
OPSP	Operation Spare Part	运行备件
OPSS	Orifice Plate Specification Sheet	孔板规格表
OPT	Output Transformer	输出变压器
OPTRAN	Operational-Transient	运行瞬态
OQA	Operation Quality Assurance	运行质量保证
OQAM	Orientation of QAM	质量保证手册指南
OQAP	Operations Quality Assurance Programme	运行质量保证大纲
OQC	Outside Quality Control	外部质量控制
OR	Open Report/Operating Range/Operation Record/Operational Requirements/Outside Radius/Overload Relay	公开报告/运行(工作)范围/运行记录/运行要求/外围半径/过载继电器
ORNL	Oak Ridge National Laboratory	橡树岭国家实验室(美国)
ORR	Operations Readiness Review	运行准备就绪审查

英文简称	英文全称	中文名称
OS	Oil Switch/Operating System/Overall-Shield/Overseas/Oversize	油开关/操作系统/全屏蔽/海外/超大尺寸,超差
OSA	Optical Society of America	美国光学学会
OSART	Operational Safety Review Team/Operational Safety Assessment Review Team	(核电厂)运行安全评议(检查)组/运行安全评审团(IAEA)
OSF	Onsite Facility	就地设施
OSG	Operating Support Group	运行支持组
OSHA	Occupation Safety and Health Administration	职业性安全和保健管理局(美国)
OSIP	Operational Safety Indicator Programme	运行安全指标大纲
OSOR	On Site Organization Regulation	现场组织管理条例
OST	Office of Science and Technology	科学与技术局
OT	Oil Turbine/Operating Temperature/Operating Time/Output Transformer/Outstanding Task/Overtime	油透平/运行温度/运行时间/输出变压器/遗留项,甩项/加班
OTI	Oil Temperature Indication	油温指示器
OTL	Operating Time Log/Output Transformerless	运转时间记录/无输出变压器,输出变压器不可用
OTO	Out-to-Out	外廓尺寸
OTP	Operator Test Procedure	操作员试验程序
OTPT	Ownership and Terminal Point Tagging	隶属关系和终端点显示牌
OTR	Overload Time Relay	过载限时继电器

英文简称	英文全称	中文名称
OTSG	Once-Through Steam Generator	直流式蒸汽发生器
OTTO	Once Through Then Out	全堆整体换料
OUSR	Operating Unit Status Report	运行机组的状态报告
OUT	Outlet	出口
OUT/S	Outside	外部
OWCC	Outstanding Work Completion Certificate	甩项完工证书
OWE	Optimum Working Efficiency	最佳工作效率
OX	Oxide	氧化物
OZ	Ounce	英两(盎司)
P	Pipe/Protection	管道/保(防)护
P & IC	Pressure & Inventory Control	压力和装量控制(系统)
P & ID	Piping and Instrumentation Diagram	管路和仪表(布置)图
P/C	Parts Catalog/Price Catagogue	零件目录/价格表
P/N	Part Number	零部件号
P3	Primavera	P3 软件
P7	Permissive Signal P7	允许信号 P7(反应堆功率>10%)
PA	Partial Admission/Power Amplifier/Protection Area/Provisional Acceptance/Public Address	部分进汽,部分认可而/功率放大器/防护区域/临时验收/有线广播,厂区播音
PAASM	Post-Accident Air Sampling and Monitoring	事故后空气取样和监测
PAC	Partial Test but Acceptance/Plasma Arc Cutting	部分需试验,但可接受/等离子弧切割
PAC	Plasma Arc Cutting	等离子弧切割

英文简称	英文全称	中文名称
PACE	Performance and Cost Evaluation/Precision Analog Computing Equipment	性能与成本评价/精密模拟计算机设备
PACP	Project Area Contingency Plan	项目区域意外事件应急预案，厂区应急预案
PACT	Production Analysis Control Technique	生产分析管理技术
PAFM	Particle-and-Force Computing Method	质点和力计算法
PAHR	Post Accident Heat Removal	事故后排热
PALS	Positioning and Location System	位置测定系统，定位系统
PAM	Post-Accident Monitoring/Pulse Amplitude Modulation	事故后监测/脉冲幅度调制
PAMI	Post-Accident Monitoring Instrumentation	事故后监测仪表
PAP	Personnel Access Passage	人员出入通道
PAS	Post-Accident Sampling System/Pressurizer Auxiliary Spray/Public Address System	事故后取样系统/稳压器辅助喷雾/有线广播系统
PAT	Production Assessment Test	产品评定试验
PAW	Plasma Arc Welding	等离子弧焊
PAX	Private Automatic Exchange	专用自动交换机
PB	Publications Bulletin/Push Button	发行季刊/按钮
PBF	Power-Burst-Facility	功率突增试验装置(美国)
PBR	Power Breeding Reactor	动力增殖反应堆

英文简称	英文全称	中文名称
PC	Partition Coefficient/Personal Computer/Pitch Circle/Pneumatic Controller/Pressure Compensator/Pressure Controller/Prestressed Concrete/Primary Containment/Processing Computer/Purification Cooler	分配系数/个人计算机/节距圆/气动调节器/压力补偿器/压力控制器/预应力混凝土/主安全壳/数据处理计算机/净化冷却器
PCB	Printed Circuit Board/Propulsion Control Box	印刷电路板/推进器操纵台
PCC	Packing Condition Certificate/Power Conditioning and Control/Primary Containment Cooling/Programme-Controlled Computer	包装状态证书/电力调节与控制/主安全壳冷却/程序控制计算机
PCCB	Product Coordination and Control Board	产品协调与控制委员会
PCCI	Pellet-Clad Chemical Interaction	芯块-包壳化学相互作用
PCCV	Prestressed Concrete Containment Vessel	预应力混凝土安全壳
PCCW	Primary Component Cooling Water System	核岛设备冷却水系统
PCD	Pitch Circle Diameter	节距圆直径
PCE	Piece	片(板),零件
PCFS	Post-EESR Change Follow-up Sheet	EESR 安装竣工状态报告(EESR)签字后变更的跟踪单
PCI	Pellet-Cladding Interaction	燃料芯块-包壳相互作用
P-CIMIS	Project-Oriented Computer Integrated Management Information System	计算机集成项目管理信息系统，基于项目的计算机集成管理信息系统

英文简称	英文全称	中文名称
PCIV	Prestressed Cast Iron Pressure Vessel	预应力铸铁压力容器
PCM	Parts and Components Manual/Planning and Contract Management/Power-Coolant- Mismatch/Pulse-Code Modulation	零(部)件手册/计划和合同管理/功率流量失配/脉冲编码调制
PCP	Primary Control Program	主控程序
PCPR	Payment Control Point Report	支付控制点报告
PCPV	Prestressed Concrete Pressure Vessel	预应力混凝土压力容器
PCR	Power Control Room/Project Completion Report	动力控制室/项目完工报告
PCRV	Prestressed Concrete Reactor Vessel	预应力混凝土反应堆容器
PCS	Primary Containment System/Primary Coolant System	主安全壳系统/主冷却剂系统
PCT	Peak Cladding Temperature	包壳峰值温度
PCV	Pressure Control Valve/Primary Containment Vessel	压力调节阀/主安全壳壳体
Pd	Paid	付讫
PD	Phase Diagram/Pitch Diameter/Potential Difference/Power Driven/Preliminary Design/Procedure/Project Document/Pulse Doppler	相位图/中径,栅格中心距/电位差/电动的,动力驱动的/初步设计/程序,规程/项目文件/多普勒脉冲
PDA	Preliminary Design Approval	初步设计批准

英文简称	英文全称	中文名称
PDC	Power Distribution Control/Programmable Digital Comparator	配电控制/可编程数字比较器
PDCV	Pump Discharge Control Valve	泵出口控制阀
PDD	Power Density Distribution/Preliminary Design Description/Project Definition Document	功率密度分布/初步设计描述/项目定义文件
PDMS	Post-Defueling Monitored Storage/Power Density Monitoring System	卸料后监控贮存/功率密度监测系统
PDP	Project Directing Plan/Preliminary Decommissioning Plan/Preliminary Design Package	工程指导计划/初步退役计划/初步设计包
PDR	Peak Dose Rate/Preliminary Design Requirements/Preliminary Design Review/Project Design Review	最高剂量率,峰值剂量率/初步设计要求/初步设计审评/项目设计审查
PDS	Parameter Display System/Plant Display System/Power Distribution System	参数显示系统/电厂显示系统/配电系统
PDS-PI	Plant Display System- Plant Information	电厂显示系统-电厂信息
PDU	Pneumatic Distribution Unit/Power Distribution Unit	气动分配器/配电装置
PE	Performance Evaluation/Permissible Error/Personal Error/Polyethylene/Potential Energy/Pressure Element	性能评价/允许误差/人因误差/聚乙烯/势能/压力元件

英文简称	英文全称	中文名称
Pe	Power (Electricity)	电功率
PEC	Photoelectric Cell/Printed Electronic Circuits	光电元件,光电池/印刷电路
PED	Pedestal	台(座)
PEF	Powerhouse Exhaust Facility	发电厂排气装置
PENET	Penetration	贯穿
PEO	Plant Equipment Operator	电厂设备操纵员
PEP	Project Execution Plan	项目执行计划
PERA	Production Engineering Research Association of Great Britain	大不列颠制造工程研究协会
PERC	Percent/Percentage	百分(比)
PERT	Program Evaluation and Review Technique	计划评审技术
PET	Performance Evaluation Test/Production Environmental Test	性能评价试验/生产环境试验
PEW	Percussion Welding	冲击焊,振动焊
PF	Pico-Farads/Power Factor/Pulse Frequency	微微法拉/功率因数/脉冲频率
PFM	Power Factor Meter	功率因数计
PFR	Prototype Fast Reactor	原型快(中子反应)堆
PFS	Program Functional Specification	程序功能说明
PGA	Peak Ground Acceleration/Power Generator Assembly	峰值地面加速度/发电设备
PGCC	Power Generation Control Complex	能源生产管理综合体,动力生产控制综合体

英文简称	英文全称	中文名称
PGDS	Pressurized Gas Distribution System	加压气体分配系统
PGM	Path Generation Method	通路形成法
PGP	Project General Procedure	工程总的管理程序,项目总程序
PGS	Potential Grounding Switch	电压接地开关
PGW	Pressure Gas Welding	气压焊,加压气焊
PH	Phase/Pipe Hanger	相(位);期(阶段)/管道吊架
PHD	Pipe Hanger Document	管道吊架文件
PHT	Primary Heat Transport	一回路热传输,主热传输
PHTAC	Primary Heat Transport-Auxiliary Shut Down Cooling	一回路热传输-辅助停堆冷却
PHTAD	Primary Heat Transport-Auxiliary Deuteration	一回路热传输-辅助氘化
PHTAS	Primary Heat Transport-Auxiliary Storage	一回路热传输-辅助贮存
PHTC	Pneumatic Hydraulic Test Console	气动液压试验控制台
PHTDD	Primary Heat Transport-Deuterium & De-Deut. System	一回路热传输-氘和除氘系统
PHTDL	Primary Heat Transport-Deuterium Line	一回路热传输-氘管线
PHTDP	Primary Heat Transport-Deuterium Piping	一回路热传输-氘管道
PHTFB	Primary Heat Transport-Feed & Bleed System	一回路热传输-给排水系统

英文简称	英文全称	中文名称
PHTHA	Primary Heat Transport-Hydrogen Addition Circuit	一回路热传输-加氢回路
PHTP	Primary Heat Transport Pump	主热传输泵,主泵
PHTPI	Primary Heat Transport-Pressure & Inventory	一回路热传输-压力和装载量
PHTPS	Primary Heat Transport-Purification System	一回路热传输-净化系统
PHTS	Primary Heat Transport System	一回路热传输系统
PHTST	Primary Heat Transport-Storage Transfer & Recovery Circuit	一回路热传输-贮存转移和回收回路
PHWR	Pressurized Heavy Water Reactor	加压重水(反应)堆
PI	Performance Index/Performance Indicator/Periodical Inspection/Pressure Indicator/Project Instruction/Purification Interchange	性能指数/(运行)性能指标/定期检查/压力指示器/项目指令/净化交换
PIC	Pocket Ion Chamber/Pressure and Inventory Control/Pressure Indication and Control/Process Instrumentation and Control/Program Information Center	袖珍电离室/压力和装量控制/压力指示和控制/过程仪表测量与控制/程序信息中心
PICC	People's Insurance Co. Of China	中国人民保险公司
PICE	Programmable Integrated Control Equipment	积分程序控制设备
PID	Piping and Instrumentation Diagram	管道与仪表图

英文简称	英文全称	中文名称
PIE	Parallel Interface Element/Post-Irradiation Examination/Postulated Initiating Event	并行接口部件/辐照后检验/假设始发事件
PIFS	Pressure Indicator Field Stand	就地压力指示架
PIL	Parts Installation List	部件安装清单
PIN	Position Indicator/Project Information Network	位置指示器/项目信息网
PINO	Positive Input-Negative Output	正输入-负输出
PIO	Process Input/Output	程序输入-输出，过程输入和输出
PIP	Periodic Inspection Program (Document)	定期检查大纲(文件)
PIPD	Periodic Inspection Program Document/Pre-Service Inspection Program Document	定期检查大纲文件/役前检查大纲文件
PISRC	Plant Industrial Safety & Radiation Protection Committee	电站工业安全和辐射防护委员会
PIT	Piping Integrated Team	综合管道队
PIV	Peak Inverse Voltage/Pressure Isolation Valve	峰值反向电压/压力隔离阀
PKG	Package	包(装)
PL	Panel/Parts List/Patent License/Peak Load/Pipeline/Proportional Limit	盘，屏/部件清单/专利许可证/峰值负荷/管线/比例极限
PLAT	Platinum	铂

英文简称	英文全称	中文名称
PLC	Power Line Carrier/Programmable Logic Controller	电力线线架/可编程逻辑控制器
PLD	Payload/Phase Locked Detector/Potentially Lethal Damage	有效负载/相位锁定检波器/潜在致死损伤
PLDR	Potentially Lethal Damage Repair	潜在致死损伤修复
PLEX	Plant Life Extension	核电厂寿命延长
PLI	Program Listing Index	程序表索引
PLIA	Post-LOCA Instrument Air	冷却剂丧失事故(LOCA)后测量仪表用压缩空气
PLLEX	Plant Licensing Life Extension	核电厂批准寿命延长
PLN	Plain	平的
PLOCAP	Post Loss-of-Coolant Accident Protection	冷却剂丧失事故后保护;失水事故后保护
PLT	Pilot	试验性的,典型试验的
PLU	Power Load Unbalance	电力负载不平衡
PM	Partial Melt/Permanent Magnet/Portal Monitor/Powder-Metallurgy/Project Manager/Pulse Monitor/Pump Motor	部分熔化/永久磁铁/出入口监测器/粉末冶金/项目经理/脉冲监测器/泵电机
PMAR	Preliminary Maintenance Analysis Report	初步维修分析报告
PMCR	Power Measurement and Calibration Routine	日常性功率测量和校准
PMD	Packless Metal Diaphragm	无衬垫的金属隔膜
PMETS	Probable Maximum Extratropical Storm	可能的特大热带风暴

英文简称	英文全称	中文名称
PMF	Probable Maximum Flood	可能最大洪水
PMG	Permanent-Magnet Generator	永磁发电机
PMH	Probable Maximum Hurricane	可能最大飓风
PMMA	Polymethyl Methacrylate	有机玻璃
PMOR	Pump Motor Oil Reservoir	泵电动机油箱
PMP	Probable Maximum Precipitation/Project Management Plan	最大可能降雨量/项目管理计划
PMPD	Project Management and Process Draw	工程管理和进展
PMS	Probable Maximum Seiche/Project Monitoring Scheduling	可能最大湖潮/工程监督和计划进度
PMSS	Probable Maximum Storm Surge	可能最大风暴潮
PMT	Probable Maximum Tsunami/Project Management Team	可能最大海啸/项目管理组
PMTC	Probable Maximum Tropical Cyclone	可能最大热带气旋
PMU	Pressure Measuring Unit	测压装置,测压单元
PMWR	Preventive Maintenance Work Request	预防维修工作申请
PMWST	Primary Makeup Water Storage Tank	一回路补给水贮存箱
Pn	Power (nuclear)	核功率
PNAA	Prompt Neutron Activation Analysis	瞬发中子活化分析
PNEUM	Pneumatic	气压(动)的

英文简称	英文全称	中文名称
PNL	Panel	盘，屏
PNO	Preliminary Notification of Occurrence	事件初步通知
PNP	Prototype Nuclear Process Heat Plant	原型核供热厂
PNRA	Prompt Nuclear Reaction Analysis	瞬发核反应分析
PNS	Planning Network & Schedule	计划网络和进度
PNSC	Plant Nuclear Safety Committee	电站核安全委员会
PNSRC	Plant Nuclear Safety Review Committee	电厂核安全审评委员会
PO	Polarity/Power Oscillator/Power Output/Purchase Order	极性/功率振荡器/功率输出/定购单
PO/TO	Pole-One End/Threaded-Other End	一端有支撑的，另一端有螺纹的
POB	Post-Office Box	邮政信箱
POC	Port of Call	停靠港
POD	Port of Debarkation	卸载港
POE	Port of Entry/Pressure (Pnematically) Operated Equipment	进口港/气动装置
POL	Provisional Operating License	临时运行许可证(执照)
POLYE	Polyethylene	聚乙烯
POLYS	Polystyrene	聚苯乙烯
POQAP	Plant Operational Quality Assurance Procedure	电厂运行质量保证程序
PORC	Plant Operations Review Committee	电厂运行审评委员会

英文简称	英文全称	中文名称
PORV	Pilot Operated Relief Valve/Power Operated Relief Valve	先导式释放阀/电动释放阀
POS	Position/Positioning/Pressure (Pneumatically) Operated Switch	位置/定位/气动开关
POT	Potentiometer/Pre-Operational (System) Test	电位计/运行前(系统)试验
PP	Peak Power/Peak Pressure/Polypropylene/Power Plant/Process Procedure	峰值功率/峰压力/聚丙烯/发电厂/工艺程序
PPA	Preliminary Pile Assembly	初级反应堆装置(美国)
PPB	Parts per Billion	十亿分之(几),十亿分率
PPC	Plant Process Condition/Program Planning and Control	电厂工作状态/程序设计与控制
PPL	Production Punch List	生产性消缺单,生产流程打孔通过单
PPM	Parts per Million/Procedure Preparation Manual	百万分之(几),百万分率/程序编制手册
PPO	Power Plant Operator	电厂操纵员
PPS	Plant Performance Specification/Plant Protection System/Preferred Power Supply/Primary Protection System	电站性能规格书/电厂保护系统/优先保证电源/主保护系统
PQA	Procurement Quality Assurance	采购质量保证
PQC	Production Quality Control	生产质量控制
PQOM	Plant Quality Organization Manual	电站质量管理手册

英文简称	英文全称	中文名称
PQTR	Personnel Qualification Training Requirements	人员资格评定培训要求
PR	Performance Ratio/Pressure Ratio/Procedure/Productivity Ratio/Progress Report/Public Relations/Purchase Requisition	性能比/压力比/程序/生产率/进展报告/公共关系/采购申请
PRA	Probabilistic Risk Analysis/Probabilistic Risk Assessment	概率风险分析/概率风险评价
PRAISE	Piping Reliability Analysis Including Seismic Event	包括地震在内的管系可靠性分析
PRC	Pressure Record and Control/Pressure Retaining Components	压力记录和控制/承压部件
PRCF	Plutonium Recycle Critical Facility	钚再循环临界装置(美国)
PRCS	Primary Reactor Coolant System	(反应)堆主冷却剂系统
PRESS	Pressure	压力
PRIM	Primary	最初的,初级的;首要的,主要的
PRIS	Power Reactor Information System	动力堆信息中心(国际原子能机构)
PRISM	Power Reactor Inherently Safe Module	动力(反应)堆固有安全模块(单元)
PRM	Power Range Monitor/Project Review Meeting	功率区段监测器/项目评审会
PRO	Procedure	程序
PROM	Programmable Read-Only Memory	可编程只读存储器

英文简称	英文全称	中文名称
PROSPER	Peer Review of Operational Safety Performance Experience	运行安全性能经验同行评议
PROT	Prototype	原型;样机
PROX	Proximity	近似,接近
PRP	Preliminary Remediation Plan/Primary Radiolytic Product	初步补救计划/主要辐照分解产物
PRS	Probabilistic Risk Statement	概率风险陈述
PRTR	Plutonium Recycle Test Reactor	钚再循环试验堆(美国)
PRU	Pneumatic Regulation Unit	气动调节装置
PRV	Peak Reverse Voltage/Pressure-Reducing Valve/Pressure Regulating Valve/Pressure Relief Valve	峰值反向电压/减压阀/调压阀/泄压阀
PRX	Pressure Regulation Exhaust	调压排气(装置,口)
PS	Planning & Schedule/Polystyrene/Power Supply/Pressure Switch/Prestress/Process Specifications/Product Specification/Program Specification/Project Schedule/Project service/Pull Switch/Pump Suction	计划与进度/聚苯乙烯/电源/压力开关/预应力/工艺规程/产品规格书/程序说明/项目进度/工程服务/拉线开关/泵吸入口
PSA	Probabilistic Safety Analysis/Probabilistic Safety Assessment	概率安全分析/概率安全评价
PSC	Parallel-to-Serial Converter/Pressure System Control/Prestressed Concrete	并联-串联变换器/压力系统控制/预应力混凝土
PSCD	Probability of Severe Core Damage	严重堆芯损伤概率

英文简称	英文全称	中文名称
PSD	Position-Sensitive Detector/Power Spectra Density	位置灵敏探测器/功率谱密度
PSDD	Preliminary System Design Description	初步系统设计描述
PSE	Pressurized Subcritical Experiment/Project Support Equipment	加压次临界试验(美国)/工程辅助设备
PSH	Pump Suction Head	泵吸入压头
PSI	Pre-Service Inspection	役前检查
PSI(G)	Pounds per Square Inch (Gauge)	磅/平方英寸(表压)
PSL	Power Supply List/Process Safety Limit	电源清单/工艺安全限值
PSMS	Power Shape Monitor System	功率分布轮廓监测系统
PSP	Plasma Spraying	等离子喷涂
PSRMS	Personal Salary and Record Management System	人员工资与档案管理系统
PSRV	Pressurizer Safety Relief Valve	稳压器安全释放阀
PSS	Pressure Suppression System/Process Sampling System/Program Specification-Specification/Pump Specification Sheet	抑压系统/工艺取样系统/程序说明-规格书/泵规格表
PSSCD	Probability of Subsequent Severe Core Damage	继发的严重堆芯损坏概率
PSV	Pressure Safety Valve/Primary Safety Valve	压力安全阀/主安全阀
PSWS	Potable and Sanitary Water System	饮用水和卫生用水系统

英文简称	英文全称	中文名称
PT	Performance Test/Periodic Test/Point/Potential Transformer/Power Tilt/Pressure Transducer/Pressure Tube/Proof Test	性能试验/定期试验/点/电压互感器/堆芯(象限)功率倾斜/压力变送器,压力传感器/压力管/验证试验
PTBPD	Poison Tank-Ball Position Detection System	毒物箱-球型液位检测系统
PTC	Part Through Crack/Plant Training Committee/Positive Temperature Coefficient/Power Test Criteria	非贯穿裂纹,未穿透裂纹/电厂培训委员会/正温度系数/动力试验规范,功率试验准则
PTFE	Polytetrafluoroethylene	聚四氟乙烯
PTR	Pool Test Reactor/Pressure Tube Rupture	池式试验堆(加拿大)/压力管破裂
PTS	Periodic Test System/Pneumatic Test Set/Pressurized Thermal Shock	定期试验系统/气动测试设备/受压热冲击
PTSL	Pressure Tube Seal and Latch	压力管密封和压紧
PU	Polyurethane/Pump Unit	聚氨基甲酸酯/泵组
PUD	Plant Unit Description	电厂机组说明
PUMPH	Pumphouse	泵房
PUR	Purification	净化
PV	Pilot Valve/Plug Valve/Peak Voltage/Pressure Vessel	先导阀/柱塞阀/峰值电压/压力容器
PVA	Polyvinyl Alcohol	聚乙烯醇
PVC	Polyvinyl Chloride	聚氯乙烯
PVDC	Polyvinyl Dichloride	聚二氯乙烯

英文简称	英文全称	中文名称
PVF	Polyvinyl Fluoride	聚氟乙烯
PVME	Polyvinyl Methyl Ether	聚乙烯基甲基醚
PVQAB	Pressure Vessel Quality Assurance Board	压力容器质量保证局(英国)
PVT	Pressure-Volume-Temperature	压力-体积-温度
PVTS	Pressure Vessel Thermal Shock	压力容器热冲击
PW	Pipe Weld	管子焊接
PW	Projection Welding	凸焊
PWGR	Pressurized Water Graphite Reactor	压水石墨(反应)堆
PWHT	Post-Weld Heat Treatment	焊后热处理
PWP	Peak Working Pressure/Process Welding Procedures	最大工作压力,峰值工作压力/工艺焊接程序
PWR	Power/Pressurized Water Reactor	功率;电力/压水(反应)堆
PWS	Plant Watch Supervisor	电厂监控长
PWT	Power Transducer	功率变送器
PWW	Precision Wire Wound	精密导线绕制
Q&A	Question and Answer	问题和回答
QA	Quality Assurance	质量保证
QADP	Quality Assurance Data Package	质量保证文件数据包
QAM	Quality Assurance Manual	质量保证手册
QAP	Quality Assurance Plan/Quality Assurance Program	质量保证计划/质保大纲
QAPR	Quality Assurance Program Review	质量保证大纲审查

英文简称	英文全称	中文名称
QASR	Quality Assurance Surveillance Report	质量保证监督报告
QC	Quality Control	质量控制
QCD	Quality Control Division	质量控制科
QCE	Quality Control Engineer	质量控制工程师
QCI	Quality Control Inspector	质量控制检查员
QCK	Quick	快速
QCM	Quality Control Manual	质量控制手册
QCO	Quality Control Organization	质量控制组织
QCS	Quality Control Standards	质量控制(管理)标准
QD	Quick Disconnect	快速断开
QEP	Quality Engineering Plan	质量工程计划
QF	Quality Factor	品质(质量)因数
QI	Quality Inspection	质量检查
QM	Quality Manual	质量手册
QNPC	Qinshan Nuclear Power Company	秦山核电公司
QOD	Quick-Opening Device	快速断路(开启)装置
QOR	Quality Observation	质量观察
QOTR	Quality Observation Trace Report	质量观察跟踪报告
QOV	Quick Open Valve	快开阀
QP	Quality Plan/Quality Procedure/Quality Program	质量计划/质量程序/质量大纲
QPCL	Quality Program Category List	质量大纲分类清单
QPD	Quantized Probability Design	量化概率设计

英文简称	英文全称	中文名称
QPL	Qualified Parts List	合格零件一览表
QPM	Quality Plan Management	质量计划管理
QPP	Quality Project Plan	质量项目计划
QPT	Qinshan Project Team	秦山项目队,秦山工程队
QR	Quality Related	与质量有关的,质量相关的
QRC	Quick Reaction Capability	快速反应能力
QS	Quality Surveillance/Quality Sur-veyor	质量监督/质量监察员
QSR	Quality And Safety Related	与质量及(核)安全有关的
QSP	Quality Surveillance Plan	质量监督计划
QT	Quart/Qualification Test	夸脱(容量单位)/质量评定试验
QTP	Qualification Test Procedure	质量鉴定试验程序
QUAL	Qualified/Qualification/Qualita-tive	合格的,鉴定过的/资格,合格性,鉴定/定性的
QUANT	Quantitative	定量的
QUP	Quantity per Unit Pack	每包的数量
QWA	Qinshan Work Authorization	秦山工程审批
R	Resistor	电阻(器)
R&D	Research & Development	研究与开发,研发
R&F/H	Reactor & Fuel Handling	反应堆与燃料操作
R&QC	Reliability and Quality Control	可靠性与质量控制
R. P.	Repairing Part	修复部件
RPS	Reference Power Station	参考电厂
R/A	Rack Assembly/Radius of Ac-tion	齿条(导轨)装置/作用半径

英文简称	英文全称	中文名称
R/B	Release to Birth Ratio	裂变气体释放量与产生量之比
R/B	Reactor Building	反应堆厂房
R/O	Reverse Osmosis	反渗透
R/R	Remove and Replace	拆卸与更换
RA	Reliability Analysis/Re-Qualification Assessment	可靠性分析/重新鉴定评估
RAB	Restrict Area Boundary	限制区边界
RABNVS	Reactor Auxiliary Building Normal Ventilation System	反应堆辅助厂房正常通风系统
RABSRVS	Reactor Auxiliary Building Switchgear Room Ventilation System	反应堆辅助厂房开关间通风系统
RACE	Rapid Automatic Checkout Equipment	快速自动检查设备
RAD	Radiation/Radiation Absorption Dose	辐射/辐射吸收剂量
Radwaste	Radioactive Waste	放射性废物
RAM	Random-Access Memory/Reliability-Availability-Maintainability	随机存取存储器/可靠性、可利用率和可维修性
RAP	Recommendation & Authorization to Purchase	采购建议和授权
RAPAT	Radiation Protection Advisory Team	辐射防护咨询组
RAPP	Registered Air Parcel Post	航空挂号邮包
RAR	Reasonably Assured Reserves/Reasonably Assured Resources	合理保证储量/合理保证的资源
RAS	Report Audit Summary	监察报告摘要

英文简称	英文全称	中文名称
RASM	Request for Authorization by the Station Manager	电站经理批准的申请
RASP	Reactor Analysis Support Package	反应堆分析支持包
RAX	Remote Access Computing System	遥控计算系统
RB	Reactor Building/Roller Bearing	反应堆厂房/滚柱轴承
RB	Resistance brazing	电阻钎焊
RBA	Rescue Breathing Apparatus	救急氧气设备
RBBAS	Reactor Building-Breathing Air System	反应堆厂房-通气系统
RBCT	Reactor Building-Cable Tray	反应堆厂房-电缆托(桥)架
RBCW	Reactor Building Chilled Water System	反应堆厂房冷冻水系统
RBD	Reliability Block Diagram	可靠性方块图
RBE	Radiation Biological Effectiveness/Relative Biological Effectiveness	辐射生物效应/相对生物效应
RBEP	Reactor Building-Embedded Part	反应堆厂房-预埋件
RBH	Relative Biological Hazard	相对生物危害
RBHPR	Reactor Building-High Pressure Recirculation	反应堆厂房-高压再循环
RBHPT	Reactor Building-High Pressure Trip System	反应堆厂房-高压脱扣系统
RBIA	Reactor Building-Instrument Air	反应堆厂房-仪表空气
RBIT	Reactor Building-Instrument Tubing	反应堆厂房-仪表管

英文简称	英文全称	中文名称
RBMCG	Reactor Building-Moderator Cover Gas	反应堆厂房-慢化剂覆盖气体
RBSAS	Reactor Building-Service Air System	反应堆厂房-厂用压缩空气系统
RBSW	Reactor Building-Service Water	反应堆厂房-厂用水
RBV	Reactor Building-Ventilation	反应堆厂房-通风
RBVS	Reactor Building Ventilation System	反应堆厂房通风系统
RC	Reinforced Concrete/Remote Control/Restriction Orifice	钢筋混凝土/遥控，远距离控制/节流孔板
RCB	Reactor Containment Building	反应堆安全壳厂房
RCC	Radiochemical Center/Radiochemical Co.	放射化学中心(英国)/放射化学公司(加拿大)
RCCC	Recommended Construction Contract Change	建造合同变更建议
RCCP	Reinforced Concrete Culvert Pipe	钢筋混凝土涵管
RCE	Remote Control Equipment	遥控设备
RCEEA	Radio Communication and Electronic Engineering Association	无线电通信和电子学工程协会(英国)
RCFC	Reactor Containment Fan Cooler	反应堆安全壳风冷器
RCG	Radioactivity Concentration Guide	放射性浓度指南
RCGM	Reactor Cover Gas Monitor	反应堆覆盖气体监测器
RCGVS	Reactor Coolant Gas Vent System	反应堆冷却剂放气系统

英文简称	英文全称	中文名称
RCICS	Reactor Coolant Inventory Control System/Reactor Core Isolation Cooling System	反应堆冷却剂装量控制系统/反应堆堆芯隔离冷却系统
RCM	Reliability-Centered Maintenance	以可靠性为核心的维修
RCR	Reverse Current Relay	逆流继电器
RCS	Reaction Control System/Reactor Coolant System/Remote Control System/Room Cleanup System	反应控制系统/反应堆冷却剂系统/远距离控制系统/室内清洁系统
RCS-NAA	Radiochemical Separation-Neutron Activation Analysis	放化分离-中子活化分析
RCTR	Reactor	反应堆
RCU	Reactivity Control Unit/Reactor Control Unit	反应性控制装置/反应堆控制装置
RCW	Recirculating Cooling Water	再循环冷却水
RD	Rupture Disc	爆破盘
RD/HD	Reducing Head	缩(异)径封头
RDAR	Reliability Design Analysis Report	可靠性设计分析报告
RDB	Reference Data Base/Research and Development Board	参考数据库/研究发展局(美国)
RDH	Radioactive Drain Header	放射性疏水总管
RDI	Read Digital Input	数字量读入
RDM	Rod Drive Mechanism	控制棒驱动机构
RDO	Rebars Delivery Order	钢筋发料单
RDROP	Rod Drop Test	落棒试验
RDS	Reactor Diagnostic System/Registered Design Submission	反应堆诊断系统/登记了的设计提交文件

英文简称	英文全称	中文名称
RDT	Resistance Temperature Detector	电阻温度探测器
RE	Radiation Effects/Rare Earth/Reliability Evaluation/Resident Engineer/Reynold's Number	辐射效应/稀土元素/可靠性评价/常驻工程师/雷诺数
REAC	Radiological Emergency Assessment Center	辐射应急评估中心
REACT	Reactivity	反应性
REB	Roentgen-Equivalent Biological	生物伦琴当量
REC	Reconciliation Statement/Recorder	调解声明,协议书/记录器
RECEP	Receptacle	容器;插座;贮存所
RECIR	Recirculating/Recirculation	(再)循环
RECMF	Radio and Electronic Component Manufacturers Federation	无线电和电子元器件商联盟
RECOV	Recovery	恢复;回收;补偿
RECOVER	Remote Continual Verification System	远距离连续核实系统,遥控持续核查系统
RECT	Rectifier	整流器
REDUC	Reducing	减(缩)小;降低,简化
REF	Reference	参考
REFL	Reference Line	基准线
REFP	Refurbishment Package	整修包,整治包
REG	Regulating/Regulator	调节/调节器
REHSL	Rehearsal	演习
REL	Rate of Energy Loss/Relief	能量损失率/释放,卸压
REM	Removable	可移(动)的

英文简称	英文全称	中文名称
REMAS	Radiation Effects Machine Analysis System	辐射效应计算机分析系统
RENDA	Requests for Neutron Data Measurement	对中子数据测量的要求
REP	Reactor Establishment Permit	反应堆装置许可证
REPU	Reprocessed Uranium	后处理铀,堆后铀
REQ	Request for Quotation	报价申请,要求报价
RER	Radiation Effects Reactor	辐射效应(反应)堆
RERTR	Reduced Enrichment for Research and Test Reactors	被降低的研究堆和试验堆的燃料富集度
RESFLD	Residual Field	剩余磁场
RESIS	Resistant/Resistance	抗,耐,防/阻力;电阻
RET	Reliability Evaluation Test	可靠性评价试验
RETMA	Radio, Electronics and Television Manufacturers Association	无线电、电子学和电视机制造商协会(美国)
RETNG	Retaining	承压的,保持
REV	Revision	修正;修订版本
RF	Radial Flow/Radio Frequency/Reduction Factor	径向流/射频/减容因子
RFC	Radio Frequency Current/Release for Construction	射频电流/发放建造许可证
RFEL	Reactor Fuel Element Laboratory	反应堆燃料元件实验室
RFF	Reset Flip-Flop	触发器复位
RFH	Reactor and Fuel Handling	反应堆和燃料装卸
RFM	Reactive Factor Meter	无功功率因数计
RFP	Reactor Feed water Pump	反应堆给水泵

英文简称	英文全称	中文名称
RFPY	Reactor Full Power Year	(反应)堆满功率年
RFSP	Reactor Fueling Simulation Program	反应堆换料模拟程序
RFT	Radio Frequency Transformer	射频变压器
RFW	Reserve Feed Water	备用给水
RG	Regulatory Guide/Reticulated Grating/Reverse Grade	管理导则/网状格栅/反向梯度
RH	Relative Humidity/Right Hand/Rockwell Hardness	相对湿度/右手/洛氏硬度
RHE	Recuperative Heat Exchanger	再生式热交换器,同流换热器
RHM	Roentgen per Hour at One Meter	距辐射源 1 m 处每小时的伦琴数
RHR	Reactor Heat Removal/Reheater/Residual Heat Removal	反应堆传热/再热器/余热排出
RHRS	Residual Heat Removal System	余热排出系统
RI	Radioactive Isotope/Receiving Inspection/Report of Investigation/Rubber Insulation	放射性同位素/接收检查/调查报告/橡皮绝缘
RIA	Radioimmunoassay/Reactivity-Initiated Accident	放射免疫分析/反应性事故
RIBA	Royal Institute of British Architects	英国皇家建筑师学会
RIC	Representative in Charge	总代表
RICM	Radiation Isolating Control Monitor	辐射隔离控制监测器
RIH	Reactor Inlet Header	反应堆进口集管

英文简称	英文全称	中文名称
RIL	Research Information Letter	研究信息通信
RINPO	Research Institute of Nuclear Power Operation	核动力运行研究所
RIP	Reactor Internal Pump	反应堆内置泵
RIPS	Radioactive Isotope Power Supply	放射性同位素电池
RIR	Research and Isotope Reactor	研究和制备同位素的反应堆
RISA	Radioisotope Smoke Alarm	放射性同位素烟雾报警器
RL	Room List	房间清单
RLP	Radioisotope Laboratories Protocol	放射性同位素实验室规章
RLSE	Release	释放；出版发行
RLWY	Railway	铁路
RLY	Relay	继电器
RM	Radiation Monitor/Raw Material/Reactivity Mechanism/Reference Manual/Reference Material/Routine Maintenance	辐射监测仪/原(材)料/反应性机理/参考手册/参考材料，参考资料/例行维修
RMA	Radio Manufacturers' Association	无线电制造商协会
RMF	Reactivity Measurement Facility	反应性测量装置，反应性测量仪
RMM	Read-Mostly Memory	主读存储器
RMR	Reactor for Metallurgical Research	冶金研究用反应堆
RMS	Radiation Monitoring System/Remote Monitoring System/Root Mean Square	辐射监测系统/远距离监测系统/均方根
RMT	Read Magnetic Tape	读磁带

英文简称	英文全称	中文名称
RMWS	Reactor Makeup Water System	反应堆补水系统
RNB	Return to Nucleate Boiling	返回泡核沸腾
RND	Round	圆形的;环绕
RNG	Ring	圆环
RNPP	Reference Nuclear Power Plant	参考核电厂
RNPSH	Required Net Positive Suction Head	要求的净正吸入压头
RO	Reactor Operator/Restriction Orifice/Reverse Osmosis	反应堆操纵员/节流孔板/反渗透
ROH	Reactor Outlet Header	反应堆出口集管
ROL	Roller	滚轴;滚轧机
ROM	Read Only Memory	只读存储器
ROMIS	Remotely Operated Miniature Inspection System	遥控微型检查系统
ROP	Record of Production	生产记录
ROPT	Regional Overpower Protection Trip	区域超功率保护停堆
ROS	Read Only Storage	只读存储器
ROT	Rotary	转动,旋转的
ROV	Remote-Operated Vehicle	遥控小车
ROW	Roll Welding	热轧焊
RP	Radiation Protection/Reactor Power/Real Property/Rehearsal Port	辐射防护/反应堆功率/固定资产/演习通道
RPDB	Reliability Parameters Data Bank	可靠性参数数据库
RPM	Revolutions per Minute	每分钟转数

英文简称	英文全称	中文名称
RPS	Reactor Protection System	(反应)堆保护系统
RR	Radionuclide Release/Recirculation Ratio/Relative Risk/Research Reactor/Review Report	放射性核素释放/再循环倍率/相对风险/研究(反应)堆/审查报告
RRS	Radiation Research Society/Reactor Regulating System/Required Response Spectrum	辐射研究学会/反应堆调节系统/要求的响应谱
RS	Rapidly Solidified/Reactive Sputtering/Resistor/Rotary Switch	快速固化的/活性溅射/电阻(器)/旋转开关
RS	Resistance Soldering	电阻钎焊
RSAC	Reactor Safety Advisory Committee	反应堆安全咨询委员会
RSB	Reactor Service Building	反应堆辅助厂房
RSC	Reactor Safeguards Committee/Restart Capability	反应堆安全保障委员会/重新启动能力
RSCS	Rate Stabilization Control System	速率稳定控制系统,稳速控制系统
RSD	Relative Standard Deviation	相对标准偏差
RSEW	Resistance Seam Welding	电阻缝焊
RSEW-HF	High-frequency Resistance Seam Welding	高频电阻缝焊
RSEW-I	Induction Resistance Seam Welding	感应电阻缝焊
RSEW-MS	Mash Seam Welding	压平缝焊
RSN	Reactor Safety Note	反应堆安全备忘录,反应堆安全便函

英文简称	英文全称	中文名称
RSP	Radwaste Solidification Plant/Recommended Service Parts/Rotating Stream Plate	放射性废物固化车间，放射性废物固化装置/建议备件/旋流塔板
RSPL	Recommend Spare Parts List	推荐的备件清单，建议备件清单
RSS	Remote Shutdown System/Reserve Shutdown System	遥控停堆系统/备用停堆系统
RSSF	Retrievable Surface Storage Facility	可回取地面贮存设施(美国)
RSST	Reserve Station-Service Transformer	备用厂用变压器
RST	Recommend Special Tool/Reinforcing Steel/Restart	推荐的专用工具/补强钢(筋)/重新启动，再启动
RSTNT	Restraint	约束，制约；阻尼
RSV	Reheat Stop Valve	再热截止阀
RSW	Raw Service Water/Recirculated Service Water	厂用(原)水/再循环厂用水
RSW	Resistance Spot Welding	接触点焊
RSWF	Radioactive Scrap and Waste Facility	放射性废物压缩减容装置(美国)
RT	Radiographic Test/Room Temperature	射线探伤/室温
RTC	Reheat Temperature Control	再热温度控制
RTD	Resistance Temperature Detector/Resistive Thermal Detector	电阻温度计/电阻式热探测器
RTE	Recycle Test Element	再循环试验元件

英文简称	英文全称	中文名称
RTI	Relative Toxicity Index/Rotor Temperature Indicator	相对毒性指数/转子温度指示器
RTM	Real-Time Monitor/Running Time Meter	实时监视器/运转计时器
RTN	Return	返回;复原
RTO	Reactor Thermal Output	反应堆热输出
RTOS	Real-Time Operating System	实时操作系统
RTR	Resonance Test Reactor	共振试验(反应)堆(美国)
RTU	Remote Terminal Unit	远程终端装置
RU	Recombiner Unit	复合单元
RUHS	Reserve Ultimate Heat Sink	备用最终热阱
RUM	Remote Underwater Manipulator	遥控水下机械手
RUPT	Rupture	破(断)裂
RV	Reactor Vessel/Release Valve/Relief Valve/Regulating Valve	反应堆容器/排气阀,排放阀/卸压阀/调节阀
RVA	Reactive Volt-Ampere Meter	无功伏安计
RVIC	Return Valve Insulating Cabinet	回流阀隔离室
RVLIS	Reactor Vessel Level Instrumentation System	反应堆压力容器水位测量系统
RVRSE	Reverse	反向的;逆流的
RW	Raw Water/Repair Welding	原水/补焊
RW	Resistance Welding	电阻焊
RWCS	Reactor Water Cleanup System	反应堆冷却剂净化系统
RWCU	Reactor Water Cleanup	反应堆冷却剂净化
RWL	Remaining Work List	剩余工作清单

英文简称	英文全称	中文名称
RWM	Rod Worth Minimizer	控制棒价值最小化装置
RWMP	Radioactive Waste Management Program	放射性废物管理大纲
RWP	Radiation Work Permit/Reusing Water Pool	辐射工作许可/复用水池
RW-PC	Pressure Controlled Resistance Welding	压力控制电阻焊
RWPS	Radioactive Waste Processing System	放射性废物处理系统
RWST	Recirculation Water Storage Ta-nk	再循环水贮存箱
RY	Reactor-Years	(反应)堆年
S	Soldering	软钎焊
S	Stall/Service Air Station	停转,拖延/压缩空气站
S&I	Surveillance and Inspection	监督与检查
S. I. C	Specific Inductive Capacity	电容率,介电常数
S/C	Short Circuit	短路
S/N	Serial Number/Signal-to-Noise Ratio/Shipping Note	序号/信号-噪声比/装船通知
S/P	Single Pole/Snout Plug/Spare Parts/Static Pressure	单极/嘴塞/备件/静压力
S/T	Straight Thread	直螺纹
SA	Sectional Area/Situation Aware-ness/Specific Activity/Sub-As-sembly Drawing	截面积/情况认知/(放射性)比活度/组件装配图

英文简称	英文全称	中文名称
SAB	Safety Analysis Basis (Document)/Station Advisory Bulletin	安全分析基础(文件)/电站警示公告
SADL	Safety Analysis Data List	安全分析数据清单
SAF	Safety	安全;安全设备
SAFR	Sodium Advanced Fast Reactor	先进钠冷快堆
SAFSTOR	Safe Storage	安全贮存
SAIM	System Analysis and Integration Model	系统分析与综合模型
SAL	Specific Activity Limit	比放射性活度(限值)
SAR	Safety Analysis Report/Safety Assessment Report	安全分析报告/安全评估报告
SAS	Safety Assessment System/Secondary Alarm Station/Small-Angle Scattering/Statistical Analysis System/Surface Analysis Station	安全评价系统/辅助警报站/小角度散射/统计分析系统/表面分析站
SAT	Simulator Acceptance Test/System Acceptance Test	模拟机验收试验/系统验收试验
SATCOM	Satellite Communications Agency	卫星通信机构
SAW-S	Series SAW	横列双丝埋弧焊
SB S/B	Service Building/Stuffing Box/Switch Board	辅助厂房/填料箱/开关板
SBC	Standard Buried Collector	标准埋层集极(集成电路)
SBCP	Service Building-Condensate Piping	辅助厂房-凝结水管
SBDS	Service Building-D_2O Sampling	辅助厂房-重水取样

英文简称	英文全称	中文名称
SBDSS	Service Building-D_2O Supply System	辅助厂房-重水供应系统
SBGTS	Standby Gas Treatment System	备用气体处理系统
SBLC	Standby Liquid Control System	备用液体控制系统
SBLOCA	Small Break Loss-of-Coolant Accident	小破口失水事故
SBMP	Service Building-Moderator Deuterium	辅助厂房-慢化剂氘
SBP	Service Building-Panel	辅助厂房-控制盘
SBR	Skirtboard-Rubber	侧壁橡胶
SBV	Steam By Volume	容积含汽率
SBW	Steam By Weight	重量含汽率
SBWR	Small (Simplified) BWR	小型(简化)沸水(反应)堆
SC	Safety Class/Short Circuit/Shutdown Cooler/Stores Clerk	安全级/短路/停堆冷却器/库房管理员
SCA	Scope Change Application/Secondary Control Area/Servo Corporation of America	范围变更申请/二级控制区,次要控制区/美国伺服系统公司
SCAR	Significant Corrective Action Request	重大纠正运行要求
SCC	Security Control Center/Station Control Computer/Stress Corrosion Cracking/System Control Center	安全控制中心/电站控制计算机/应力腐蚀裂纹/系统控制中心
SCD	Severe Core Damage	严重堆芯损伤
SCEP	Safety Culture Enhancement Programme	安全文化增强计划

英文简称	英文全称	中文名称
SCF	Standard Cubic Foot/System Commissioning Files/Service Control Fabrication	标准立方英尺/系统调试文件/制造质量检查部(法国)
SCH	Schedule/Socket Head	计划;日程,进度;作业(表)/插座头
SCHST	Schedule-Standard	作业表-标准
SCHWR	Steam Cooled Heavy Water Reactor	蒸汽冷却重水堆
SCHXH	Schedule-Extra Heavy	作业表-过重
SCHXS	Schedule-Extra Strength	作业表-过重
SCI	Society of Chemical Industry/Stress Corrosion Index	化学工业学会(英国)/应力腐蚀指数
SCIC	Semiconductor Integrated Circuit	半导体集成电路
SCIF	Standard Thermal Column Irradiation Facility	标准热柱辐照设施
SCL	Screed Check List/System Classification List	找平层检查单/系统分类清单
SCN	Stock Code Number	存货编码号
SCO	Senior Control Operator/Set Time Counter	控制室高级操作员/置位时间计数器
SCP	Standard Commissioning Procedure	标准调试程序
SCR	Silicon-Controlled Rectifier/Software Change Request/Summary Commissioning Report	可控硅整流器/软件变更申请/概要调试报告

英文简称	英文全称	中文名称
SCS	Secondary Containment System/Shutdown Cooling System/Stabilization Control System/Suit Communication System/System Completion Schedule	二次安全壳系统/停堆冷却系统/稳定性控制系统/气衣通信系统/系统完工计划
SCSM	Structural Carbon Steel Medium	结构用中碳素钢
SCTL	Sodium Component Test Loop	钠设备试验回路
SCU	Storage Control Unit/System Control Unit	存储控制装置/系统控制装置
SCUAE	State Committee on the Utilization of Atomic Energy	国家原子能利用委员会(前苏联)
SCWR	Supercritical Water Reactor	超临界水慢化反应堆
SCWS	Stator Cooling Water System	定子冷却水系统
SD	Shut Down	停堆(役);关闭
SDA	Supply/Demand Adjustment	供求调节
SDB	State Development Bank of China	中国国家开发银行
SDC	Shutdown Cooling (Pump)	停堆冷却(泵)
SDCE	Society of Die Casting Engineers	压铸工程师协会(美国)
SDCS	Shutdown Cooling System	停堆冷却系统
SDD	Software Design Description	软件设计说明
SDE	Site Design Earthquake	厂址设计地震
SDG	Safety Design Guide/Standby Diesel Generator	安全设计导则/备用柴油发电机
SDH	Software Development Handbook	软件开发手册

英文简称	英文全称	中文名称
SDM	Safety Design Matrix (Model)/Shutdown Margin/Station Data Manual/System Design Manual	安全设计模型/停堆裕(深)度/电站数据手册/系统设计手册
SDP	Safety Design Principles	安全设计原则
SDR	S/F Discharge Room/Safety Design Requirement/Sodium-D_2O Reactor	乏燃料卸料间/安全设计要求/钠冷重水(反应)堆
SDS	Safety Depressurization System/Shutdown System/Software Design Specification	安全卸压系统/停堆系统/软件设计说明
SDS1	Shutdown System No. 1	1 号停堆系统
SDS2	Shutdown System No. 2	2 号停堆系统
SDV	Steam Discharge Valve	排汽阀
SDVS	Safety Depressurization and Vent System	安全卸压和排气系统
SE	Solder End/Special Equipment/Standard Error/System Engineer/System Engineering	焊端/专用设备/基本(标准)误差/系统工程师/系统工程
SEB	Source Evaluation Board	原始资料评价委员会(英国)
SEC	Second/Station Engineering Changes	秒/电站工程设计变更
SECS	Space Environmental Control System	空间环境控制系统
SECT	Section	剖(截)面;段,节
SEE-IN	Significant Event Evaluation and Information Network	重要事件评价和通报网络(核安全)

英文简称	英文全称	中文名称
SEIC	Solar Energy Information Center	太阳能资料中心
SEL	Selector	选择器
SEL	Signal Engineering Laboratories	信号工程实验室
SEPAR	Separator	分离器
SEPC-NP	Shandong Electric Power Construction Corporation	山东核电工程公司
SEPP	Standard Erection Procedure Package	标准安装程序包
SER	Safety Evaluation Report/Series/Significant Event Report/Site Evaluation Report	安全评价报告/系列/重大事件报告/现场评价报告
SERL	Service Electronics Research Laboratory	检修用电子仪器研究实验室
SERV	Service	服务;作业
SES	Supplier Evaluation Sheet	供应商评审表
SESA	Society for Experimental Stress Analysis	实验应力分析协会
SETR	Service Engineering Trouble Report	维修工程故障报告
SETP	System Engineering Test Program	系统工程试验程序
SF	Scale Factor/Signal Frequency/Spent Fuel/Standard Frequency/Supply Fan	比例系数/信号频率/标准频率/乏燃料/送风扇
SFB	Spent Fuel Bay	乏燃料池

英文简称	英文全称	中文名称
SFC	Specific Fuel Consumption	比燃料消耗
SFCI	Specialized Fuel Channel Installation	专门燃料通道安装
SFCR	Single Fuel Channel Replacement	单个燃料通道更换
SFCS	Spent Fuel Cooling System	乏燃料冷却系统
SFD	Severe Fuel Damage	燃料严重损坏
SFDRB	Spent Fuel Discharge Reception Bay	乏燃料卸料接收池
SFE	Society of Fire Engineers	消防工程师协会
SFIM	Surplus Free Issue Material	免费调拨的剩余材料
SFPAVS	Spent Fuel Pool Area Ventilation System	乏燃料池区域通风系统
SFPCCS	Spent Fuel Pool Cooling and Cleanup System	乏燃料水池冷却及净化系统
SFRS	Search for Random Success	随机结果的查找
SFTAA	Spent Fuel Transfer Auxiliary Air	乏燃料转运辅助压缩空气
SG	Screen Grid/Specific Gravity/Standby Generator/Steam Generator/Strain Gauge	帘栅极/比重/备用发电机/蒸汽发生器/应变仪
SGAFWS	Steam Generator Auxiliary Feed Water System	蒸汽发生器辅助给水系统
SGBD	Steam Generator Blowdown	蒸汽发生器排污
SGFLP	Steam Generator Feed Line Low Pressure	蒸汽发生器给水低压力

英文简称	英文全称	中文名称
SGFPT	Steam Generator Feed Water Pump Turbine	蒸汽发生器给水泵汽机
SGLC	Steam Generator Level Control (System)	蒸汽发生器液位控制(系统)
SGM	Station General Manager	电站总经理
SGMW	Steam Generator Make-up Water (System)	蒸汽发生器补给水(系统)
SGOG	Steam Generator Owner's Group	蒸汽发生器业主集团
SGP	Stepback Guard Program	快速落棒功率监视程序
SGPC	Steam Generator Pressure Control	蒸汽发生器压力控制
SGR	Sodium Graphite Reactor	钠(冷)石墨(反应)堆
SGTS	Standby Gas Treatment System	备用气体处理系统
SGW	Security Guard Window	安全防护窗
SH	Schedule/Service Hours/Super-Heater	进度/运行小时/过热器
SH/P	Shield Plug	屏蔽塞
SHD	Shield/Shielded/System History Docket	屏蔽/屏蔽的/系统历史记录
SHF	Super-High Frequency/System History File	超高频/系统历史文件
SHGS	Sodium Hypochlorite Generation Station	次氯酸钠发生站
SHLDR	Shoulder	台,肩
SHP	Super-High Pressure	超高压
SHT	Solution Heat Treatment	固溶热处理

英文简称	英文全称	中文名称
SHTL	Shuttle	气压传送,往复式
SI	Safety Injection/Seal-In/Site Instruction/Système Internationale d'Unités	安全注入/密封,封入/现场指示书/(法)国际单位制
SIB	Solid Integration Building/Station Information Bulletin	刚性组合体/电站情报公告
SIC	Semiconductor Integrated Circuit	半导体集成电路
SID	Sudden Ionosphere Disturbance/System Installation Document	电离层的突然扰动/系统安装文件
SIE	Selective Ion Electrode/Solvent Isotope Effect	离子选择电极/溶剂同位素效应
SIG	Signal	信号
SIGS	Stellar Inertial Guidance System	天体惯性制导系统
SIML	Similar	类似的,相似的
SING	Single	单(一)的
SINH	Hyperbolic Sine	双曲正弦
SIP	Safety Injection Pump	安全注射泵
SIR	Safe Integral Reactor/Single Item Record/Straight Individually Replaceable	安全一体化反应堆/单项记录/可直接单独更换
SIS	Safeguards Information System/Safety Injection Signal/Safety Injection System/Station Information System	保障信息系统/安全注射信号/安全注射系统/电站综合信息系统
SITP	System (Subsystem) Integration Test Plan	系统(子系统)综合试验计划

英文简称	英文全称	中文名称
SIWT	Safety Injection Water Tank	安全注射水箱
SJ	Slip Joint	滑动接头
SJAE	Steam Jet Air Ejector	蒸汽喷射抽气泵
SK	Sketch	草图,简图
SL	Safety License/Signal Lamp	安全许可证/信号灯
SLA	Service Level Agreement	服务水准协议
SLAC	Stanford Linear Accelerator Center	斯坦福直线加速器中心
SLAR	Spacer Location and Reposition	定位圈定位和复位
SLCS	Standby Liquid Control System	备用液体控制系统
SLD	Shipping List of Drawings/Solid	图纸装箱单/减速/固体(态)
SLG	Sealing	密封
SLO	Slow	慢
SLR	Service Loading Report	使用负载报告
SLS	Speed Limiting Switch	限速开关
SLTD	Slotted	开槽的
SLTR	Service Life Test Report	使用寿命试验报告
SM	Security Monitor/Sheet Metal/Synchronous Motor	安全监控器/金属薄板/同步电动机
SMAC	Shielded Metal Arc Cutting	焊条电弧切割
SMACNA	Sheet Metal and Air-Conditioning Contractor's National Association	金属板材和空调承包商全国协会(美国)
SMAW	Shielded Metal Arc Welding	焊条电弧焊
SMB	Safety Margin Basis	安全裕度基准
SMI	Standard Measuring Instrument	标准测量仪

英文简称	英文全称	中文名称
SMLS	Seamless	无缝的
SMM	Surplus Miscellaneous Material	其他剩余材料
SMSS	Surplus Material Shipped to Site	运往现场的剩余材料
SMSUT	Special Maintenance & Set-up Tool	专门维修和安装工具
SN	Shipping Notice/Sine of Amplitude	装船通知单/振幅的正弦
SNAP	System for Nuclear Auxiliary Power	核辅助动力系统
SNERDI	Shanghai Nuclear Engineering Research & Design Institute	上海核工程研究设计院
SNG	Substitute Natural Gas	代用天然气
SNIF	Standard Neutron Irradiation Facility	标准中子辐射设施
SNL	Standard Nomenclature List	标准术语表
SNSE	Society of Nuclear Scientists and Engineers	核能科学家和工程师协会
SNSS	Senior Nuclear Shift Supervisor	核电厂高级值长
SO	Security Officer/Shut-off	安全员/关闭;停机(堆)
SOA	State-of-the-Art(s)	现代技术发展水平,现代知识水平,工艺现状
SOAR	State of the Art Reports	技术现况报告
SOAV	Solenoid Operated Air Valve	电磁操纵空气阀
SOC	Socket	塞孔,插座

英文简称	英文全称	中文名称
SOE	Sequence of Events	事件顺序
SOER	Significant Operating Experience Report	重要运行经验报告
SOF	System Operational Flow Sheet	系统运行流程图
SOH	Scheduled Outage Hours	计划停役时数
SOL	Solenoid	电磁线圈
SONOAN	Sonic Noise Analyzer	噪声分析器
SOP	Standard Operating Procedure	标准操作程序
SOR	Shutoff Rod	停堆棒
SORL	Simulation of Reactor Operation	反应堆运行模拟
SOV	Shut-off Valve/Solenoid Operated Valve	截流(止)阀/电磁控制阀
SP	Shield Plug/Site Plan	屏蔽塞/现场平面图
SP/C	Spare Parts/Consumables	备件/消耗品
SPAR	Space Power Advanced Reactor	先进空间动力(反应)堆(美国)
SPARC	Suppression Pool Aerosol Removal Cycle	抑压池气溶胶排除循环
SPC	Steam Pressure Control/Stored Program Control/System Power Control	蒸汽压力控制/存储程序控制/系统电源控制器
SPCAN	Sulzer Pump Canadian	加拿大苏尔寿泵业公司
SPCS	Suppression Pool Cooling System	抑压池冷却系统
SPCU	Suppression Pool Cleanup	抑压池净化

英文简称	英文全称	中文名称
SPD	Self-Powered Detector/Self-Powered Flux Detector/Software Problem Definition	自给能探测器/自给能通量探测器/软件问题定义
SPDS	Safety Parameter Display System	安全参数显示系统
SPDT	Single-Pole Double-Throw Contact	单刀双掷闸刀
SPE	Special-Purpose Equipment	专用设备
SPEC	Scram Prevention Evaluation Checklist	防止紧急停堆评定清单
SPGR	Specific Gravity	比重
SPH	Space Heater/Standards and Procedures Handbook	空间加热器/标准和程序手册
SPL	Spare Parts List	备品备件清单
SPMO	Site Project Management Organization	(AECL)现场项目管理机构
SPPO	Senior Power Plant Operator	电厂高级操纵员
SPRG	Spring	弹簧
SPS	Secondary Protection System/Standard Pipe Size/System Performance Specification (Manufacturer's)	二次保护系统/标准管径/系统性能说明(制造商的)
SPSB	Shenzhen Power Supply Bureau	深圳供电局
SPT	Standard Penetration Test/Under Standard Pressure and Temperature	标准贯穿试验/在标准气压和温度下
SQ	Significant Quantity/Square	显著量/方形
SQ/HD	Square Head	方头

英文简称	英文全称	中文名称
SQA	Site Quality Assurance	现场质量保证
SQAP	Software Quality Assurance Program	软件质量保证大纲
SQCL	Seismically Qualified Component List	抗震设备清单
SQM	Site Quality Manager	现场质量经理
SQP	Software Qualification Plan	软件评定计划
SQR	Seismic Qualification Report	抗震评定报告
SQR	Software Qualification Report	软件评定报告
SQRT	Square Root	平方根
SQUG	Seismic Qualification Utility Group	抗震条件评定单位集团
SR	Safety Related/Safety Relief/Scanning Radiometer/Short Radius/Specific Resistance/Split Ring/Stress Report/Supply Request	安全相关的/安全卸压、安全释放/扫描式辐射仪/短半径/电阻率/开口环/应力报告/供应要求书
SRBT	Single Rod Burst Test	单棒破裂试验
SRC	Safety Review Committee/Spinning Reserve Capacity	安全评审委员会/空转备用容量
SRCAS	Safety-Related Compressed Air System	安全相关压缩空气系统
SRD	Safety and Reliability Directorate	安全与可靠性管理局(美国)
SRDL	Signal Research Development Laboratory	信号研发实验室

英文简称	英文全称	中文名称
SRE	Safety-Related Event/Sodium Reactor Experiment	安全相关事件/钠冷反应堆试验
SRIAR	Scientific Research Institute for Atomic Reactors	原子反应堆科学研究院
SRM	Source Range Monitor/Standard Reference Material/Standard Repair Manual/Status Review Meeting/Stuck Rod Margin	源区监测器/标准参考材料/标准检修手册/状态评审会/卡棒裕度
SRN	Software Release Notice	软件释放通知
SRP	Scram Rod Program/Single Rotating Plug/Standard Review Plan	紧急停堆棒控程序/单旋塞/标准审查大纲
SRPV	Steel Reactor Pressure Vessel	反应堆钢制压力容器
SRR	Software Review Report	软件审查报告
SRS	Scientific Research Society of America/Software Requirement Specification/Standard Response Spectrum	美国科学研究学会/软件要求规格书/标准响应谱
SRST	Safety Related System Test	安全相关系统试验
SRTF	Spent Resin Tank/Standard RTD Transmitter/System Reaction Time/Site Radwaste Treatment Facilities	废树脂箱/标准电阻温度计传感器/系统反应时间/现场废物处理设施
SRV	Safety Relief Valve/Steam Reject Valve	安全卸压阀/排汽阀
SRWPS	Solid Radioactive Waste Processing System	放射性固体废物处理系统

英文简称	英文全称	中文名称
SRWSV	Solid Radioactive Waste Storage Vault	放射性固体废物储存库
SS	Seismic Support/Shear Strength/Shift Supervisor/Specification Sheet/Stainless Steel/Superheated Steam	抗震支撑/抗剪强度/值长/规格书单页/不锈钢/过热蒸汽
SSAFV	Steam Seal Auxiliary Feed Valve	蒸汽密封辅助进汽阀
SSAS	Small-Sample Assay System	小样品测定系统
SSBR	Safety System Buffer Relay	安全系统缓冲继电器
SSCV	Spherical Steel Containment Vessel	球形钢安全壳
SSD	Solid State Detector/Station Status Document/Status Source Document/Subsoil Drain/System Status Display	固体探测器/电站状态文件/状态原始文件/地下排水/系统状态显示
SSETF	Solid Storage Engineering Test Facility	固体(放射性废物)贮存工程试验设施
SSFI	Safety System Functional Inspection	安全系统功能检查
SSFV	Steam Seal Feed Valve	蒸汽密封进汽阀
SSI	Secondary Side Inspection/Soil-Structure Interaction	二次侧检查/土壤-结构相互作用
SSP	Special Serveillance Plan/System Surveillance Plan	专门监督计划/系统监督计划
SSPB	Swedish State Power Board	瑞典国家电力局

英文简称	英文全称	中文名称
SSPC	Steel Structure Painting Commission (America)	钢结构油漆委员会(美国)
SSRM	Secondary System Radiation Monitor	二回路系统辐射监测器
SSRS	Safeguards Status Report System	核保障现况报告系统
SSS	Silicon Symmetrical Switch/Site Security System/Special Safeguards Study/Special Safety System	硅对称开关/厂区保安系统/特殊核保障研究/专设安全系统
SSSA	Safe Shutdown State Achieved	达到的安全停堆状态
SSSD	Safe Shutdown State Definition	安全停堆状态定义
SSSF	Solid Storage Surveillance Facility	固体(放射性废物)贮存监视设施
SSSP	Safe Shutdown State Plan	安全停堆状态计划
SST	Safety System Test/Siting Source Term/Station Service Transformer/System Service Transformer	安全系统试验/厂址源项/厂用辅助变压器/系统辅助变压器
SSTP	Subsystem Test Procedure	子系统试验程序
SSTR	Subsystem Test Report	子系统试验报告
SSTU	Seamless Steel Tubing	无缝钢管
SSU	Semi-Conductor Storage Unit	半导体存储器
SSW	Solid State Welding	固态焊
SSW	Standby Service Water	备用厂用水
SSWP	System Specific Walk down Procedure	系统专用巡检程序

英文简称	英文全称	中文名称
ST	Shunt Trip/Special Tool/Speed Transducer/Standard Drawing/Startup Transformer	并联跳闸/专用工具/速度变送器/标准图纸/启动变压器
STA	Shift Technical Advisor	值班技术顾问
STB	Stabilization Treatment	稳定化处理
STD	Safety Topic Discussion/Standard/Standard Test Dose	安全专题讨论/标准/标准试验剂量
STEP	Source Term Experimental Program	源项实验大纲
STES	Shield Tank & End Shield	屏蔽箱与端屏蔽
STI	Station Instruction	电站指令
STL	Steel	钢
STN	Station	站
STOR	Storage	贮存
STP	Shielded Transfer Platform/Standard Temperature and Pressure/Station Program	屏蔽运输平台/标准温度和压力/电站大纲
STR	Straight/Subsystem Test Report	直线,直的/子系统试验报告
STS	Special Treatment Steel/Static Transfer Switch	特殊处理钢/静态切换开关
SUCT	Suction	抽吸
SUPP	Support	支持(撑,架)
SUPPR	Suppressor	抑制(阻尼)器
SUPT	Superintendent	监管人,负责人,主管
SUSH	Set Up Sheet	配置图
SUT	System Under Test	正在试验的系统

英文简称	英文全称	中文名称
SV	Safety Valve/Slide Valve/Solenoid Valve/Specific Volume/Stop Valve	安全阀/滑阀/电磁阀/比容/截流(止)阀
SVR	Software Verification Report	软件验证报告
SVS	Software Validation Specification	软件确认说明
SVVP	Software Verification and Validation Plan	软件验证和确认计划
SVVR	Software Validation and Verification Report	软件确认和验证报告
SW	Safety Switch/Salt Water/Secondary Winding/Socket Weld/Specific Weight/Spot Weld	安全开关/盐水/次级线圈/承插焊接/比重/点焊/
SW	Stud Arc Welding	螺栓电弧焊
SWGLK	Swagelok	接头
SWGR	Switchgear	开关装置，开关设备
SWH	Switch	开关
SWIL	Switch Indicator Light	开关指示灯
SWMS	Solid Waste Management System	固体废物处理系统
SWO	Stop Work Order	停工令
SWP	Safe Working Pressure/Software Work Practices/Steam Working Pressure	安全工作压力/软件工作实践/蒸汽工作压力
SWRT	Sodium Water Reaction Test	钠水反应试验
SWU	Separative Work Unit	分离功单位
SWYD	Switch Yard	开关站,升压站
SY	Synchronizing	同步

英文简称	英文全称	中文名称
SYM	Symbol	符号
SYN	Synchronous/Synthetic	同步的/人造的
SYS	System	系统，体系
T	Transformer	变压器
T&H	Temperature and Humidity	温度和湿度
T&M	Testing and Maintenance	试验与维修
T/B	Turbine Building	汽轮机厂房
T/O	Turnover	移交
TA	Tangent Angle/Tantalum/Task Analysis/Technical Adaptation/Technical Assistance/Thermal Analysis	正切角/钽/任务分析/技术适应性/技术援助/热分析
TAA	Technical Assistance Administration	技术援助局
TAB	Turbine Auxiliary Bay	汽轮机辅助隔间
TAC	Technical Assistance Committee/Time-to-Amplitude Converter/Total Acceptance/Turbine Automatic Control	技术援助委员会/时-幅变换器/完全接受/汽轮机自动控制
TAD	Technical Acceptance Data	技术验收日期
TAFG	Two-Axis Free Gyro	双自由度陀螺仪
TAG	Technical Advisory Group	技术顾问组(美国)
TAP	Tapping	抽头，分路
TARS	Three-Axis Reference System	三轴坐标系统
TAVS	Turbine Area Ventilation System	汽轮机区域通风系统
TB	Tee-Bend/Terminal Board	三通/接线板

英文简称	英文全称	中文名称
TB	Torch Brazing	火焰钎焊,吹管硬焊
TBA	To Be Advised	待告知
TBCCWS	Turbine Building Component Cooling Water System	汽轮机厂房设备冷却水系统
TBD	Technical Basis Document	技术基础文件
TBE	Total Binding Energy	总结合能
TBEDT	Turbine Building Equipment Drain Tank	汽轮机厂房设备疏水箱
TBF	Travelling Belt Filter	移动带过滤器
TBNAA	Total-Body Neutron Activation Analysis	全身中子活化分析
TBO	Time Between Overhaul/Turbine Building Operator	大修间隔期/汽轮机厂房操纵员
TBR	Thorium Breeder Reactor/Total Breeding Ratio	钍增殖反应堆/总增殖比
TBS	To Be Specified	待定
TC	Technical Committee/Temperature Coefficient/Thermocouple/Time Closing/Trip Coil	技术委员会/温度系数/热电偶/定时闭合/脱扣线圈
TC	Thermal Cutting	热切割
TCA	Tank-type Critical Assembly/Temporary Control Alterations/Thermal Critical Assembly	罐式临界装置/临时性控制变更/热临界装置
TCAB	Twin Carbon Arc Brazing	双碳弧钎焊
TCBV	Temperature Coefficient of Breakdown Voltage	击穿电压温度系数

英文简称	英文全称	中文名称
TCC	Thiokol Chemical Corporation	聚硫橡胶化学公司
TCD	Technical Coordination Department	技术协调部
TCIC	Total Capital Investment Cost	总基建费,总投资成本
TCO	Test Control Officer	试验控制人员,试验管理人员
TCR	Test Clarification Request/Temperature Coefficient of Resistance	试验澄清要求/电阻温度系数
TCS	Tritium Cleanup System/Turbine Control System	氚纯净系统/汽轮机调节系统
TCV	Temperature Control Valve/Throttle Control Valve/Turbine Control Valve	温度控制阀/节流阀/汽轮机控制阀
TCW	Turbine Cooling Water	汽轮机冷却水
TD	Technical Description/Technical Document/Tendering Documents/Tendering Drawing/Thermal Diffusion/Time Delay	技术描述/技术文件/招标文件/投标图纸/热扩散/时间延迟,延时
TDC	Technical Data Center/Time Delay Closing	技术资料中心/延时闭合
TDH	Total Dynamic Head	总动压头
TDL	Technical Document List	技术文件清单
TDM	Time Division Multiplex	时分多路传输
TDO	Time Delay Opening	延时断开
TDR	Time Delay Relay/Turnover Date Review	延时继电器/交工日期审查
TDS	Technical Data System	技术数据系统

英文简称	英文全称	中文名称
TE	Tangent Elevation/Technical Evaluation/Temperature Element/Turbine End	仰角/技术评估/温度元件/汽轮机端
TED	Thermal Expansion Device/Thermal Expansion Difference/Track Etch Device	热膨胀装置/热膨胀差/径迹蚀刻装置
TEEAR	Test Equipment Error Analysis Report	试验设备误差分析报告
Teflon	Polytetrafluorethylene	特氟隆(聚四氟乙烯的商名)
TEL	Total Exhaust Loss	总排汽损失
TELE	Telephone	电话
TEMA	Tube Heat Exchanger Manufacturers Association	管式热交换器制造商协会
TEMP	Temperature	温度
TEMP DIFF	Temperature Difference	温差
TEN	Total Except Noted	完全接受,注明者除外
TERMS	Terminals	终端
TES	Telemetering Evaluation Station	遥测结果评价站
TFD	Test Flow Diagram/Traveling Flux Detector	试验流程图/移动式注量率探测器
TFE	Tetrafluooethylene	四氟乙烯
TFR	Trouble and Failure Report	故障事故报告
TFT	Thin-Film Technique	薄膜工艺
TG	Turbine Generator	汽轮发电机
TGC	Turbine Generator Controller	汽轮发电机控制系统
TGM	Turbine Generator Monitoring	汽轮发电机监测

英文简称	英文全称	中文名称
TGOP	Turning Gear Oil Pump	盘车油泵
TGT	Target	目标
TH	Thermometer	温度计
THCF	Transient Hot Channel Factor	瞬态热通道因子
THD	Thread	螺纹
THK	Thick/Thickness	厚/厚度
THP	Thrust Horsepower	推进马力
THS	Total Heating Surface	总受热面，总发热面
THSP	Thermal Spraying	热喷涂
THTF	Thermal Hydraulic Test Facility	热工水力试验装置(美国)
THTR	Thorium High-Temperature Reactor	钍高温(反应)堆
THUD	Thorium-Uranium Deuteriumr	钍-铀-氘系统
THV	Throttle Valve	节流阀
TIAM	Tritium in-Air Monitoring	空气中氚的监测
TIC	Technical Information Center/Temperature Indicating Controller/Temperature of Initial Combustion	技术信息中心/温度指示控制器/燃烧起始温度
TID	Technical Information Division/Technical Information Document	技术情报司/技术资料文件
TIG(AW)	Tungston-Inert-Gas Arc Welding	钨极惰性气体保护(电弧)焊
TIM	Time Meter	计时器
TIP	Traversing In-core Probe	移动式堆内探头
TIS	Test Instrumentation System	测试仪表系统
TK	Tank	箱

英文简称	英文全称	中文名称
TL	Task List/Totalizer	任务清单/加法器
TLC	Transferable Loan Certificate/Trip Logic Cubicles	可转让贷款证/脱扣逻辑柜
TLD	Thermoluminescent Detector/Thermoluminescent Dosimeter	热释光探测器/热释光剂量计
TLN	Test Listing Number	试验清单编号
TLOFW	Total Loss of Feed Water	给水总损失
TM	Technical Manual/Technical Memorandum/Temperature Meter/Training Manual/Twisting Moment	技术手册/技术备忘录/温度计/培训手册/扭矩
TMA	Thermomechanical Analysis	热机械(性能)分析
TMCA	Titanium Metals Corporation of America	美国钛合金公司
TMD	Transient Mass Distribution	瞬态质量分布
TMI	Three Mile Island/Total Metallic Impurities	三哩岛/总金属杂质
TN	Technical Note	技术说明,技术函件
TNR	Thermonuclear Reactor	热核反应堆
TNS	Tank Nitrogen Supply	罐装氮气供应
TO	Technical Order/Time Opening/Turnover	技术命令/定时断开(开启)/移交
TOB	Takeover for Blocking	模块式接受(管)
TOC	Timing Operations Center/Top-blown Oxygen Converter/Total Organic Carbon	计时操作中心/氧气顶吹转炉/有机碳总量

英文简称	英文全称	中文名称
TOD	Time of Delivery/Total Oxygen Demand	交货时间/总需氧量
TOI	Temporary Operation Instruction	临时运行指令
TOM	Takeover for Maintenance	维修移交
TOP	Temporary Operating Procedure	临时运行程序
TOTO	Takeover for Temporary Operation	临时运行移交
TP	Technical Procedure/Temperature Probe/Test Pressure/Test Procedure/Trap/Twisted Pairs	技术规程/温度探头/试验压力/试验规程/捕集器/双芯绞合线
TPA	Thimble Plug Assembly	阻力塞组件
TPFL	Two-Phase Flow Loop	两相流回路
TPL	Training Plan	培训计划
TPRD	Tapered	锥形的
TPS	Tagged Photon Spectrometer	标记光子谱仪
TPSD	Turnover Package Scope Definition	移交包范围定义
TQC	Total Quality Control	全面质量控制
TQNPC	Third Qinshan Nuclear Power Company Ltd.	秦山第三核电有限公司
TQNPP	Third Qinshan Nuclear Power Plant	秦山第三核电厂
TQP	Typical Quality Plan	典型质量计划
TR	Technical Regulations/Technical Report/Temperature Recorder/Test Report/Test Requirement/Thermal Relay/Time Relay/Transmitter Receiver	技术管理/技术报告/温度记录仪/试验报告/试验要求/热继电器/定时继电器/收发两用机

英文简称	英文全称	中文名称
TRANS	Transport	传输(送)
TRAS	Test Report Analysis Sheet	试验报告分析卡
TRC	Temperature Record and Control	温度记录和控制
TRD	Triad	三价,三元,三合一的
TREAT	Transient Reactor Test Facility	瞬态反应堆试验装置(美国)
TRIPR	Training, Research and Isotope Production Reactor	培训、研究和同位素生产(反应)堆(美国)
TRM	Turbine Run-up Monitor	汽轮机冲转监测器
TRP	Technical Review Package	技术审查包
TRR	Technical Review Response	技术审查答复
TRS	Test Response Spectrum	试验响应谱
TRU	Turbine Run-up Program	汽轮机冲转程序大纲
TRVM	Transistorized Voltmeter	晶体管伏特计
TS	Technical Specification/Tensile Strength/Tool Steel/Transient State/Turbine System	技术规格(说明)书/抗拉强度/工具钢/瞬态/汽轮机系统
TS	Torch soldering	火焰钎焊
TSA	Technical Safety Appraisal/Technical Service Agreement	技术安全评价/技术服务协议
TSCOP	Telescopic	望远镜式,套筒式,拉杆式
TSE	Test Support Equipment/Test System Flow-Sheet/Time Since Exposure	试验辅助设备/试验系统流程图/照射后时间
TSI	Turbine Supervisory Instrumentation	汽轮机监视仪表
TSPL	Throttle Steam Pressure Limiter	节流蒸汽压力限制器

英文简称	英文全称	中文名称
TSS	Terminal Solid Solubility/Total Suspended Solid	固体极限溶解度/悬浮固体物总量
TSSA	Technical Standards and Safety Authority	技术标准和安全局
TSSHD	Terminal Solid Solubility for Hydride Dissolution	氢化物极限固(体)溶(解)度
TSTS	Thrust Structure Test Stand	推力结构试验台
TSW	Technical Scope of Work/Test Switch	工作技术范围/试验开关
TTC	Terminating Toll Center	终端长途电话局
TTF	Thermal Transient Facility	热瞬态装置
TTI	Tender Technical Information	投标技术资料
TTO	TQNPC Toronto Office	TQNPC 多伦多代表处
TTR	Thermal Test Reactor/Toronto Technical Report	热中子试验反应堆(美国)/多伦多技术报告
TU	Thermal Unit/Tube	热量单位/管
TV	Television/Test Voltage	电视/试验电压
TVL	Tenth-Value Layer	十分之一值层
TVR	Television Recording	电视录像
TW	Thermit Welding	热剂焊
TW	Total Weight	总重
TWG	Test Working Group	测试工作组
TX	Twin Axial	双轴
TY	Type	型式
TYP	Typical	典型的

英文简称	英文全称	中文名称
U. p.	Under Proof	在试验中，待验证
U/C	Unclassified	未分类的；不保密的
UAL	Unit Area Loading	单位面积负荷
UATI	Union of International Engineering Organizations	国际工程组织联合会
UCCRL	Union Carbide and Carbon Research Laboratories	碳及碳化物联合研究实验室
UCL	Upper Confident Limit/Upper Control Limit	置信度上限/控制上限
UDC	Upper Dead Center	上死点
UEC	Unit Energy Cost	单位能量成本
UED	United Electrodynamics, Inc	美国电动力学公司
UES	Unexpected Event Sheet	意外事故单
UEW	United Electrical, Radio and Machine Workers of America	美国电气、无线电和机器工人联合会
UGRR	Underground Railroad	地下铁路，地铁
UHF	Ultra High Frequency	超高频
UHS	Ultimate Heat Sink	最终热阱
UHV	Ultra-High Vacuum/Ultra-High Voltage	超高真空/超高电压
UIE	Union Internationale D'Electrothermie	国际电热协会
UKAEA	United Kingdom Atomic Energy Authority	英国原子能管理局
ULF	Ultra-Low Frequency	超低频率
ULL	Unit Local Loading	单位局部加载

英文简称	英文全称	中文名称
UM	User Manual	用户手册
UN	United Nations	联合国
UNAEC	United Nations Atomic Energy Commission	联合国原子能委员会
UNCHE	United Nations Conference on the Human Environment	联合国人类环境会议
UNCOL	Universal Computer Oriented Language	通用计算机语言
UNCTAD	United Nations Conference on Trade and Development	联合国贸易和发展会议
UNDP	United Nations Development Programme	联合国开发计划署
UNIDO	United Nations Industrial Development Organization	联合国工业发展组织
UNISIST	World Scientific Information System	世界科学情报系统
UNITAR	United Nations Institute for Training and Research	联合国训练研究所
UNIV	Universal	通用的;万能的
UNPOC	Utility Nuclear Power Oversight Committee	电力公司核电监督委员会(美国)
UNS	Upper Neutron Shielding	上部中子屏蔽
UNSCCUR	United Nations Scientific Conference on the Conservation and Utilization of Resources	联合国资源保存和利用科学会议
UNSCEAR	United Nations Scientific Committee on the Effects of Atomic Radiation	联合国原子辐射效应科学委员会

英文简称	英文全称	中文名称
UNTAA	United Nations Technical Assistance Administration	联合国技术援助署
UOP	Universal Oil Products Company	美国环球石油产品公司
UP	Ultimate Pressure	最终压力
UPD	Uniform Probability Design	均匀概率设计
UPGR	Upgrade	升级，提浓
UPO	Undistorted Power Output	不失真功率输出
UPR	Unit Power Regulator	机组功率调节器
UPTF	Upper Plenum Test Facility	上腔试验装置
UPV	Unfired Pressure Vessel	非受火压力容器
URC	Ultrasonic Resin Cleaner	超声波树脂净化器
US AEC	United States Atomic Energy Commission	美国原子能委员会
US ERDA	United States Energy Research and Development Administration	美国能源研究与发展署
US NRC	United States Nuclear Regulatory Commission	美国核管理委员会
USBS	United States Bureau of Standards	美国标准局
USCEA	United States Committee for Energy Awareness	美国能源了解委员会
USDC	United States Department of Commerce	美国商业部
US-EXIM	Export-Import Bank of United States	美国进出口银行
USGPM	United State Gallons per Minute	(美国)加仑/分钟

英文简称	英文全称	中文名称
USHE	Upstream Heat Exchanger	逆流式热交换器
USI	Unresolved Safety Issue	未解决的安全问题
USP	United States Patent	美国专利
USPO	US Patent Office	美国专利局
USS	United States Standards	美国标准
UST	Unit Service Transformer	机组辅助变压器,厂用变压器
USW	Ultrashort Wave/Uranium Separative Work	超短波/铀分离功
USW	Ultrasonic Welding	超声波焊
UTC	United Technology Corporation	联合技术公司
UTIL	Utility	公用事业单位,电力公司
UTS	Ultimate Tensile Strength	极限抗拉强度
UUS	Ultrasonic Soldering	超声波钎焊
UV	Ultraviolet/Under Voltage	紫外线/电压不足
UVC	Universal Contact	万能接头
UW	Ultrasonic Wave/Underwater	超声波/水下
UW	Upset Welding	电阻对焊
UW-HF	High-Frequency UW	高频电阻对焊
UW-I	Induction UW	感应电阻对焊
V	Valve/Velocity/Voltage/Volts	阀门/速度/电压/伏(特)
V/L	Vapor-Liquid Ratio	汽液比
VA	Variable Area/Volt-Ampere	可变区域/伏-安
VAC	Volts-Alternating Current	伏(交流)
VACU	Vacuum	真空
VAN	Variable Area Nozzle	可变截面喷管

英文简称	英文全称	中文名称
VAP	Vapor/Vehicle Access Passage	蒸汽/车辆出入口
VARH	Volt Amp Reactive Hours	无功伏安小时
VATE	Versatile Automatic Test Equipment	多功能自动测试设备
VB	Valve Body	阀体
VC	Variable Capacitor/Volume of Compartment	可变电容器/舱室容积
VCG	Vertical Location of the Center of Gravity	重心的垂向位置
VCL	Vertical Center Line	垂直中心线
VCS	Volume Control System	容积控制系统
VCT	Volume Control Tank	容积控制箱
VDC	Volts Direct Current	伏(直流电)
VDI	Verein Deutscher Ingenieure	德国工程师协会
VDR	Voltage Dependent Resistor	压敏电阻
VDRIF	Variable Dose Rate Irradiation Facility	可变剂量率辐照设施
VENT	Ventilation	通风
VF	Velocity Factor/Video Frequency	速度系数/视频
VFC	Valid for Construction	适用于建造的
VFCR	Vendor Field Change Request	供货商现场变更申请
VFD	Variable-Frequncy Drive	变频启动
VFDA	Vertical Flux Detector Assembly	轴向注量率探测组件
VFGS	Variable Frequency Generating Set	变频发电机组

英文简称	英文全称	中文名称
VFO	Variable Frequency Oscillator	变频振荡器
VFT	Valid for Test/Verified for Test	适用于试验的/用于试验已验证过的
VFTO	Very Fast Transient Overvoltage	快速瞬变过电压
VH	Vent Hole/Vickers Hardness	排气孔/维氏硬度,Vickers 硬度
VHF	Very High Frequency	甚高频
VHL	Very Heavy Lift	特大型吊装
VHTR	Very High Temperature Gas Cooled Reactor	超高温气冷堆
VI	Volume Indicator	音量计
VIB	Vibration	振动
VL	Valve List/Variable Loss	阀门清单/可变损耗
VLF	Very Low Frequency	甚低频
VLV	Valve	阀门
VM	Validation Matrix/Velocity Meter/Voltmeter	确认矩阵/速度计/伏特计
VMR	Variable Moderator Reactor	可变慢化剂反应堆
VMS	Velocity Measuring System/Ventilation Management System	速度测量系统/通风管理系统
VO	Variation Order	变更令
VOL	Volume	体(容)积,音量
VOM	Volt-Ohm Meter	伏欧计
VOTA	Vibration Open Test Assembly	振动开口检验组件
VP	Vapor Pressure	蒸汽压
VPL	Valve Position Limit	阀位限制,阀位限值

英文简称	英文全称	中文名称
VPSP	Vacuum Plasma Spraying	真空等离子喷涂
VR	Variable Resistor/Verification Report/Voltage Regulator	可变电阻/验证报告/调压器
VRL	Vertical Reference Line	垂直基准线
VRS	Vapor Recovery System	蒸汽回收系统
VS	Valve Specification	阀门规格书
VSO	Very Stable Oscillator	极稳定振荡器
VSS	Valve Specification Sheet/Vapour Suppression System	阀门技术规格单/蒸汽抑压系统
VT	Vacuum Tube/Visual Inspection/Voice Tube/Voltage Transformer	真空管/目视检查/音频管/变压器
VTR	Validation Test Report/Video Tape Recorder	确认试验报告/磁带录像机
VTVM	Vacuum-Tube Voltmeter	真空管伏特计
VV	Vice Versa	反之亦然
VVVF	Variable Voltage and Variable Frequency	变压和变频
VX	Vapor Extractor	抽汽器
W	Watt/Weight/Wide/Width/Work	瓦/重量/宽的/宽度/功
W/A	With Attachment	带附件
W/B	Waybill	运货单
W/C	Water-Cement Ratio	水灰比
W/O	Weight Percent/Without	重量百分比/没有
W/P	Weather or Water Proof/Witness Point	防风雨或防水/见证点

英文简称	英文全称	中文名称
W/T	Wall Thickness	壁厚
WAMAP	Waste Management Advisory Programme	(放射性)废物管理咨询计划
WAN	Wide Area Network	宽域网络,宽带网
WANO	World Association of Nuclear Operators	世界核营运者协会
WAP	Work Activity Plan	工作活动计划
WAS	Withdrawal Authorization Sheet	撤销授权书
WAZ	Weld-Affected Zone	焊缝影响区
WBC	Whole Body Counter	全身计数器
WBNS	Water Boiler Neutron Source	水锅炉中子源
WBR	Water Boiler Reactor/Whole Body X-Irradiation	水锅炉反应堆/X射线全身照射
WBS	Work Breakdown Structure	工作分解结构
WCA	Whole Core Accident/Work Control Area	全堆芯事故/工作控制区
WD	Wiring Data/Wiring Diagram	接线数据/接线图
WDB	Wiring Database	接线数据库
WDCS	Wetwell Drywell Cooling System	湿井干井冷却系统
WDR	Water Displacer Rod	挤水棒
WE	Western Electric	西部电气公司
WEC	Westinghouse Electric Corporation	西屋电气公司(美国)
WESF	Waste Encapsulation and Storage Facility	放射性废物封装贮存设施
WF	Work File	工作文件

英文简称	英文全称	中文名称
WFP	Waste Fixation Program	废水固化计划
WFQC	Welding Follow-up and Quality Control	焊接质量跟踪管理
WG	Water Gauge	水表,水标
WGS	Work Group Supervisor	工作组主管
WH	Watt Hours/White	瓦时/白色
WHE	Water Hammer Eliminator	水锤效应消除器
WHM	Watt-Hour Meter	瓦时计
WI	Working Instruction	工作指南
WIS	Weld Information System	焊接信息系统
WL	Water Line/Wave Length/Wind Load/Wiring List	水线/波长/风荷载/接线清单
WLDMT	Weldment	焊件
WM	Water Meter/Watt-Meter/White Metal/Wire Mesh	水量计,水表/瓦特计/白金属/金属丝网
WMIS	Waste Management Information System	废物管理信息系统
WMP	Waste Management Plan	废物管理计划
WO	Work Order	工作指令
WP	Welding Part/Witness Point/Work Permit/Work Plan/Water-Proof/Working Pressure	焊接部件/见证点/工作票/工作计划/防水/工作压力
WPC	World Power Conference	世界电力会议
WPEGDC	Work Permit and Equipment Guarantee During Commissioning	调试期间的工作许可和设备保修

英文简称	英文全称	中文名称
WPRM	Weekly Progress Review Meeting	每周进度评审会
WR	Work Request	工作申请
WRENDA	Work Request List for Neutron Data Measurements	中子数据测量工作申请单
WRN	Work Request Notice	(合同外)工作要求单
WRSDC	Work Request System During Commissioning	调试期间的工作申请系统
WRT	Wrought	锻造的,精炼的
WS	Wave Soldering	波峰钎焊
WS	Wetwell Spray	湿井喷淋
WSL	Work Section Leader	施工队长,工长
WSP	Working Steam Pressure	工作汽压
WSR	Weekly Summary Report	每周总结报告
WSW	Wet Solid Waste	湿固体废物
WT	Warning Tag/Watertight/Weight	警告标志/水密的/重量
WTI	Winding Temperature Indicator	绕组温度指示器
WTP	Water Treatment Plant	水处理厂
WV	Working Voltage	工作电压
WW	Wire Wound	线绕的
WWA	Wolsong Work Authorization	月城(核电厂)工作授权
WWER	Water-Water Energetic Reactor	水-水动力堆
XFER	Transfer	转移(换,让)
XHV	Extreme High Vacuum	极高真空
XMTR	Transmitter	变换(送)器

英文简称	英文全称	中文名称
XPS	Expanded Polystyrene X-Ray Laser	多孔聚苯乙烯X射线激光器,发泡聚苯乙烯X射线激光器
XTLO	Crystal Oscillator	晶体振荡器
XUV	Extreme Ultra-Violet	超紫外
Y	Young's Modulus	弹性模量
YD	Yard	场区,场地
YP	Yield Point	屈服点
YS	Yield Strength	屈服强度
Z-2	Zircaloy-2	锆-2合金
ZAM	Zinc Alloy for Antifriction Metal	锌基轴承抗磨合金
ZCRJ	Zero Clearance Rolled Joint	零间隙滚轧接合
ZCU	Zone Controller Unit	区域控制单元
ZEA	Zero Energy Assembly	零功率装置
ZEBRA	Zero Energy Breeder Reactor Assembly	零功率增殖反应堆装置
ZEEP	Zero Energy Experimental Pile	零功率实验性堆
ZETA	Zero Energy Thermonuclear Assembly	零功率热核装置
ZETR	Zero Energy Thermal Reactor	零功率热反应堆
ZIRC	Zirconium	锆
ZN	Zinc	锌
ZOE	Zero Oxide Eau	零功率氧化铀重水反应堆(法国)
ZP	Zero Power	零功率
ZPA	Zero Period Acceleration	零周期加速度

英文简称	英文全称	中文名称
ZPPT	Zero Power Physics Testing	零功率物理试验
ZPR	Zero Power Reactor	零功率反应堆
ZPT	Zero Power Test	零功率试验

Ⅲ. AP1000 核电厂系统缩写

英文简称	英文全称	中文名称
ADS	Automatic Depressurization System	自动泄压系统
ASS	Auxiliary Steam Supply System	辅助蒸汽供应系统
BDS	Steam Generator Blowdown System	蒸汽发生器排污系统
CAS	Compressed and Instrument Air System	压缩空气和仪表空气系统
CCS	Component Cooling Water System	设备冷却水系统
CDS	Condensate System	凝结水系统
CES	Condenser Tube Cleaning System	冷凝器管清洁系统
CFS	Turbine Island Chemical Feed System	常规岛化学药剂供给系统
CGS	Generator Stator Cooling Water System	发电机定子冷却水系统
CMS	Condenser Air Removal System	冷凝器排气系统
CNS	Containment System	安全壳系统
CPS	Condensate Polishing System	凝结水精处理系统
CVS	Chemical and Volume Control System	化学和容积控制系统
CWS	Circulating Water System	循环水系统
DAS	Diverse Actuation System	多样化驱动系统，多样化触发系统
DDS	Data Display and Processing System	数据显示和处理系统
DOS	Standby Diesel and Auxiliary Boiler Fuel Oil System	备用柴油机和辅助锅炉燃油系统
DPS	Domestic Water Yard Piping System	厂区生活水分配系统
DRS	Storm Drain System	雨水排放系统

英文简称	英文全称	中文名称
DTS	Demineralized Water Treatment System	除盐水处理系统
DWS	Demineralized Water Transfer and Storage System	除盐水传输和储存系统
ECS	Main AC Power System	主交流电电源系统
EDS	Non Class 1E DC and UPS System	非1E级直流电和不间断电源系统
EDS1	Non Class 1E DC and UPS System-Load Group 1	非1E级直流电和不间断电源第1负荷组
EDS2	Non Class 1E DC and UPS System-Load Group 2	非1E级直流电和不间断电源第2负荷组
EDS3	Non Class 1E DC and UPS System-Load Group 3	非1E级直流电和不间断电源第3负荷组
EDS4	Non Class 1E DC and UPS System-Load Group 4	非1E级直流电和不间断电源第4负荷组
EDSS	Non Class 1E DC and UPS System-Spare	非1E级直流电和不间断电源备用组
EFS	Communication System	通信系统
EGS	Grounding and Lightning Protection System	接地和避雷保护系统
EHS	Special Process Heat Tracing System	特殊工艺加热跟踪系统
ELS	Plant Lighting System	电厂照明系统
EMS	Medium Voltage AC Distribution System	常规岛中压交流配电系统
EPS	Electrical Protection System	常规岛电气保护系统
EQS	Cathodic Protection System	阴极保护系统
EVS	Low Voltage AC Distribution System	常规岛低压交流配电系统

英文简称	英文全称	中文名称
FHS	Fuel Handling and Refueling System	装换料系统
FPS	Fire Protection System	消防系统
FWS	Main and Startup Feedwater System	主给水及启动给水系统
GPS	GPS Clock System	GPS 时钟系统
GSS	Gland Seal System	轴封系统
HCS	Generator Hydrogen and CO_2 System	发电机氢气和二氧化碳系统
HDS	Heater Drain System	加热器疏水系统
HSS	Hydrogen Seal Oil System	氢气密封油系统
IDS	Class 1E DC and UPS System	1E 级直流电和不间断电源系统
IDSA	Class 1E DC and UPS System-Division A	1E 级直流电和不间断电源分系统 A
IDSB	Class 1E DC and UPS System-Division B	1E 级直流电和不间断电源分系统 B
IDSC	Class 1E DC and UPS System-Division C	1E 级直流电和不间断电源分系统 C
IDSD	Class 1E DC and UPS System-Division D	1E 级直流电和不间断电源分系统 D
IDSS	Class 1E DC and UPS System-Spare	1E 级直流电和不间断电源备用分系统
IIS	Incore Instrumentation System	堆内仪表系统
IWS	Industrial Water System	工业水系统
LHS	Electric Hydraulic Fluid System	(汽轮机)电动液压油系统
LOS	Main Turbine and Generator Lube Oil System	主汽轮机和发电机润滑油系统

英文简称	英文全称	中文名称
LPS	Lube Oil Purification System	(汽轮机)润滑油净化系统
MES	Meteorological and Environmental Monitoring System	气象和环境监测系统
MHS	Mechanical Handling System	机械操作系统
MSS	Main Steam System	主蒸汽系统
MTS	Main Turbine System	主汽轮机系统
NSSS	Nuclear Steam Supply System	核蒸汽供应系统
OCS	Operation and Control Centers System	运行和控制中心系统
OWS	Opened Cycle Cooling Water System	开式循环冷却水系统
PCS	Passive Containment Cooling System	非能动安全壳冷却系统
PGS	Plant Gas System	电厂气体系统
PLS	Plant Control System	电厂控制系统
PMS	Protection and Safety Monitoring System	保护和安全监控系统
PPS	Industrial Water Yard Piping System	厂区生产水管网系统
PSS	Primary Sampling System	一回路取样系统
PWS	Potable Water System	生活水系统
PXS	Passive Core Cooling System	非能动堆芯冷却系统
RCS	Reactor Coolant System	反应堆冷却剂系统
RDS	Gravity and Roof Drain Collection System	重力和屋顶排水收集系统
RMS	Radiation Monitoring System	辐射监测系统
RNS	Normal Residual Heat Removal System	正常余热导出系统
RWS	Raw Water System	原水系统
RXS	Reactor System	反应堆系统
SDS	Sanitary Drainage System	生活污水排放系统
SES	Plant Security System	电厂保安系统

英文简称	英文全称	中文名称
SFS	Spent Fuel Pool Cooling System	乏燃料池冷却系统
SGS	Steam Generator System	蒸汽发生器系统
SJS	Seismic Monitoring System	地震监测系统
SLS	Seal Water System	密封水系统
SMS	Special Monitoring System	特殊监测系统
SPS	Sanitary Water Yard Piping System (including treatment sys. and facilities)	厂区生活污水排放系统(包括生活污水处理系统和设施)
SRS	Moisture Separator and Reheater (MSR) System	汽水分离再热器(MSR)系统
SSS	Secondary Sampling System	二回路取样系统
STS	Simulator Training System	模拟机培训系统
SWS	Service Water System	厂用水系统
TBS	Turbine Bypass System	汽轮机旁路排放系统
TCS	Turbine Building Closed Cooling Water System	汽轮机厂房闭式冷却水系统
TDS	Turbine Island Vents, Drains and Relief System	常规岛排气、疏水和泄压系统
TES	Turbine Extraction Steam System	汽轮机抽汽系统
TIS	Turbine Supervisory Instrument System	汽轮机监督仪表系统
TOS	Main Turbine Control and Diagnostics System	主汽轮机控制和诊断系统
TPS	Water Treatment Plant System	水处理厂系统
TVS	Closed Circuit TV System	闭路电视系统
VAS	Radiologically Controlled Area Ventilation System	放射性控制区通风系统

英文简称	英文全称	中文名称
VBS	Nuclear Island Nonradioactive Ventilation System	核岛非放射性通风系统
VCS	Containment Recirculation Cooling System	安全壳再循环冷却系统
VDS	Demineralized Water Treatment Building HVAC System	除盐水厂房暖通空调系统
VES	Main Control Room Emergency Habitability System	主控制室应急可居留系统
VFS	Containment Air Filtration System	安全壳空气过滤系统
VGS	Auxiliary Boiler Building HVAC System	辅助锅炉房暖通空调系统
VHS	Health Physics and Hot Machine Shop HVAC System	保健物理和热机车间暖通系统
VLS	Containment Hydrogen Control System	安全壳氢气控制系统
VNS	Switchyard Control Building HVAC System	开关控制楼通风空调系统
VOS	Oil Treatment Facility Room Ventilation System	油处理室通风系统
VPS	Pump House Building Ventilation System	泵站楼通风系统
VQS	Chlorination Workshop HVAC System	加氯车间通风和空调系统
VRS	Radwaste Building HVAC System	放射性废物厂房通风和空调系统
VTS	Turbine Building Ventilation System	汽轮机厂房通风系统
VUS	Containment Leak Rate Test System	安全壳泄漏率试验系统
VVS	Waste Water Treatment Plant Ventilation	废水处理车间通风
VWS	Central Chilled Water System	中央冷冻水系统

英文简称	英文全称	中文名称
VXS	Annex/Aux building Nonradioactive Ventilation System	附属/辅助厂房非放射性通风系统
VYS	Hot Water Heating System	热水系统
VZS	Diesel Generator Building Heating and Ventilation System	柴油发电机厂房暖通系统
WDS	Sea Water Desalinized System	海水淡化系统
WGS	Gaseous Radwaste System	放射性废气系统
WIS	Circulate Water Reagent Injection System	循环水次氯酸钠制备和加药系统
WLS	Liquid Radwaste System	放射性废液系统
WRS	Radioactive Waste Drain System	放射性废物排出系统
WSS	Solid Radwaste System	放射性固体废物系统
WWS	Waste Water System	废水系统
ZAS	Main Generator System	主发电系统
ZBS	Transmission Switchyard and Offsite Power System	输电开关站和厂外电源系统
ZOS	Onsite Standby Power System	现场备用电源系统
ZVS	Excitation and Voltage Regulation System	励磁和电压调节系统

Ⅳ. 设备监造专用缩写

英文简称	英文全称	中文名称
ACC	Acceptable	接受，可接受
ADL	Applicable Document List	适用文件清单
AEN	Acceptable Except Noted Comment	除了提到的意见之外，可接受
AR	Action Report	（监造）行动报告
ATP	Authorization to Proceed	开工令，授权开工
BEA	Bid, Evaluate and Award	招标，评标与定标
B/L	Bill of Lading	（海运）提单
CAB	Certified as Built	完工确认的，完工鉴定的
CAE	Certified as Executed	经实践检验的
CAR	Corrective Action Request	纠正行动要求
CCC	China Compulsory Product Certificate	中国强制性产品认证
CEAR	Construction & Erection All Risks	建安一切险
CFC	Certified for Construction	可用（于建造）的批准文件版本
CIF	Cost, Insurance and Freight	到岸价（包括保险费、运费）
CIP	Cost, Insurance Paid to	运费、保险费付至（指定目的地）
C&F	Cost and Freight	成本加运费，运费在内价
CFR	Cost and Freight	成本加运费
CPT	Carriage Paid to	运费付至（指定目的地）
CMTR	Certified Material Test Report	材料鉴定试验报告
CR	Communication Report	通信联络单，联系单

英文简称	英文全称	中文名称
D	Daily Record	设备监造日志
DAF	Delivered At Frontier	边境(指定地点) 交货
DBA	Design Basis Accident	设计基准事故
DCD	Design Control Document	设计控制文件
DCP	Design Change Proposal	设计变更申请
DCR	Design Change Request	设计变更申请
DDU	Delivered Duty Unpaid	未完税交货
DDP	Delivered Duty Paid	完税交货
DEQ	Delivered EX Quay	目的港码头交货
DES	Delivered EX Ship	目的港船上交货
DES	Description	叙述性文件,说明书
DID	Design Interface Document	设计接口文件
DOC	Document Charges	文件费
DR	Deviation Report	偏差报告
DRCF	Document Review Comment Form	文件审查意见单
DRCS	Document Review Comment Sheet	文计审查意见单
DSF	Document Submittal Form	文件提交形式
DT	Dye Penetrant Testing	渗透检验
EBP	Evaluated Bid Price	评标价
ECS	Equipment Expediting Communication Sheet	设备催交联络单
EDCR	Engineering & Design Coordinate Report	工程设计协调报告
EFI	Engineering Field Instruction	工程现场指令
ELFR	Elementary List for Receipt	初步验收清单
EOMR	End of Manufacturing Report	完工报告
EOC	Effectiveness of Contract	合同生效

英文简称	英文全称	中文名称
EPL	Elementary Packaging List	初步装运清单
EPT	End of Performance Test	性能试验结束
ESP	Equipment Supervision Program	设备监督大纲
EUR	European Utility Requirements	欧洲电力公司要求文件
EXW	EX WORKS	工厂交货
FAC	Final Acceptance Certificate	最终验收证书
FAS	Fee Alongside Ship	船边交货价
FCA	Free Carrier	交至承运人
FCD	First Concrete Date	第一罐混凝土浇灌日期
FDD	Final Delivery Date	最后交货期
FMEA	Failure Mode and Effect Analysis	故障模式与效应分析
FOB	Fee on Board	离岸价、船上交货价
FSAR	Final Safety Analysis Report	最终安全分析报告
FSSD	Final Scheduled Shipment Date	最后计划的装运日期
H	Hold Point	停工待检点
ICM	Interface Control Manual	接口控制手册
IED	Index of Engineering Document	工程文件索引
IPB	Isolated Phase Busduct	隔离相母线槽
IPR	Intellectual Property Right	知识产权
IQP	Initial Quality Plan	初始质量计划
ISI	In-Service Inspection	在役检查
ITP	Inspection Test Plan	检查试验计划
L	Letters	信函
L/C	Letter of Credit	信用证
M	Memorandum	备忘录

英文简称	英文全称	中文名称
M/T	Measurement Ton	吨尺码(即货物收费以体积计费)
MPR	Month Progress Report	月进度报告
NCR	Non Conformance Report	不符合项报告
NDE	Non Destructive Examination	无损检验
NOC	No Comment	没意见
O/F	Ocean Freight	海运
ON	Observation Notice	观察意见单
OMG	Operation and Maintenance Guide	运行维修指南
OMPM	Overall Manufacturing Progress Meeting	总体制造进度会
PAC	Provisional Acceptance Certificate	临时验收证书
PAN	Pre-Award Notification	预中标通知
PRE	Preliminary	(文件)初稿
PSA	Probabilistic Safety Assessment	概率安全评价
PSS	Procurement Support Services	采购支持服务
QA	Quality Assurance	质量保证
QC	Quality Control	质量控制
QOR	Quality Observation Report	质量观察报告
QP	Quality Plan	质量计划
QSCS	Quality Surveillance Communication Sheet	质量监督联络单
QSDR	Quality Surveillance Deficiency Report	质量监督缺陷报告
QSL	Qualified Supplier List	合格供货商清单
QTAR	Quality Tendency Analysis Report	质量趋势分析报告
RFC	Unacceptable and Return for Correction	不接受并退回修改

英文简称	英文全称	中文名称
RT NDT	Reference Temperature of Nil Ductility Transition	无延性转变参考温度,脆性转变参考温度
SAR	Safety Analysis Report	安全分析报告
SEO	Stone & Webster Site Engineering Office	石伟公司现场工程办公室
SET	Consortium Site Engineering Team	西屋联队现场工程队
SG	Surveillance Guide	监督导则
SP	Spare Part	备件
SR	Specific Report	专题报告
SSP	Specific Supervision Plan	专项监督计划
SWN	Stop Work Notice	停工令
TBD	To Be Determined	待决定
TBS	Turbine Bypass System	汽轮机旁路系统
TSS	Technical Support Services	技术支持服务
URD	User Requirement Document	电力公司要求文件,用户要求文件
W	Witness Point	见证点
W/M	Weight or Measurement Ton	以重量吨和体积折算吨的高者计费
W/T	Weight Ton	吨重量(即货物收费以重量吨数计费)
WN	Witness Notification	出席见证点通知单
WTR	Welding Technical Report	焊接技术报告

V. 法规、标准缩写

英文简称	英文全称	中文名称
AFCEN		法国核岛设备设计和建造规则协会标准
AFNOR	Association Francaise de Normalisation	法国标准化协会标准
AMS	American Material Standard	美国材料标准
ANS	American Nuclear Society	美国核协会标准
ANSI	American National Standard Institute	美国国家标准学会标准
ASME	American Society of Mechanical Engineers	美国机械工程师协会标准
ASME-PTC	American Society of Mechanical Engineers-Performance Test Code	美国机械工程学会-性能测试规范
ASTM	American Society for Testing and Material	美国材料与试验协会标准
BS	British Standard	英国标准
CB		中国船舶行业标准
CFR	Code of Federal Regulations	美国联邦法规
EJ		中国核工业行业标准
FWPCA	Federal Water Pollution Control Act	美国联邦水污染控制法
GB	People's Republic of China National Standards	中华人民共和国国家标准
GJB		中国国家军事工业标准
HAD	National Nuclear Safety guidelines	中国核安全导则
HAF	National Nuclear Safety Codes	中国核安全法规

英文简称	英文全称	中文名称
HAF003	Nuclear Power Plant Safety Requirements for Quality Assurance	核电厂质量保证安全规定
IAEA	Code 50-C/SG-Q	IAEA制定的《核电厂及其他核设施安全质量保证法规和导则》
IEEE	Institute of Electrical and Electronics Engineers	美国电气与电子工程师协会标准
INCOT-ERMS	International Rules for the Interpretation of Trade Terms	国际贸易术语解释通则
ISO	International Standardization Organization	国际标准化组织标准
ISO 6215		核电厂质量保证
JB,NJ		中国机械部标准
JIS	Japanese Industrial Standards	日本工业标准
NEC	National Electrical Code	美国国家电气法规
NEPA	National Environmental Policy Act	美国国家环境政策法令
NF	Norm de France	法国国家标准
RCC-C		法国压水堆PWR核电厂燃料组件设计和制造规范
RCC-E	PWR Nuclear Islands Electrical Equipment Design and Construction Rules	法国压水堆核岛电气设备设计建造规范
RCC-G		法国压水堆核电厂土建工程的设计和建造规范
RCC-I	Pressurized Water Reactor Nuclear Power Plant Design and Construction of Fire Safety Rules	法国压水堆核电厂防火安全设计和建造规范

英文简称	英文全称	中文名称
RCC-P	PWR Nuclear Power Plant System Design and Construction Regulations	法国压水堆核电厂系统设计和建造规范
RCC-M	Pressurized Water Reactor Nuclear Island Construction Machinery and Equipment Design Rules	法国压水堆核岛建造机械设备设计规程
RRC-EV	Règle de Réalisation et Contrôle Eau/Vapeur	法国汽/水控制实施规程
RRC-TA	Règle de Réalisation et Contrôle Turbine et Alterateur	法国汽轮机与发电机控制实施规程
RSE-M	Règle de Surveillance en Exploitation des Matériels Mécaniques	法国材料性能监督检测规程
SL,DL		中国水利、电力行业标准
SH		中国石油化工行业标准
SY		中国石油行业标准
US NRC 10CFR 50, Appendix B	U. S. Nuclear Regulatory Commission	美国核管会联邦法规《生产和使用核设施的许可证审批》的附录 B《核电厂和燃料后处理厂的质量保证大纲》
YS		中国有色冶金行业标准
ZB		中国国家专业标准

Ⅵ. 组织与机构缩写

英文简称	英文全称	中文名称
AACC	American Automatic Control Council	美国自动控制委员会
AAE	American Association of Engineers	美国工程师协会
AAEC	Australian Atomic Energy Commission	澳大利亚原子能委员会
AASM	Association of American Steel Manufacturers	美国钢铁制造商协会
ABMA	American Boiler Manufacturers Association	美国锅炉制造商协会
ACI	American Concrete Institute	美国混凝土学会
ACM	Association for Computing Machinery	计算机协会(美国)
ACMUI	Advisory Committee on Medical Uses of Isotopes	同位素医用咨询委员会
ACRS	Advisory Committee on Reactor Safeguards	反应堆保障咨询委员会(美国)
AEA	Atomic Energy Authority	原子能管理局(英国)
AEC	Atomic Energy Commission	原子能委员会(美国)
AECB	Atomic Energy Control Board	原子能管理局(加拿大)
AECL	Atomic Energy of Canada Limited	加拿大原子能有限公司
AERB	Atomic Energy Regulatory Board	原子能管理局(印度)
AERE	Atomic Energy Research Establishment	原子能研究中心(英国)
AES	American Electromechanical Society American Electronical Society	美国机电学会 美国电子学会
AESC	American Engineering Standards Committee	美国工程标准委员会
AFBMA	Antifriction Bearing Manufacturers Association	减磨轴承制造商协会
AFI	Air Filter Institute	空气过滤器学会

英文简称	英文全称	中文名称
AGMA	American Gear Manufacturers Association	美国齿轮制造商协会
AGN	Aerojet-General Nucleonics	喷气发动机通用核子学公司
AIA	American Institute of Architects Authorized Inspection Agency	美国建筑学会 指定的检查机构
AICE	American Institute of Chemical Engineers	美国化学工程师学会
AIEE	American Institute of Electrical Engineers	美国电气工程师学会
AIF	Atomic Industrial Forum	原子工业论坛(美国)
AIIE	American Institute of Industrial Engineers	美国工业工程师学会
AISC	American Institute of Steel Construction	美国钢结构协会
AISI	American Iron and Steel Institute	美国钢铁学会
AMCA	Air Movement and Conditioning Association	美国通风与空调协会
AMCA	Aeromechanic and Control Association	空气动力和调节协会
AMS	American Metallurgical Society	美国冶金学会
AMTBA	American Machine-Tool Builders' Association	美国机床制造业协会
ANI	American Nuclear Insurance	美国核保险公司
ANL	Argonne National Laboratory	美国阿贡国家实验室
ANS	American Nuclear Society	美国核协会
ANSI	American National Standards Institute	美国国家标准学会
ANSTO	Australian Nuclear Science and Technology Organization	澳大利亚核科学与技术机构
API	American Petroleum Institute	美国石油学会
APS	American Physical Society	美国物理学会
ARI	Air-conditioning and Refrigeration Institute	空调和制冷学会

英文简称	英文全称	中文名称
ASA	Acoustics Society of America/American Standard Association/Atomic Scientists' Association	美国声学学会/美国标准协会/原子能科学工作者协会(英国)
ASC	American Standards Committee	美国标准委员会
ASCE	American Society of Civil Engineers	美国土木工程师学会
ASEA	American Society of Engineers and Architects	美国工程师及建筑师学会
ASHRAE	American Society of Heating , Refrigeration and Air Condition Engineers	美国采暖、冷冻及空调工程师学会
ASHVE	American Society of Heating and Ventilating Engineers	美国采暖与通风工程师学会
ASII	American Science Information Institute	美国科学情报研究所
ASIS	American Society for Industrial Security	美国工业安全学会
ASLAB	Atomic Safety & Licensing Appeal Board	原子安全和审批上诉委员会(美国)
ASLB	Atomic Safety & Licensing Boards	核安全和审批委员会(美国)
ASLBP	Atomic Safety and Licensing Board Panel	核安全和审批委员会专门小组(美国)
ASM	American Society for Metals	美国金属学会
ASME	American Society of Mechanical Engineers	美国机械工程师学会
ASNT	American Society for Nondestructive Testing	美国无损探伤学会
ASQC	American Society for Quality Control	美国质量控制学会
ASST	American Society for Steel Treating	美国钢处理学会
ASTM	American Society for Testing Materials	美国材料试验学会

英文简称	英文全称	中文名称
AWS	American Welding Society	美国焊接学会
AWWA	American Water Works Association	美国水工协会
BESA	British Engineering Standards Association	英国工程标准协会
BGMA	British Gear Manufacturers Association	英国齿轮制造商协会
BIC	Bureau of International Cooperation	国际合作局
BINE	Beijing Institute of Nuclear Engineering	北京核工程研究院
BIRE	British Institution of Radio Engineers	英国无线电工程师协会
BISRA	British Iron and Steel Research Association	英国钢铁研究协会
BNES	British Nuclear Energy Society	英国核能学会
BNFL	British Nuclear Fuels Ltd.	英国核燃料有限公司
BPVC	Boiler and Pressure Vessel Committee	锅炉及压力容器委员会
BRDC	Bureau of Research and Development Center	研究及发展中心局
BSI	British Standards Institute	英国标准学会
BUSTDS	The National Bureau of Standards	国家标准局(美国)
BWI	Babcok & Wilcox International	B&W 国际公司
CAGI	Canadian Air and Gas Institute	加拿大空气和天然气学会
CEN	European Committee for Standardization	欧洲标准化委员会
CFHI	China First Heavy Industries Co., Ltd.	中国一重集团有限公司
CGNPC	China Guangdong Nuclear Power Corporation	中国广东核电集团公司
CIETAC	China International Economic and Trade Arbitration Commission	中国国际经贸仲裁委员会
CMAA	Crane Manufacturing Association of American	美国起重机制造商协会
CNECC	China Nuclear Engineering & Construction Corporation	中国核工业建设集团公司

英文简称	英文全称	中文名称
CNEGC	China National Erzhong Group Co.	中国二重集团公司
CNEIC	China Nuclear Energy Industry Corporation	中国原子能工业公司
CNNC	China National Nuclear Corporation	中国核工业集团公司/中国核工业总公司
CNPE	China Nuclear Power Engineering Company	中国核电工程公司
CNTIC	China National Technical Import & Export Corporation	中国技术进出口公司
CPI	China Power Investment Corporation	中国电力投资集团公司
CSEO	The Consortium Site Engineering Organization	西屋现场工程设计联队
DEMA	Diesel Engine Manufacturers Association	柴油发动机制造商协会
DFB	Dongfang Boiler (Group) Co. Ltd.	东方锅炉集团有限公司
DIN	Deutsches Institut fur Normung	德国标准化委员会
DOE	Department of Energy	能源部(美国)
DOOSAN	Doosan Heavy Industries & Construction Co. LTD.	斗山重工建设有限公司(韩国)
ECEPDI	Eastern China Electric Power Design Institute	华东电力设计院
EDF	Electricite de France	法国电力公司
EMD	Curtiss-Wright Electro-Mechanical Corporation	美国 Curtiss-Wright 电机公司(屏蔽电机泵制造商)
EPRI	Electric Power Research Institute	电力研究院(美国)
FEMA	Federal Emergency Management Agency	联邦应急管理机构
FM	Factory Mutual Engineering and Research Corporation	工厂共同设计与研究公司

英文简称	英文全称	中文名称
FRA	Framatome	法玛通公司(法国)
GE	General Electric Co.	通用电气公司(美国)
GYNPC	Guangdong Yangjiang Nuclear Power Company	广东阳江核电公司
HBC	Harbin Boiler Co.，Ltd.	哈尔滨锅炉集团有限公司
HPEC	Harbin Power Equipment Company Ltd.	哈尔滨动力设备股份有限公司
IAEA	International Atomic Energy Agency	国际原子能机构
ICEA	Insulated Cable Engineers Association	绝缘电缆工程师协会
IEC	International Electrotechnical Commission	国际电工委员会
IEEE	The Institute of Electrical and Electronic Engineering	电气与电子工程协会(美国)
IES	Illumination Engineering Society	照明工程学会(美国)
INEL	Idaho National Engineering Laboratory	爱达荷国家工程实验室
ISA	Instrument Society of American	美国仪表学会
ISA	International Standardization Association	国际标准化协会
ISO	International Organization for Standardization	国际标准化组织
JPMO	Joint Project Management Organisation	联合项目管理组
JSA	Japan Standard Association	日本标准化协会
JSW	Japan Steel Work	日本制钢所
JTG	Joint Test Group	联合试验组
KEDO	Korean Peninsula Energy Development Organization	朝鲜半岛能源开发组织

英文简称	英文全称	中文名称
KEPCO	Korea Electric Power Corporation	韩国电力集团
KEPRI	Korean Electric Research Institute	韩国电力研究院
KHIC	Korean Heavy Industries Corporation	韩国重工业公司
KHNP	Korea Hydro & Nuclear Power Co. Ltd	韩国水力核电有限公司(原 KEPCO)
KINS	Korean Institute of Nuclear Safety	韩国核安全研究所
KNFC	Korean Nuclear Fuel Company	韩国核燃料公司
KOPEC	Korea Power Engineering Company	韩国电力工程公司
MHI	Mitsubishi Heavy Industries, LTD.	三菱重工业有限公司(日本)
MSS-SP	Manufacturers' Standardization Society	阀门及配件工业制造商标准化协会(美国)
NAE	National Academy of Engineering	美国工程院
NAS	National Academy of Sciences	美国科学院
NBS	National Bureau of Standards	美国国家标准局
NCRP	National Council on Radioactive Protection	国家辐射防护委员会(美国)
NDRC	National Defense Research Committee	国防部研究委员会(美国)
NDRC	National Development and Reform Commission	国家发展与改革委员会
NEI	Nuclear Energy Institute	核能研究所(美国)
NEMA	National Electrical Manufacturers Association	国家电子制造商协会
NFPA	National Fire Protection Association	国家防火协会(美国)
NIAC	Nuclear Industry Assessment Committee	核工业评估委员会

英文简称	英文全称	中文名称
NIST	National Institute of Standards and Technology	国家标准化及技术研究院
NNSA	National Nuclear Safety Administration	国家核安全局
NPD	Nuclear Power Division	核电部
NPIC	Nuclear Power Institute of China	中国核动力院
NRC	Nuclear Regulatory Commission	核管理委员会(美国)
NUMARC	Nuclear Management and Resources Council	核管理和资源委员会
NUPIC	Nuclear Procurement Issues Committee	核的采购事物委员会
OSHA	Occupational Safety and Health Administration	职业安全健康管理局
PAEC	Pakistan Atomic Energy Commission	巴基斯坦原子能委员会
PaR Nuclear		西屋分包商(生产环吊，换料机)
PRC	People's Republic of China	中华人民共和国
PUC	Public Utility Commission	美国公用事业委员会
SAC	Standardization Administration of PRC	中国国家标准化管理委员会
SDNPC	Shandong Nuclear Power Company Ltd.	山东核电有限公司
SEO	Site Engineering Office	现场工程管理办公室
SEPA	State Environment Protection Administration (China)	国家环保总局(中国)
SBW	Shanghai Boiler Works, Ltd.	上海锅炉集团有限公司
SIMENS		德国西门子电气公司
SMACNA	Sheet Metal and Air-Conditioning Contractor's National Association	全国板材及空调承包商协会

英文简称	英文全称	中文名称
SMNPC	Sanmen Nuclear Power Company Ltd.	三门核电有限公司
SMNPP	Sanmen Nuclear Power Plant	三门核电厂
SNERDI	shanghai Nuclear Engineering Research & Design Institute	上海核工程研究设计院
SNPTC	State Nuclear Power Technology Company	国家核电技术公司
SPMO	Site Project Management Organization	现场项目管理组织
SQSB	State Quality Supervision Bureau	国家质量监督局
SQUG	Seismic Qualification Group	地震评定小组
SSPC	Steel Structures Painting Council	钢结构喷漆委员会
SSW	Shaw-Stone and Webster	肖-石伟工程公司
SWA	Stone and Webster Asia, Inc.	石伟亚洲公司
SWI	Stone & Webster International Engineering	石伟国际工程公司
TEMA	Tube Heat Exchanger Manufacturers Association	管式热交换器制造商协会
TEPCO	Tokyo Electric Power Company	东京电力公司
UPCO	(The SPX-Copes Vulcan subcontractor)	生产爆破阀的分供货商
WANO	World Association of Nuclear Operators	世界核电营运者协会
WEC	Westinghouse Electric Company	西屋电气公司
WEC	Consortium	西屋联合体,西屋联队
WIPICO	Westinghouse Industry Products International Company, LLC	西屋工业产品国际有限公司
WOG	Westinghouse Owners Group	西屋业主集团
ZTPC	Zhejiang Thermal Power Construction Company	浙江火电建设公司

附录 常见计量单位符号及换算关系

Unit Symbols 单位符号	Name of Unit 单位名称	Name of Quantity 量的名称	Conversion Relation 换算关系	Remarks 备注
A	ampere 安[培]	electric current 电流	—	许用*
Å	ångström 埃	length 长度	1 Å=0.1 nm $=10^{-10}$ m	非许用*
a	year 年	time 时间	—	许用
at	technical atmosphere 工程大气压	pressure 压强,压力	1 at=98 066.5 Pa	非许用
atm	standard atmosphere 标准大气压	pressure 压强,压力	1 atm=101 325 Pa	非许用
b	barn 靶恩	cross section 截面	1 b=10^{-28} m^2	非许用
bar	bar 巴	pressure 压强,压力	1 bar=100 kPa	非许用
Bq	becquerel 贝可[勒尔]	activity [放射性]活度	1 Bq=1 s^{-1}	许用
Btu	British thermal unit 英热单位	heat 热, quantity of heat 热量	1 Btu=1 055.056 J	非许用
C	coulomb 库仑	quantity of electricity 电荷[量]	1 C=1 A·s	许用
℃	degree Celsius 摄氏度	temperature 温度	1 ℃=K−273.15	许用

* 表中许用或非许用是我国国家标准规定的。

续表

Unit Symbols 单位符号	Name of Unit 单位名称	Name of Quantity 量的名称	Conversion Relation 换算关系	Remarks 备注
cal	calorie 卡[路里]	heat 热量	$1\ cal_{IT}$(1 cal)=4.186 8 J； $1\ cal_{th}=4.184\ 0$ J	非许用
cd	candela 坎[德拉]	luminous intensity 发光强度	—	许用
Ci	curie 居里	activity [放射性]活度	$1\ Ci=3.7\times10^{10}$ Bq	非许用
d	day 天,(日)	time 时间	1 d=86 400 s	许用
dB	decibel 分贝	level difference 级差	无量纲	许用
dyn	dyne 达因	force 力	$1\ dyn=10^{-5}$ N	非许用
erg	erg 尔格	work 功	$1\ erg=10^{-7}$ J	非许用
eV	electronvolt 电子伏[特]	energy 能[量]	$1\ eV=1.602\ 177\times10^{-19}$ J	许用
F	farad 法[拉]	capacitance 电容	1 F=1 C/V	许用
℉	degree Fahrenheit 华氏度	temperature 温度	$1\ ℉=\frac{9}{5}K-459.67$ $=\frac{9}{5}℃+32$	非许用
fermi	fermi 费米	length 长度	1 fermi=1 fm $=10^{-15}$ m	非许用
ft	foot 英尺	length 长度	1 ft=0.304 8 m	非许用
G,Gs	gauss 高斯	magnetic flux density 磁通量密度	$1\ G\triangleq10^{-4}$ T	非许用

续表

Unit Symbols 单位符号	Name of Unit 单位名称	Name of Quantity 量的名称	Conversion Relation 换算关系	Remarks 备注
Gb	gilbert 吉伯	magnetic potential difference 磁位差	1 Gb=0.795 8 安匝	非许用
Gy	gray 戈[端]	absorbed dose 吸收剂量	1 Gy=1 J/kg	许用
H	henry 享[利]	inductance 电感	1 H=1 V・s/A	许用
h	hour[小]时	time 时间	1 h=3 600 s	许用
ha	hectare 公顷	area 面积	1 ha=10^4 m^2	许用
HP,hp	horsepower [英制]马力	power 功率	1 英制马力= 745.700 W	非许用
Hz	hertz 赫[兹]	frequency 频率	1 Hz=1 s^{-1}	许用
in	inch 英寸	length 长度	1 in=0.025 4 m	非许用
J	joule 焦[耳]	energy, work, heat 能量，功，热	1 J=1 N・m =1 W・s	许用
K	kelvin 开[尔]文	thermodynamic temperature 热力学温度	K=℃+273.15	许用
kg	kilogram 千克， (公斤)	mass 质量	—	许用
M	克分子浓度	concentration of gram molecule 克分子浓度	1 M≙1 mol/L	非许用

续表

Unit Symbols 单位符号	Name of Unit 单位名称	Name of Quantity 量的名称	Conversion Relation 换算关系	Remarks 备注
kgf	kilogram-force 千克力(公斤力)	force 力	1 kgf= 9.806 65 N	非许用
L,l	litre 升	volume 容积,体积	$1\ L=10^{-3}\ m^3$	许用
lb	pound 磅	mass 质量	1 lb=0.453 592 kg	非许用
lbf	pound-force 磅力	force 力	1 lbf=4.448 22 N	非许用
lm	lumen 流明	luminous flux 光通量	1 lm=1 cd・sr	许用
lx	lux 勒[克斯]	illuminance 光照度	$1\ lx=1\ lm/m^2$	许用
l. y.	light year 光年	length 长度	1 l. y. =9.460 53× 10^{15} m	非许用
m	metre 米	length 长度	—	许用
mile	mile 英里	length 长度	1 mile= 1 609.344 m	非许用
min	minute 分	time 时间	1 min=60 s	许用
mmHg	conventional millimetre of mercury 约定毫米汞柱	pressure 压强	1 mmHg= 133.322 4 Pa	非许用
mmH_2O	conventional millimetre of water 约定毫米水柱	pressure 压强	$1\ mmH_2O=$ 9.806 65 Pa	非许用
mol	mole 摩[尔]	amount of substance 物质的量	—	许用

续表

Unit Symbols 单位符号	Name of Unit 单位名称	Name of Quantity 量的名称	Conversion Relation 换算关系	Remarks 备注
mol/L	mole per litre 摩[尔]每升	concentration of substance 物质的浓度	1 mol/L= 10^3 mol/m^3	许用
Mx	maxwell 麦克斯韦	magnetic flux 磁通量	1 Mx≙10^{-8} Wb	非许用
N	newton 牛[顿]	force 力	1 N=1 kg·m/s^2	许用
N	equivalent concentration 当量浓度	concentration of substance 物质的浓度	1 *N*≙(1 mol/L)× 离子价数	非许用
n mile	nautical mile 海里	length 长度	1 n mile=1 852 m	许用
Np	neper 奈培	level difference 级差	1 Np=8.685 909 dB	许用
nt	nit 尼特	luminance 光亮度	1 nt=1 cd/m^2	非许用
Oe	oersted 奥斯特	magnetic field strength 磁场强度	1 Oe≙ (1 000/4π) A/m	非许用
P	poise 泊	dynamic viscoity [动力]黏度	1 P=10^{-1} Pa·s	非许用
Pa	pascal 帕[斯卡]	pressure, stress 压力,压强,应力	1 Pa=1 N/m^2	许用
phot	phot 辐透	illuminance 光照度	1 phot=10^4 lx	非许用
R	röntgen 伦琴	exposure 照射量	1 R=2.58× 10^{-4} C/kg	非许用
°R	degree Rankine 兰氏度	temperature 温度	1 °R=$\frac{5}{9}$K	非许用

续表

Unit Symbols 单位符号	Name of Unit 单位名称	Name of Quantity 量的名称	Conversion Relation 换算关系	Remarks 备注
r	revolutuion 转	number of revolution 转数	无量纲量	许用
rad，rd	rad 拉德	absorbed dose 吸收剂量	1 rad=10^{-2} Gy	非许用
rad	radian 弧度	[plane]angle [平面]角	1 rad=1 m/m=1	许用
rem	rem 雷姆	dose equivalent 剂量当量	1 rem=10^{-2} Sv	非许用
S	siemens 西[门子]	conductance 电导	1 S=1 Ω^{-1}	许用
s	second 秒	time 时间	无量纲量	许用
sr	steradian 球面度	solid angle 立体角	无量纲量	许用
St	stokes 斯[托克斯]	kinematic viscosity 运动黏度	1 St=10^{-4} m^2/s	非许用
Sv	sievert 希[沃特]	dose equivalent 剂量当量	1 Sv=1 J/kg	许用
T	tesla 特[斯拉]	magnetic flux density 磁通量密度	1 T=1 Wb/m^2	许用
t	tonne 吨	mass 质量	1 t=10^3 kg	许用
tf	ton-force 吨力	force 力	1 tf=9.806 65×10^3 N	非许用
Torr	torr 托	pressure 压力，压强	1 Torr=133.332 4 Pa	非许用
u	atomic mass unit 原子质量单位	mass 质量	1 u=1.660 540×10^{-27} kg	许用

续表

Unit Symbols 单位符号	Name of Unit 单位名称	Name of Quantity 量的名称	Conversion Relation 换算关系	Remarks 备注
V	volt 伏[特]	electric potential, electric difference, electromotive force 电位，电压，电动势	1 V=1 W/A	许用
V·A	volt-ampere 伏安	apparent power 视在功率	1 V·A=1 W	许用
var	var 乏	reactive power 无功功率	1 var=1 W	许用
W	watt 瓦[特]	power 功率	1 W=1 J/s	许用
Wb	weber 韦[伯]	magnetic flux 磁通量	1 Wb=1 V·s	许用
γ	gamma 伽马	magnetic induction 磁感应强度	1 $\gamma=10^{-9}$ T	非许用
μ	micron 微米	length 长度	1 μ=1 μm=10^{-6} m	非许用
Ω	ohm 欧[姆]	resistance 电阻	1 Ω=1 V/A	许用
(°)	degree 度	[plane]angle [平面]角	1°=(π/180) rad	许用
(′)	minute[角]分	[plane]angle [平面]角	1′=(1/60)°	许用
(″)	second[角]秒	[plane]angle [平面]角	1″=(1/60)′	许用